ELECTROMAGNETIC INVERSE PROFILING
Theory and numerical implementation

ELECTROMAGNETIC INVERSE PROFILING
Theory and numerical implementation

Proefschrift

ter verkrijging van de graad van doctor aan de Technische Universiteit Delft, op gezag van de Rector Magnificus, prof.dr. J.M. Dirken, in het openbaar te verdedigen ten overstaan van een commissie aangewezen door het College van Dekanen op dinsdag 12 mei 1987 te 16.00 uur

door

ANTONIUS GREGORIUS TIJHUIS

Natuurkundig doctorandus

geboren te Oosterhout N.B.

CRC Press
Taylor & Francis Group
Boca Raton London New York

CRC Press is an imprint of the
Taylor & Francis Group, an **informa** business

CRC Press
Taylor & Francis Group
6000 Broken Sound Parkway NW, Suite 300
Boca Raton, FL 33487-2742

Visit the Taylor & Francis Web site at
http://www.taylorandfrancis.com

and the CRC Press Web site at
http://www.crcpress.com

In memory of my mother
To my father

FOREWORD

I am pleased to have read this monograph that addresses the important research issues of direct scattering transient problems and applications to inverse profiling.

Over the years the singularity expansion method (SEM) has been the subject of ongoing controversy regarding the completeness of the natural mode expansion stemming from the vanishing or nonvanishing contributions from the closing contour at infinity. This issue is now completely resolved for the cases of a lossy dielectric slab and a radially inhomogeneous cylinder by emphasizing the inadequacy of field descriptions solely in terms of mode contributions. To avoid direct numerical integration on the closing contours, two independent approaches have been implemented. For the homogeneous slab case, decomposition is made for the total Laplace transformed field which brings out the branch cut contribution as the required entire function augmentation of the SEM expansion. The pulse-truncation approach, which was independently proposed by Morgan in 1984 for the study of the transient response of three-dimensional conducting targets, is applied to the inhomogeneous lossy slab and cylinder cases.

The direct-scattering problem is also studied with time-domain techniques by numerically solving the source-type integral equation that governs the transient scattering of electromagnetic fields by one- and two-dimensional objects. Since in numerical solutions, each field value is computed from field values at previous instants, the computational errors due to discretization accumulate, and as a consequence the solution obtained may become unstable. The author derives two stability criteria for the discretization of the space-time integral which guarantee that instability can be controlled by reducing the discretization steps. These stability conditions are sufficient but not necessary since in some cases the marching-on-in-time method may yield stable results even though the criteria are violated.

The inverse-scattering problems are investigated for electromagnetic systems identification with the Prony-type method, and for determining an unknown susceptibility and/or conductivity profile of a one-dimensional, inhomogeneous, and lossy dielectric slab from reflection data in the frequency domain and in the time domain. Different inversion schemes and many related issues are studied. These are all important research topics at inception and will undoubtedly lead to further advances and discoveries.

A.G. Tijhuis has published many technical papers and made significant contributions in this new and exciting field of research. This book is certainly a timely contribution to the archival literature. I take great pleasure in writing this foreword.

J.A. KONG
MASSACHUSETTS INSTITUTE OF TECHNOLOGY
CAMBRIDGE, MASSACHUSETTS

PREFACE

This monograph overviews the author's research on the direct scattering of electromagnetic waves by one- and two-dimensional objects and its use for one-dimensional inverse profiling. Having investigated these problems for a number of years, the author has obtained a series of results both in method and in application to specific problems. The systemization of these results, the majority of which has earlier been published as individual research papers, is the basis of the present monograph. Compared with the original research papers, most of the unresolved issues have been cleared up, and a number of new numerical techniques and results are presented and discussed. This book aims at an audience of engineers, mathematicians, physicists and geophysicists interested in the theoretical description of physical phenomena. Special care was taken to make the presentation self-contained and understandable for graduate students. In fact, part of the material has already been used as main text in an advanced course on direct and inverse scattering given at the Department of Electrical Engineering of the Delft University of Technology.

In this monograph, several techniques are presented for the solution of transient electromagnetic direct-scattering problems. These problems are solved indirectly via a Fourier or Laplace transformation to the real- or complex-frequency domain, as well as directly in the time domain. Both the analytical and the computational aspects of each solution technique are discussed. For each technique, representative numerical results of its application to at least one example problem are presented and discussed. Special attention is devoted to the physical interpretation of the theoretical and numerical results.

For the one-dimensional case, it is also described how the special features of the techniques can be utilized to solve the inverse problem of determining obstacle properties from the scattered field excited by a known incident field. Both the problems of "identification" and of "inverse profiling" are addressed. In this part of the analysis, the same kind of analytical, numerical and physical considerations are emphasized as in the direct-scattering part. Special attention is devoted to the band-limiting effects that may arise due to possible restrictions in the frequency spectrum of the incident field, and due to approximation errors inherent in the solution procedures employed.

A more detailed survey of the direct- and inverse-scattering problems that have been resolved, and of the algorithms applied to arrive at the solutions, is given in Chapter 1. In addition, it is described how the presentation of these problems is organized.

The author would like to thank all persons who, directly or indirectly, have contributed to the work described in the present monograph. In particular he is indebted to: Maria Ammerlaan, for her expert typing of this and previous manuscripts; Mrs. S. Massotty for her continual help in improving the English text; Mr. P.J.J. van Daalen and Mr. B. Zorn for their invaluable advice concerning programming matters; the former research students Mr. C.P.L. Cames van Batenburg, Mr. P.C. Kempen, Mr. R.E.C. Koster, Mr. B.L. Michielsen, Mr. R.M. van der Weiden, and Mr. C. van der Worm for their enthusiastic participation in and substantial contributions to parts of this study; his colleagues at the Laboratory of Electromagnetic Research of the Department of Electrical Engineering of the Delft University of Technology for creating the kind of stimulating atmosphere in which research can flourish; Professor A.T. de Hoop of the Delft University of Technology and Professor J. Boersma of the Eindhoven University of Technology for their suggestions and remarks regarding some of the original research papers; and, last but not least, his thesis supervisor, Professor H. Blok, for many helpful discussions and for his permanent interest during the progress of this work.

TABLE OF CONTENTS

Introduction

Direct scattering

Identification

Inverse profiling

INTRODUCTION

1. GENERAL INTRODUCTION

The aim of *inverse scattering* is to determine physical properties of an unknown object or configuration that are not readily available from direct measurements. Typically, such direct measurements are impracticable either because of their costs or other consequences, or because the region of interest is simply inaccessible. Hence, we must probe that region indirectly by generating some wave phenomenon that can penetrate into it, and measuring the resulting scattered field. From the knowledge of that field, we then hope to recover the desired properties. In a general inverse-scattering problem, the unknown physical properties may involve geometrical aspects of the configuration, as well as the spatial distribution of certain material parameters. The probing wave may be of an electromagnetic, acoustic, or elastodynamic nature, and may be either pulsed or monochromatic.

Inverse-scattering problems can be subdivided into a number of classes, according to the kind of information that must be retrieved in the measurement. In the present monograph, we will encounter two types of inverse-scattering problems. In the first place, we will consider the *identification problem*, which amounts to establishing whether the scattering configuration is or is not identical to some fully known reference configuration. This type of problem would arise, for example, when we want to identify an unknown target from a finite class of possible targets. In the second place, we have the *inverse-profiling problem*, where the geometry of the configuration is known, and where we wish to determine the spatial distribution of one or more unknown material parameters. The designation of this class of inverse-scattering problems stems from the fact that such an unknown parameter distribution is commonly referred to as a *profile*. As indicated in the title of this monograph, we will focus our attention mainly on the inverse-profiling problem.

From the definition of an inverse-scattering problem as given above, it will be clear that such a problem cannot be resolved without a detailed understanding of the mechanisms that govern the propagation and scattering effects experienced by the probing wave inside and outside the scattering domain. Hence, a sizable part of the analysis will be devoted to investigating the so-called *direct-scattering problem* of computing the response to the probing wave in a *known* configuration. In fact, it has been the basic philosophy of this author, for one, that, by developing reliable methods for solving this direct-scattering problem, one automatically obtains useful tools for tackling its inverse counterpart.

In our applications, we will restrict ourselves to one- and two-dimensional electromagnetic scattering problems. As far as the direct problem is concerned, we consider especially transient plane-wave scattering. In the inverse case, we have limited the discussion to the class of one-dimensional problems. Thanks to these restrictions, we have been able to investigate the analytical as well as the computational aspects of each solution technique employed in full detail. Moreover, by applying different solution techniques to the same class of simple problems, we can obtain insight into the specific advantages and disadvantages of the respective procedures. The restriction to electromagnetic problems should not be considered a true objection. The extension of the discussion to acoustic and elastodynamic problems with a similar geometry is straightforward. On the other hand, the extension to more complicated geometries may be more involved. However, we have systematically formulated our solution techniques such that they seem capable of being generalized to more complicated geometries.

1.1 Direct-scattering problems

First, let us consider the direct-scattering problem in some more detail. In the first part of this monograph, we will describe three different methods for solving this problem. As an illustration, a schematic view of these methods is presented in Figure 1.1.1. This figure shows that we can distinguish between *frequency-domain* techniques and *time-domain* techniques. In frequency-domain techniques, the transient response of the scattering obstacle is represented as a Laplace inversion integral over a Bromwich contour in the right half of the time-Laplace domain (complex s-plane). Such a representation has the advantage that in the equation(s) governing its constituents (the Laplace-transformed field quantities), the partial

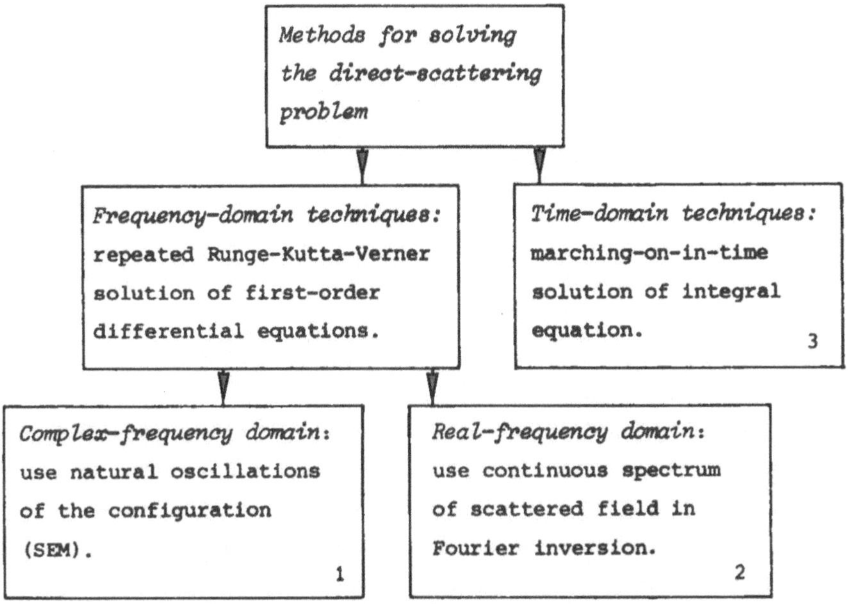

Figure 1.1.1 Schematic view of the methods for solving transient electromagnetic direct-scattering problems that are discussed in this monograph.

time differentiation from the corresponding time-domain equations is replaced by a scalar multiplication with the complex frequency s. In Chapter 2, two different methods will be discussed to evaluate the Laplace inversion integral.

In the *singularity expansion method,* commonly referred to as SEM, this integral is evaluated by contour deformation. This results in a field representation in terms of residual contributions from the singularities in the left half of the complex s-plane, i.e. poles and, possibly, branch points. Each pole is then identified as a natural frequency associated with a natural oscillation of the scatterer. The natural frequencies, in particular, are characteristic of the scattering configuration. The validity of the modal representation obtained depends on the possible presence of a contribution from the closing contour in the left half of the complex s-plane.

For each space point in the configuration, a finite initial instant can be indicated after which the modal representation holds. Before that instant, there may be a time interval where a nonvanishing electromagnetic field cannot be represented by modal contributions only. The main attraction of the singularity expansion method is that it leads in an elegant way to a profound insight into the physical aspects of the process of scattering by an obstacle.

The second technique that can be applied to evaluate the Laplace inversion integral is its *direct computation* with the aid of an *FFT algorithm*. One advantage of this technique is that it yields the excited field at all instants.

In the third method of solution, we solve the transient direct-scattering problem directly in the time domain using source-type integral equations for the electromagnetic field. In particular, we use one property that these equations have in common, namely that the scattered field is expressed in terms of one or more integrals of field values at previous instants. In the *marching-on-in-time* method, the integral equations are discretized in space and time. This results in a system of linear equations for the approximate field values at the discrete space-time points that can be solved by a step-by-step updating procedure. The most important difficulty in applying this method is the accumulation of discretization errors in the time-recursive solution procedure, which may lead to exponentially increasing instabilities. In Chapter 3, the method will be formulated in a general form, and it will be demonstrated how the instabilities can be avoided.

Each of the three methods outlined above will be applied to a set of one- and two-dimensional transient electromagnetic scattering problems. Figure 1.1.2 presents a schematic view of these problems and the methods that have been employed to resolve them. In this figure, two points may require further explanation. First, the application of the SEM to a homogeneous, lossy dielectric slab has been included since, for this configuration, the integrand of the Laplace inversion integral is known in closed form. This enables us to explain the general features of the method in an orderly manner. Second, the only two-dimensional problem that has been analyzed with

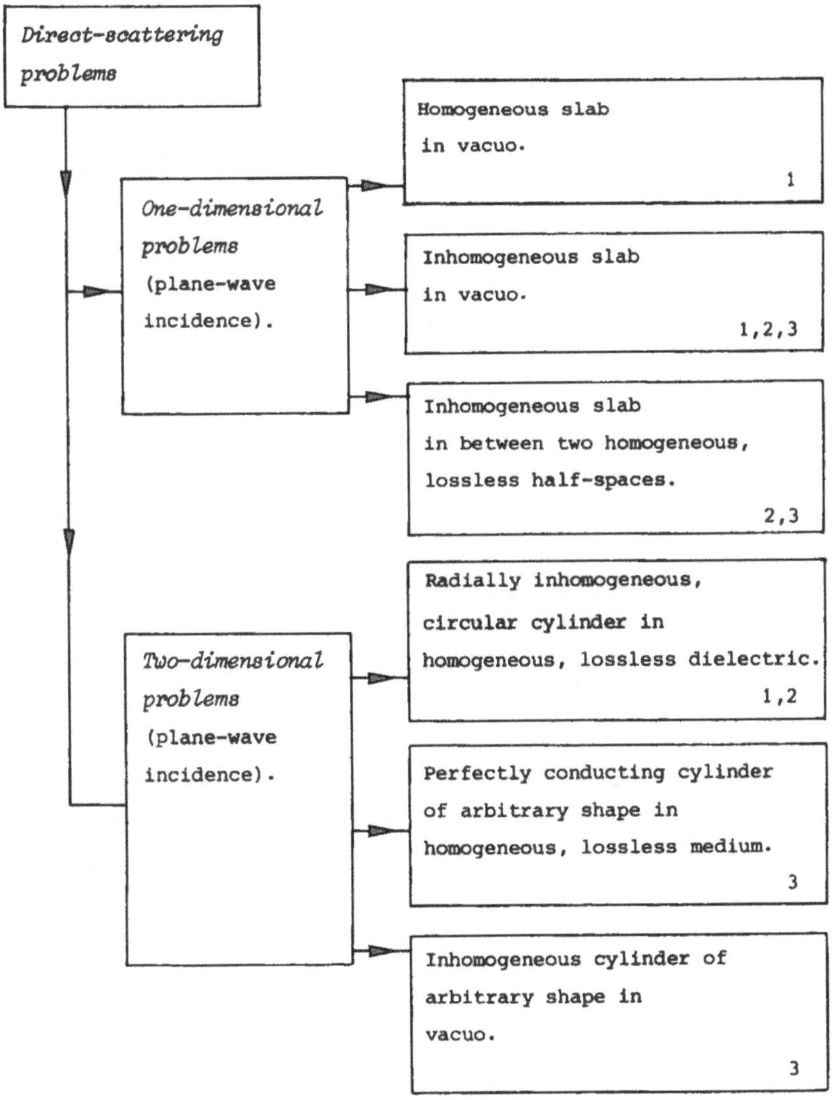

Figure 1.1.2 Schematic view of the direct-scattering problems that are investigated in Chapters 2 and 3. Unless otherwise indicated, the scattering obstacles consist of a lossy dielectric medium with permittivity $\varepsilon(\underline{x}) \geq \varepsilon_0$, conductivity $\sigma(\underline{x}) \geq 0$, and permeability $\mu(\underline{x}) = \mu_0$, where ε_0 and μ_0 denote the permittivity and permeability in vacuo, respectively. The numbers in the lower right corners of the rectangles relate to the methods employed to resolve the respective problems; they correspond to those in Figure 1.1.1.

the aid of frequency-domain techniques is the scattering by a radially inhomogeneous, lossy dielectric circular cylinder. The reasoning behind this restriction is that the circularly symmetric problem can be decomposed into a series of one-dimensional scattering problems, each of which is amenable to the solution procedure developed for the slab problem. Thus, we are able to come up with conditions for closing the Bromwich contour in the left half-plane as well as with an accurate numerical solution for the frequency-domain direct-scattering problem. Moreover, the symmetry in the configuration facilitates the physical interpretation of the two types of ordering observed in the natural-frequency spectrum. For the marching-on-in-time method, which is more of the numerical "brute-force" type, no restrictions need be imposed on the symmetry of the configuration.

1.2 Inverse-scattering problems

In the second and third parts of this monograph, we investigate how the knowledge gathered in solving the direct-scattering problem can be used to tackle the inverse-scattering problem. In view of the complexity of this problem, we have chosen to restrict ourselves to considering the one-dimensional configuration of an inhomogeneous, lossy dielectric slab embedded in a vacuum. For this configuration, we assume that we know the reflected and/or transmitted fields resulting from an incident pulse of finite duration at one or two directions of incidence, or, alternatively, the corresponding real-frequency spectra. We will investigate both the identification problem of recovering the natural frequencies and the inverse-profiling problem of reconstructing the susceptibility and/or the conductivity profile of the unknown slab.

A schematic view of the methods employed to resolve these problems is presented in Figure 1.2.1. As indicated in this figure, we start off by briefly analyzing the identification problem. To this end, we try to determine natural-mode parameters from the known reflected or transmitted fields in the slab configuration at hand. As mentioned in Section 1.1, this configuration can then be identified from the natural frequencies obtained. The determination of the modal param-

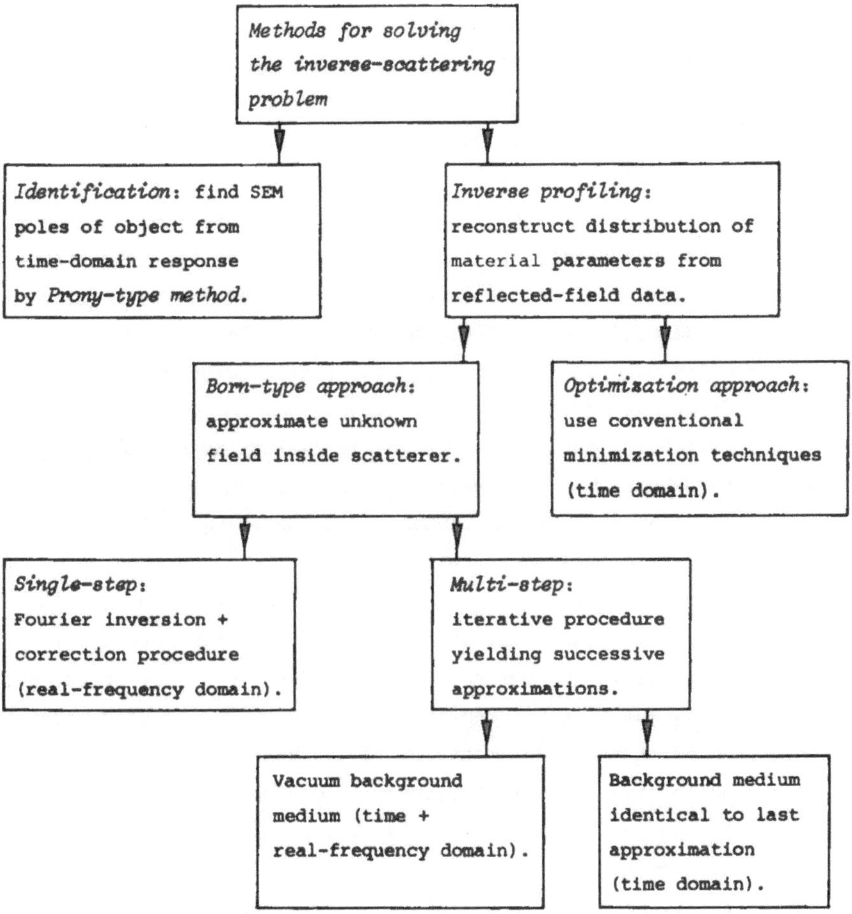

Figure 1.2.1 Schematic view of the methods for solving one-dimensional inverse-scattering problems that are discussed in this monograph. Where relevant, it has been indicated whether time-domain or frequency-domain reflected-field data have been used.

eters will be discussed in Chapter 4. A preprocessing procedure that adapts the known time signals to the limitations of *Prony's algorithm* will be proposed. This procedure is based on the *a priori* knowledge that we are dealing with a slab configuration, and results in a time sequence from which the desired natural frequencies can indeed be recovered. We will not consider the possible reconstruction of the slab configuration from the modal parameters obtained.

Instead, we consider in Chapters 5 and 6 the reconstruction of this configuration directly from the known reflected fields. In these chapters, we discuss the reconstruction of the susceptibility profile and/or the conductivity profile from the given time or frequency behavior of the reflected field at one or two directions of incidence. Both the situations where one and two profiles are unknown will be considered. It will be argued that we can reconstruct a single profile from the given reflected field at a single direction of incidence, provided that the other profile is known. For the reconstruction of both profiles, reflected-field data at two independent directions of incidence are necessary.

As Figure 1.2.1 shows, we can basically follow two different approaches to solving the inverse-scattering problem. The most direct one is the so-called *optimization approach*, where we apply a conventional minimization procedure to find the smallest possible value of a so-called *cost function*. In our case, this cost function consists of the integrated squared deviation between the known actual field and the field resulting from the incident pulse in a known reference configuration, and of small regularization terms that supply missing information about the local behavior of the unknown profile(s). This information is missing from the given reflection data because of the band limitation in the incident field. The principal advantage of the optimization approach is that we can use *all* the reflected-field data available. The principle disadvantage is that, owing to the general nature of the minimization methods employed, rather a large number of approximate direct-scattering problems must be resolved numerically to obtain an acceptable profile reconstruction. As a consequence, the optimization approach is relatively time-consuming. For this reason, we have restricted its application to that domain for which the fastest direct-scattering algorithm was available, i.e. the time domain. This is not an undue restriction, since, by Parseval's theorem, any frequency-domain integrated squared error can be reduced to a time-domain equivalent.

More dedicated procedures for solving the inverse-profiling problem result from the so-called *Born-type approach*. The basic idea in this approach is to draw up a source-type integral representation for the difference between the actual reflected field and the reflected field excited in some known *background medium*. In such a representation, the integration runs over the interior of the scattering obstacle. The integrand is a product of a known Green's function, a contrast function representing the difference in material parameters between the actual configuration and the background medium, and the total field inside the actual configuration. Of these three quantities, both the contrast function and the actual total field inside the obstacle are unknown. In the integral representation obtained, we apply the *Born-type* approximation of replacing the unknown total field by the field that would be excited in some *reference medium*. It should be stressed that this reference medium may differ from the background medium. The Born-type approximation reduces the integral representation for the reflected-field difference to an approximate linear integral equation of the first kind for the unknown contrast function. Solving this integral equation in a least-squares sense then yields an approximation to the unknown profile(s).

In our work, the Born-type approximation is used in two ways. In the first one, we attempt to reconstruct the unknown susceptibility and/or conductivity in a single step. In this case, both the background medium and the reference medium are taken as a vacuum. This results in an identification of the frequency-domain reflected field in terms of spectral components of the unknown profile(s). Hence, the configuration can be reconstructed by a straightforward Fourier inversion. For small values of the susceptibility and the conductivity in the unknown configuration, this procedure directly yields an acceptable reconstruction. For larger values, non-linear correction procedures must be applied.

In our second Born-type method for solving the inverse-scattering problem, the relevant approximation is incorporated in an iterative procedure. In each iteration step, the reference configuration is taken identical to the estimate of the unknown configuration obtained in the previous step. Thus, we alternately obtain successive approximations to the unknown total field and the unknown material properties inside the scattering obstacle. The reconstructive capacity of the iterative procedure depends greatly on the particular choice of the background medium made in drawing up the integral representation for the reflected field. As indicated in Figure 1.2.1, two possible choices will be considered. The most simple formulation and numerical implementation are obtained when we choose the background medium to be a vacuum. For that choice, however, the reconstructed profiles suffer from a high-frequency band limitation which is inherent in the Born-type approximation. This inherent band limitation manifests itself independently of the one imposed by a possible restriction in the frequency content of the incident pulse; it shows up even for wide-band reflected-field data. With our second choice, we remove the band limitation caused by the error in the Born-type approximation at the cost of a more involved formulation and numerical implementation. In each iteration step, we now take the background medium identical to the reference medium, i.e. the previously obtained approximate reconstruction. When the reflected-field data are given in the frequency domain, this modification still does not suffice to remove the band limitation. When the data are given in the time domain, however, an appropriate time window may be imposed. On the one hand, this time window leaves sufficient reflected-field information to obtain a complete reconstruction. On the other hand, it removes a sufficient fraction of the Born error to make the inherent band limitation in the reconstruction disappear as the iterative procedure converges.

As in the optimization approach, we may have an additional imperfection in the recovered profile(s) due to a possible restriction in the frequency content of the incident field. As mentioned earlier in this subsection, such an imperfection can only be relieved by supplying suitable *a priori* information about the local behavior of the unknown configuration. In fact, the choice of the proper Born-type method may depend on the amount of reflected-field information available.

The inverse-profiling part of this monograph is organized as follows. In Chapter 5, we first consider some general aspects of the Born-type approximation. Next, we apply both the single-step and the multi-step version to two special one-dimensional frequency-domain inverse-scattering problems. In Chapter 6, we then apply the iterative Born-type approach as well as the optimization approach to some one-dimensional time-domain inverse-scattering problems.

1.3 Basic equations

An introductory chapter like the present one would not be complete without a summary of the equations that govern the behavior of the relevant wave phenomena. In this section, therefore, we summarize the equations for the wave phenomenon of interest in the present monograph, i.e. the propagation of electromagnetic waves in matter. We will restrict ourselves to giving the general time-domain version of these equations. Versions for special geometries and for frequency-domain problems will be given in the text as needed. In our formulation, we employ the International System of Units (Système International d'Unités), abbreviated SI, for expressing the relevant physical quantities.

The equations describing the behavior of electromagnetic fields in matter can be written as

$$
\begin{aligned}
-\nabla \times \underline{H}(\underline{x},t) + \underline{J}(\underline{x},t) + \partial_t \underline{D}(\underline{x},t) &= -\underline{J}^e(\underline{x},t), \\
\nabla \times \underline{E}(\underline{x},t) + \partial_t \underline{B}(\underline{x},t) &= -\underline{J}^m(\underline{x},t).
\end{aligned}
\tag{1.3.1}
$$

In this form, the electromagnetic-field equations are known as *Maxwell's equations* in matter. In (1.3.1), we encounter the physical quantities

$\underline{x}$ = Cartesian position vector (m),
t = time coordinate (s),
$\underline{E}(\underline{x},t)$ = electric-field strength (V/m),
$\underline{H}(\underline{x},t)$ = magnetic-field strength (A/m),
$\underline{J}(\underline{x},t)$ = volume density of induced electric current (A/m^2),
$\underline{D}(\underline{x},t)$ = electric-flux density (C/m^2),

$\underline{B}(\underline{x},t)$ = magnetic-flux density (T),
$\underline{J}^e(\underline{x},t)$ = volume density of external electric current (A/m^2),
$\underline{J}^m(\underline{x},t)$ = volume density of external magnetic current (V/m^2).

In addition, we have introduced the gradient operator

$$\nabla \triangleq \underline{i}_1\partial_1 + \underline{i}_2\partial_2 + \underline{i}_3\partial_3, \tag{1.3.2}$$

where $\{\underline{i}_1, \underline{i}_2, \underline{i}_3\}$ form the basis vectors of the Cartesian coordinate system at hand, and where ∂_p denotes the partial derivative with respect to the coordinate x_p ($p = 1,2,3$). Similarly, ∂_t denotes the partial derivative with respect to the time coordinate.

In the right-hand sides of Equation (1.3.1), the external or source current densities $\underline{J}^e$ and $\underline{J}^m$ constitute the active parts, which describe the action of the sources that generate the field. In the present study, where we restrict ourselves to plane-wave incidence, these current densities may be set to zero for any finite value of the position vector $\underline{x}$.

It is customary to separate off the reaction of matter to the presence of an electromagnetic field by breaking the flux densities $\underline{D}$ and $\underline{B}$ up into

$$\begin{aligned} \underline{D}(\underline{x},t) &= \varepsilon_0\underline{E}(\underline{x},t) + \underline{P}(\underline{x},t), \\ \underline{B}(\underline{x},t) &= \mu_0[\underline{H}(\underline{x},t) + \underline{M}(\underline{x},t)], \end{aligned} \tag{1.3.3}$$

where

ε_0 = permittivity in vacuum (F/m),
μ_0 = permeability in vacuum (H/m),
$\underline{P}(\underline{x},t)$ = electric polarization (C/m^2),
$\underline{M}(\underline{x},t)$ = magnetization (A/m).

The value of μ_0 is fixed by SI as $\mu_0 = 4\pi \times 10^{-7}$ H/m; the value of ε_0 follows from $\varepsilon_0 = 1/\mu_0 c_0^2$, where $c_0 = 299792458$ m/s is the speed of electromagnetic waves in vacuum.

Substitution of (1.3.3) in (1.3.1) leads to

$$
\begin{aligned}
-\nabla \times \underline{H}(\underline{x},t) + \varepsilon_0 \partial_t \underline{E}(\underline{x},t) + \underline{J}(\underline{x},t) + \partial_t \underline{P}(\underline{x},t) &= -\underline{J}^e(\underline{x},t), \\
\nabla \times \underline{E}(\underline{x},t) + \mu_0 \partial_t \underline{H}(\underline{x},t) + \mu_0 \partial_t \underline{M}(\underline{x},t) &= -\underline{J}^m(\underline{x},t).
\end{aligned}
\tag{1.3.4}
$$

In vacuum, the induced quantities $\underline{J}$, $\underline{P}$ and $\underline{M}$ vanish, and, consequently, the electromagnetic-field equations given in (1.3.4) suffice for a complete description of the electromagnetic field. In matter, the physical properties of the medium under consideration remain to be specified. To this end, Maxwell's equations in matter must be supplemented by three vectorial relations between the five field quantities $\underline{J}$, $\underline{P}$, $\underline{M}$, $\underline{E}$ and $\underline{H}$ occurring in (1.3.4). These supplementary relations are known as the *constitutive relations*; they are representative of the electromagnetic response of a piece of matter to the presence of an electromagnetic field. In that respect, they are characteristic of the material properties of the configuration, and, hence, of particular interest in inverse-scattering problems.

It lies outside the scope of the present manuscript to discuss all possible forms of these constitutive relations and their physical background. Hence, we confine ourselves to specifying the two forms that are most relevant in practical applications. For a general, inhomogeneous, linear, isotropic, locally reacting, time-invariant medium we have

$$
\begin{aligned}
\underline{J}(\underline{x},t) &= \int_0^\infty \kappa^c(\underline{x},t')\underline{E}(\underline{x},t - t')dt', \\
\underline{P}(\underline{x},t) &= \varepsilon_0 \int_0^\infty \kappa^e(\underline{x},t')\underline{E}(\underline{x},t - t')dt', \\
\underline{M}(\underline{x},t) &= \int_0^\infty \kappa^m(\underline{x},t')\underline{H}(\underline{x},t - t')dt',
\end{aligned}
\tag{1.3.5}
$$

respectively, in which

$\kappa^c(\underline{x},t)$ = conduction relaxation function (S/ms),
$\kappa^e(\underline{x},t)$ = dielectric relaxation function (s^{-1}),
$\kappa^m(\underline{x},t)$ = magnetic relaxation function (s^{-1}).

Note that, in (1.3.5), causality has been enforced by restricting the integration to the interval $0 \le t' < \infty$. This implies that the frequency-domain equivalent of the relaxation functions introduced above automatically satisfies the Kramers-Kronig relations.

Depending on the specific physical properties of the medium under consideration, it may be possible to simplify the constitutive relations specified in Equation (1.3.5). For instance, when the medium is reacting instantaneously, we have

$$\begin{aligned} \kappa^c(\underline{x},t) &\triangleq \sigma(\underline{x})\delta(t), \\ \kappa^e(\underline{x},t) &\triangleq \chi^e(\underline{x})\delta(t), \\ \kappa^m(\underline{x},t) &\triangleq \chi^m(\underline{x})\delta(t), \end{aligned} \tag{1.3.6}$$

where

$\sigma(\underline{x})$ = conductivity (S/m),

$\chi^e(\underline{x})$ = electric susceptibility (dimensionless),

$\chi^m(\underline{x})$ = magnetic susceptibility (dimensionless).

Substitution of the definitions given in (1.3.6) in the general expressions (1.3.5) yields

$$\begin{aligned} \underline{J}(\underline{x},t) &= \sigma(\underline{x})\underline{E}(\underline{x},t), \\ \underline{P}(\underline{x},t) &= \varepsilon_0\chi^e(\underline{x})\underline{E}(\underline{x},t), \\ \underline{M}(\underline{x},t) &= \chi^m(\underline{x})\underline{H}(\underline{x},t). \end{aligned} \tag{1.3.7}$$

Finally, we can now substitute the result obtained in (1.3.7) in the relations given in (1.3.3). We end up with

$$\begin{aligned} \underline{D}(\underline{x},t) &= \varepsilon(\underline{x})\underline{E}(\underline{x},t) = \varepsilon_0\varepsilon_r(\underline{x})\underline{E}(\underline{x},t), \\ \underline{B}(\underline{x},t) &= \mu(\underline{x})\underline{H}(\underline{x},t) = \mu_0\mu_r(\underline{x})\underline{H}(\underline{x},t), \end{aligned} \tag{1.3.8}$$

in which we have introduced

$\varepsilon(\underline{x})$ = (absolute) permittivity (F/m),

$\varepsilon_r(\underline{x})$ = relative permittivity (dimensionless),

$\mu(\underline{x})$ = (absolute) permeability (H/m),

$\mu_r(\underline{x})$ = relative permeability (dimensionless).

Comparing the corresponding expressions shows that

$$\varepsilon_r(\underline{x}) \triangleq 1 + \chi^e(\underline{x}),$$
$$\mu_r(\underline{x}) \triangleq 1 + \chi^m(\underline{x}), \qquad (1.3.9)$$

The special constitutive relations given in (1.3.7) and (1.3.8) hold for a wide class of materials, particularly when the electromagnetic fields do not vary too rapidly in time.

From a methodological point of view, it seems attractive not to allow an undue amount of freedom in the formulation of the constitutive relations. For instance, we must expect more severe non-uniqueness problems in reconstructing the space-time distribution of one or more relaxation functions as introduced in (1.3.5) than in reconstructing the spatial distributions of the corresponding constitutive parameters as introduced in (1.3.6). In this monograph, we have therefore restricted the considerations to instantaneously reacting dielectric materials with $\mu(\underline{x}) = \mu_0$, i.e. $\chi^m(\underline{x}) = 0$. In our numerical experiments, we have further confined the electric susceptibilities to the range $0 \leq \chi^e \leq 10$, and the conductivities to the range $0 \leq Z_0\sigma d \leq 10$, where $Z_0 = (\mu_0/\varepsilon_0)^{\frac{1}{2}}$ and where d denotes a characteristic distance across the configuration. These ranges have been chosen such that they cover the greater part of the materials for which the constitutive relations (1.3.7) and (1.3.8) are valid. It should be remarked that the restrictions imposed on χ^e and σ are not essential; there is no mathematical or numerical reason why our solution techniques could not be employed for larger values of these parameters as occurring in geophysical applications.

At the interface between two media differing in their electromagnetic properties, the electric- and magnetic-field strengths $\underline{E}(\underline{x},t)$

and $\underline{H}(\underline{x},t)$ will in general vary discontinuously. Due to this discontinuous behavior, the electric-field equations given in (1.3.1) lose their significance at the interface. Hence, these equations must be supplemented by conditions that connect the field values at both sides of the interface, the so-called *boundary conditions*.

To formulate these conditions, we consider an interface S between two domains $\mathcal{D}_1$ and $\mathcal{D}_2$ that contain matter with different material properties (see Figure 1.3.1). The properties of the media in both domains and the location of the interface are assumed to be time invariant. Moreover, it is assumed that S has everywhere a unique tangent plane, and, accordingly, a unit vector $n(\underline{x})$ normal to that plane and pointing into $\mathcal{D}_1$. Now, let $\underline{x}$ be the position vector of some point on S, and let $\{\underline{E}_1(\underline{x},t),\underline{H}_1(\underline{x},t)\}$ and $\{\underline{E}_2(\underline{x},t),\underline{H}_2(\underline{x},t)\}$ represent the limiting values of the electromagnetic field that are observed upon approaching $\underline{x}$ from $\mathcal{D}_1$ and $\mathcal{D}_2$, respectively. Then we have three types of fundamental boundary conditions according to what type of matter is present in $\mathcal{D}_1$ and $\mathcal{D}_2$.

In the general case, both an electric and a magnetic field are present in both $\mathcal{D}_1$ and $\mathcal{D}_2$. Consequently, both media are denoted as electrically and magnetically penetrable. Consistency of the integrated form of the field equations now requires that the components of both fields that are tangential to the interface be continuous

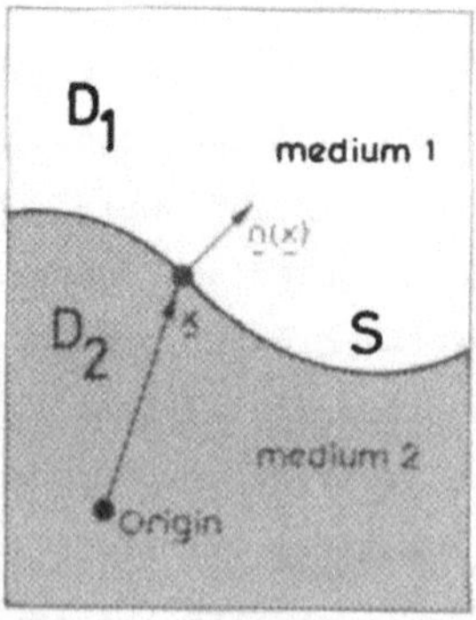

Figure 1.3.1 Interface S between two domains $\mathcal{D}_1$ and $\mathcal{D}_2$ with different material properties.

upon crossing the interface, i.e.

$$\underline{n}(\underline{x}) \times [\underline{H}_1(\underline{x},t) - \underline{H}_2(\underline{x},t)] = \underline{0},$$
$$\underline{n}(\underline{x}) \times [\underline{E}_1(\underline{x},t) - \underline{E}_2(\underline{x},t)] = \underline{0}. \qquad (1.3.10)$$

These two equations are the fundamental boundary conditions at the interface of two penetrable media.

The boundary conditions given in (1.3.10) assume a special form when one of the media cannot sustain in its interior a nonidentically vanishing electric or magnetic field, while the boundary condition of the continuity of the relevant field strength is maintained. Such a medium is denoted as electrically or magnetically impenetrable. Electrically impenetrable materials arise as limiting cases of materials whose conductivity and/or permittivity go to infinity. When $\mathcal{D}_2$ is electrically impenetrable, the second line of (1.3.10) reduces to

$$\underline{n}(\underline{x}) \times \underline{E}_1(\underline{x},t) = \underline{0}, \qquad (1.3.11)$$

for any $\underline{x}$ on the interface $\mathcal{S}$. It is not allowed to prescribe boundary conditions for the tangential part of the magnetic-field strength in this case. In fact, the tangential part of the magnetic-field strength will, in general, exhibit a discontinuity at the interface $\mathcal{S}$. Similarly, a magnetically impenetrable material arises as a limiting case of a material whose permeability goes to infinity. When such a material is present in $\mathcal{D}_2$, we have the boundary condition

$$\underline{n}(\underline{x}) \times \underline{H}_1(\underline{x},t) = \underline{0}, \qquad (1.3.12)$$

while no condition may be prescribed for the tangential part of the electric-field strength.

In combination, the electromagnetic-field relations, constitutive relations, and boundary conditions suffice for a complete description of the electromagnetic field in any configuration consisting of linear, isotropic, locally reacting, time-invariant matter.

1.4 General considerations

To conclude the introductory chapter, we mention in this section a few general considerations that have played a role throughout the present study of direct- and inverse-scattering problems. These considerations have influenced the selection of the methods of solution as well as their implementation, application and presentation.

The most important consideration in selecting the three methods employed in solving the direct-scattering problems has been that, in their formulation, the reflected field outside the scattering obstacle should in some way be connected to characteristic features of that scatterer. As explained in Section 1.3, either the natural frequencies of the configuration or the spatial distribution of relevant constitutive parameters may qualify to characterize the scattering obstacle.

A second consideration has been that we must exclude the possibility of the solution of the inverse problem being biased by numerical errors made in solving the corresponding direct-scattering problem. Two types of precautions have been taken to overcome this difficulty. The first one also concerns the selection of the methods employed to solve the direct-scattering problem at hand. We only considered methods that could be analyzed in full detail, and for which the accuracy of the results obtained can be predicted. In particular, it must be possible to indicate how, by spending some more computation time, this accuracy can be improved. Thus, we can ensure that the numerical errors made in solving the successive approximate direct-scattering problems are negligible compared to the approximation errors made in solving the inverse problem. The second precaution consists of generating the "known" reflected-field data by a direct-scattering algorithm different from the one employed in solving the inverse-scattering problem. This type of precaution arises naturally on one particular occasion, viz. in Section 6.4.

In formulating the respective solution techniques, for the direct-scattering problem as well as for the inverse-scattering problem, an important consideration has been to maintain their capability of being generalized to more complicated configurations as much as possible. To this end, we have attempted to formulate these techniques independently of the particular applications that we had in mind. In Chapter 2, this was realized by applying the singularity expansion method to a simple scattering problem for which the entire analysis can be carried out in closed form. In the remaining chapters, general formulations of the relevant solution techniques are given from which their specific advantages and disadvantages can be observed without considering individual problems in detail.

A consequence of this approach is that the conclusions obtained from the abstract formulation must be verified by computing numerical results for one or more example problems. All techniques proposed in the present monograph have been applied to at least one example problem. Details about the numerical implementation are supplied for each application. In particular, it is indicated what type of discretization is employed, and whether part of the problem can be solved by using software available from standard subroutine libraries. Of each application, representative numerical results are presented and discussed which demonstrate the specific consequences of using the method of solution in the problem at hand. Furthermore, typical accuracies and computation times are specified.

In order to enable a meaningful comparison of the computational effort required in solving a single example problem by different methods of solution, all computation times have been reduced to one type of computer. In particular, this concerns an IBM 3083/JX1 with a capacity of about 8 MIPS. Even when the computations were actually carried out on a different machine, an appropriate scaling of the computation times was enforced. All computations were performed in standard double precision, i.e. with an accuracy of 16 digits.

Some consideration should also be devoted to the application of approximation techniques. In our analysis, a number of occasions will arise where we cannot resolve (part of) the problem of interest rigorously. We mention in particular Chapter 3, where we wish to analyze a few selected eigenvalues out of an infinite set, and Chapters 2 and 5, where we need closed-form expressions for electromagnetic-field distributions that can only be determined numerically.

On such occasions, we will use approximation techniques to obtain at least a qualitative understanding of the answer to the relevant problem. Whenever the need for applying such approximation techniques arises, we will verify the trend predicted with their aid by performing suitably chosen numerical experiments.

Obviously, the considerations given above also have a bearing on the organization of the presentation. Each chapter starts with an introductory section. In that section, the existing literature is reviewed and the method or methods employed are placed within the context of that literature as well as within the framework of the present monograph. The introductory section is followed by a section which discusses the general aspects of the method or methods at hand. Subsequently, we consider the application to a number of example problems, and, in the final section, the conclusions are stated. The only chapter that does not meet this description is Chapter 6, where essentially the same method of solution is used as in Chapter 5.

DIRECT SCATTERING

2. FREQUENCY-DOMAIN TECHNIQUES

2.1 Introduction

In this chapter, we consider solving the direct-scattering problem of a pulsed, polarized electromagnetic plane wave, normally incident on a one- or two-dimensional obstacle with the aid of frequency-domain techniques. These techniques have in common the feature that the transient response of the obstacle is represented as a Laplace inversion integral over a Bromwich contour in the right half of the time-Laplace domain (s-plane). The main advantage of such a representation is that in the equation(s) governing its constituents (the Laplace-transformed field quantities), the partial time differentiation from the corresponding time-domain equation(s) is replaced by a scalar multiplication with the complex frequency s. On the other hand, the evaluation of a single time-domain response requires the solution of several frequency-domain problems. Basically, there are two different ways in which frequency-domain results can be combined into a transient response, both characterized by the way in which the Laplace inversion integral is evaluated.

In the *singularity expansion method*, commonly referred to as SEM, this integral is evaluated by contour deformation in either the right or the left half of the complex-frequency plane, according to whether the integral along the closing contour in the relevant half-plane vanishes. When the contour can be closed to the right, the field represented by the inversion integral is zero, since the Laplace transform of a causal signal is regular in the right half of the complex-frequency plane. Upon closure to the left, we end up with a field representation in terms of the contributions from the singularities in the complex s-plane, i.e. poles and, possibly, branch points. Each pole is then identified as a natural frequency s_α of the scatterer, with α being a subscript assigned according to some numbering system. The corresponding natural mode is of the form

$$\underline{E}_\alpha(\underline{x})\exp(s_\alpha t), \qquad (2.1.1)$$

with $\underline{E}_\alpha(\underline{x})$ being the natural-mode field distribution, and constitutes, along with the corresponding magnetic field, a nontrivial solution to the source-free electromagnetic-field equations. The incident field comes into the residual contribution in two ways: its spatial distribution (e.g. the angle of incidence of a plane wave) is needed, along with the object properties and $\underline{E}_\alpha(\underline{x})$, in the evaluation of the coupling coefficient D_α which determines the strength of the natural mode; the Laplace transform of its pulse shape $F(t)$ comes in as an overall multiplication factor. With these two factors, the total residual contribution from the poles assumes the form

$$\sum_\alpha F(s_\alpha)D_\alpha\underline{E}_\alpha(\underline{x})\exp(s_\alpha t). \tag{2.1.2}$$

Contributions from branch cuts in the complex-frequency plane, when present, are handled in a similar manner. It turns out that they can be incorporated in the representation (2.1.2) by allowing the index α to run over continuous ranges as well, and identifying the term $D_\alpha\underline{E}_\alpha(\underline{x})$ as being the jump discontinuity across the branch cut. It should be remarked, however, that such a spectrum is not unique: a different choice of the branch cut will generally result in a different set of discrete poles, with corresponding natural modes. We will resolve this difficulty by requiring that the Schwarz reflection principle

$$\underline{E}(\underline{x},s^*) = \underline{E}^*(\underline{x},s), \tag{2.1.3}$$

which holds in the right half of the s-plane for each $\underline{E}(\underline{x},s)$ corresponding to a real-valued causal signal, also hold for its analytical continuation into the left half-plane.

A further illustration of the SEM is given in Figure 2.1.1, which shows a flow diagram of a calculation performed according to this method. In this figure, the dependence of the relevant quantities on the object geometry, on the incident field, and on each other has been indicated. It is observed that, in the calculation of the response of more than one incident field, at most the coupling coefficients need to be recomputed.

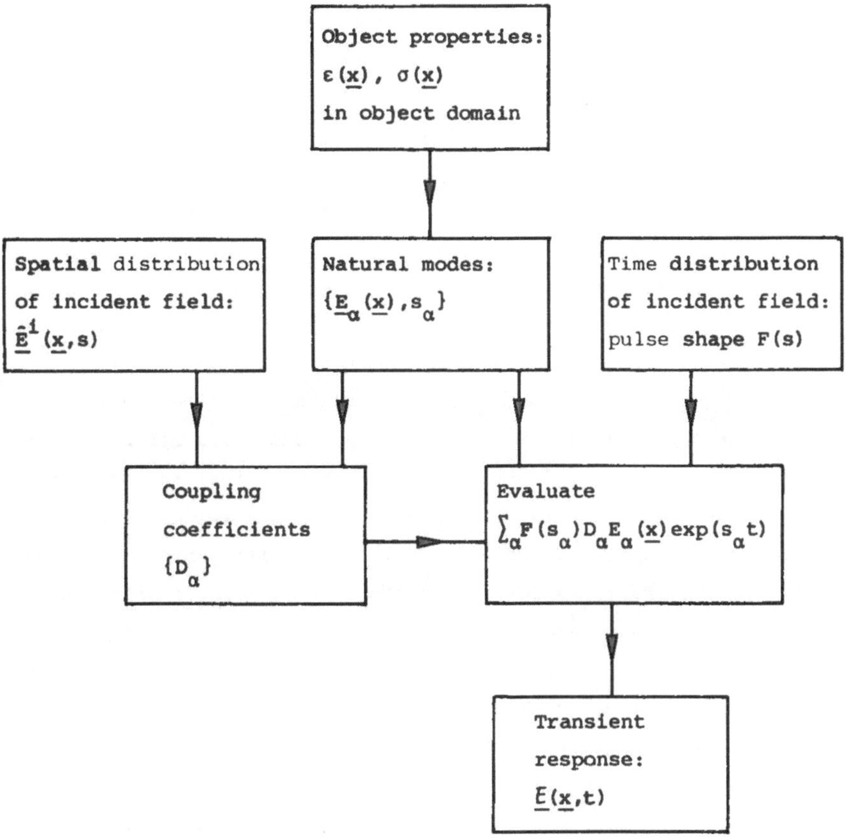

Figure 2.1.1 Flow diagram of the calculation of a transient field according to the singularity expansion method. In the diagram, $\underline{x}$ denotes a Cartesian position vector, t the time coordinate, s_α a complex natural frequency and $\underline{E}_\alpha(\underline{x})$ the field distribution of a natural mode. The summation is understood to run over both the discrete and the continuous part of the natural-frequency spectrum, when present.

Natural-mode considerations of the type sketched above have played a role earlier in the theory of forced oscillations of open resonators (see Weinstein (1969), Blok (1970)). The most simple example of such a resonator is the domain exterior to a perfectly conducting sphere, for which the natural oscillations were already analyzed by Stratton (1941). In that context, the natural modes were mainly used to explain the resonance phenomena that show up in the frequency-domain response. Similar considerations were also employed by Überall, Gaunaurd, et al. in their theory of resonance scattering, which they have applied to a wide range of electromagnetic, acoustic, and elastodynamic scattering problems (an overview was given in Flax et al. (1981)). Other references to early as well as recent work on natural resonances can be found in the extensive bibliography on the SEM and related topics compiled by Michalski (1981). The first formal treatment of a transient scattering problem with the aid of natural-mode methods was given by Baum (1971), who analyzed the plane-wave scattering by a perfectly conducting sphere. Since then, numerous papers dealing with the mathematical and physical aspects of the method as well as with its application to specific problems have appeared. We mention in particular Marin (1973), who outlined the procedure for determining the coupling coefficient D_α of a given, nondegenerate natural mode $\{\underline{E}_\alpha(\underline{x}), s_\alpha\}$ from the s-domain direct-scattering integral equation. Insight into the development and the application of the SEM can be obtained from the reviews authored by Baum (1976a,b, 1978) and from the special issues edited by Pearson and Marin (1981) and by Langenberg (1983). The latter issue includes contributions by Bollig and Langenberg and by Pearson, who surveyed work in elastodynamics and electromagnetics, respectively.

One question which has remained unresolved for a long time is that of the vanishing or nonvanishing contributions from the closing contours at infinity. These contributions are associated with the entire function that completes the Mittag-Leffler representation for the Laplace-transformed fields. Only recently (Tijhuis and Blok (1980, 1984a,b), Morgan (1984, 1985), Pearson (1984), Felsen (1984, 1985), Dudley (1985)), the inadequacy of field descriptions solely in terms of modal contributions has been emphasized widely in the literature. In order to determine whether the contribution from the closing contour vanishes in the left or the right half of the complex-frequency plane, at least a first-order high-frequency approximation to $\underline{E}(\underline{x},s)$ is needed, which holds over the entire range of arg(s). In this chapter, we consider two configurations for which such approximations are available, namely a lossy, inhomogeneous dielectric slab and a lossy, radially inhomogeneous dielectric circular cylinder, both embedded in vacuum. In these configurations, the field inside the scatterer is asymptotically represented by a first-order WKB approximation. Matching this approximation to the known representation in the surrounding medium then yields a suitable high-frequency expression. With the aid of the expressions thus obtained, it can then be analyzed whether the integral along the closing contour in the right or the left half-plane vanishes. It turns out that, for each space point, there is a finite termination instant, before which the contribution of the closing contour vanishes upon closure to the right. As argued above, this implies that the time-domain field in that interval is, apart from the incident field approaching the scatterer, zero. The termination instant for closure to the right is thus exactly the instant where the influence of the scatterer first becomes noticeable. For each space point, there is also an initial instant, after which the contribution from the closing contour in the left-plane vanishes, and, consequently, the representation (2.1.2) holds. However, there is a nonvanishing spatial region where there is a time interval in between these instants. In that time interval, we have a nonvanishing field that cannot be represented by modal contributions only. It will be shown that the length of the interval corresponds to the period during which the incident pulse is emerging into the scattering obstacle. Consequently, the computation of the integral along a closing contour can, in principle, be avoided by taking into account at each instant only that part of the incident pulse that has emerged into the obstacle. A similar approach has been proposed independently by Morgan (1984), who studied three-dimensional, perfectly conducting targets. A disadvantage of this so-called pulse-truncation approach is that the slow convergence of the resulting modified modal representation compels the application of some acceleration procedure.

For the case of the slab, the procedure outlined above can be applied directly to the relevant Laplace-transformed reflected and transmitted fields. For the case of the cylinder, the problem must first be decomposed into a series of one-dimensional scattering problems, by subjecting the time-domain fields to an angular Fourier transformation. This results in a Fourier representation that converges globally in the spatial domain, uniformly in time. The validity of the SEM can then be established by analyzing each Fourier coefficient individually. The decomposition into one-dimensional problems also supplies a consistent way to truncate the summation of the natural-mode contributions. Moreover, the modes can be subdivided into families according to their angular order.

Instead of evaluating the Laplace inversion integral by contour deformation, one can also compute it directly with the aid of a Fast-Fourier-Transform algorithm. From now on, we will designate this procedure as *direct Fourier inversion*. A flow diagram of such a computation is given in Figure 2.1.2. Comparing this figure with Figure 2.1.1 shows that, for each new choice of the incident field distribution, the total computation should be repeated. Such a situation occurs, for example, when the plane-wave response of an obstacle should be calculated for varying angle of incidence. An advantage is that the Fourier representation employed is also valid at instants where the SEM representation does not hold. For both configurations that are analyzed in this chapter with the aid of the SEM, also the direct Fourier inversion approach will be investigated.

The major attraction of the SEM is that it leads in an elegant way to a physical interpretation of the process of scattering by an obstacle. Such an interpretation is not merely restricted to the fact that the incident field excites natural modes. A more profound physical interpretation follows from analyzing the manner in which the natural frequencies corresponding to the modes are grouped in the complex-frequency plane. Such an analysis was first given by Überall and Gaunaurd (1981), for the case of radar scattering by a perfectly conducting sphere. For that configuration, it was shown that the poles can be ordered into families such that the contribution from a family of poles constitutes a creeping wave that circumnavigates the sphere during the scattering process. As such, the analysis was a natural extension to their theory of resonance scattering.

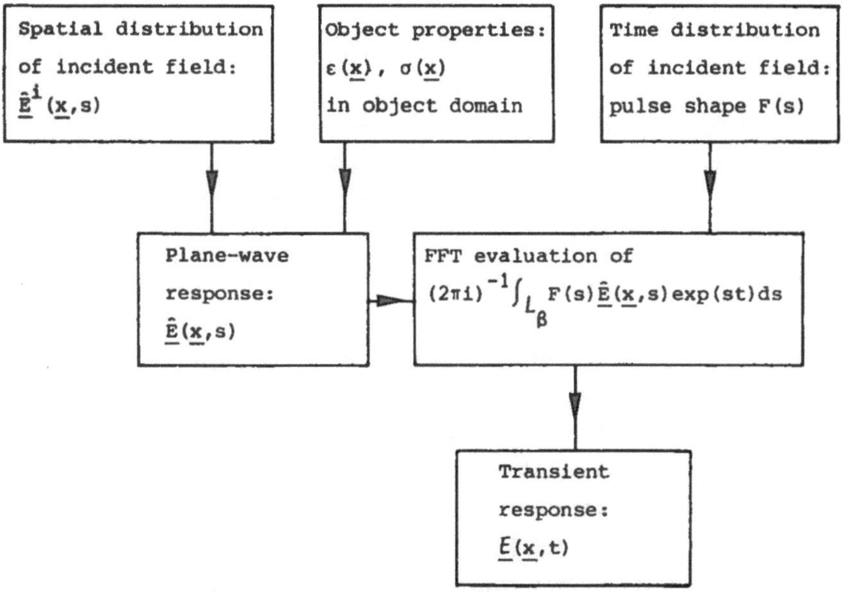

Figure 2.1.2 Flow diagram of the calculation of a transient field via Fourier transformation of the corresponding frequency-domain result. L_β denotes the Bromwich contour, other symbols as given in Figure 2.1.1.

Since then, numerous configurations have been analyzed in a similar manner (see e.g. Howell and Überall (1984), Überall and Gaunaurd (1984), Gaunaurd and Werby (1985) and references cited therein). As far as electromagnetic problems are concerned, the application has primarily been restricted to examples of rotationally symmetric, perfectly conducting targets, with or without a dielectric coating. We mention in particular the work of Heyman and Felsen (1983, 1985a), who considered a perfectly conducting cylinder. All these examples have in common the feature that the incident field does not penetrate into the interior of the scattering obstacle; i.e. that only "external" modes are excited.

For the cylinder considered in this chapter, the incident field does cause a response inside the obstacle. In order to interpret this response, we will consider the slightly more general configuration, where the surrounding medium is a homogeneous, lossless dielectric, from which a plane wave hits the cylinder. It will be shown that, for that configuration, the modes can be subdivided into whispering-gallery and creeping-wave modes, which propagate on the inside and the outside, respectively, along the cylinder boundary. The interpretation is verified by letting either the surrounding medium or the cylinder become impenetrable. It turns out that these limiting cases, in which the propagation only takes place in the complement of the impenetrable region, are included in our formulation. In the case of an impenetrable surrounding medium, which is incompatible with our choice for the incident field, this applies only to the natural modes.

A key point in the analysis is the decomposition of the harmonic response into wave constituents that circumnavigate the scattering obstacle, which is carried out before the response is continued analytically into the complex-frequency plane. One way to arrive at such a decomposition, and the most common one, starts from the separation-of-variables solution. A Watson transformation reduces the summation over the angular order occurring in this solution to a contour integral in the complex-order plane. Then, the desired decomposition is obtained by evaluating that integral by contour deformation in the complex-order plane. For a homogeneous, lossless cylinder, this procedure has been discussed in detail in the literature (see e.g. Franz (1957), Streifer and Kodis (1964a,b, 1965), and Wait (1967)). The limiting case of an impenetrable outer medium, which is important in verifying the identification of the whispering-gallery modes, is described in Wasylkiwskyj (1975) and Ishihara, Felsen and Green (1978). The only work on a radially inhomogeneous cylinder known to the author is that of Miyazaki (1981), who provided a discussion for the optical-frequency range. His approach is based on a high-frequency approximation for the integrand in the complex-order plane, which is obtained by using an asymptotic expression due to McKelvey (1959) for the field inside the cylinder. A disadvantage of this approach is that the asymptotic expression only holds when the complex order is much smaller in magnitude than the frequency. Although this restriction does not hinder the analysis of the open-waveguide modes that propagate along the axis of cylindricity (see Streifer and Kurz (1967)), it is an impediment in the present application. In particular, it prevents the estimation of the contribution from the closing contour at infinity in the complex-order plane. We avoid this difficulty by developing and interpreting a consistent decomposition for the harmonic response as expressed above. An important tool in the analysis is a new type of WKB asymptotics, which results in asymptotic expressions for the field inside the cylinder that are valid as long as at least one of the relevant parameters (the complex angular order and the complex frequency) is large in magnitude. With the aid of these expressions, asymptotic characteristic equations are derived for both types of modes. On the one hand, these equations are simple enough to allow a straightforward physical interpretation of their solutions. On the other hand, a comparison of the results obtained from these equations with numerical solutions of the exact characteristic equation shows that the approximation is surprisingly accurate. A significant characteristic of our approach is that we represent the harmonic response in terms of wave constituents that remain finite at the central axis of the cylinder. From a physical point of view, our representation has the advantage that the angular behavior is symmetric around the direction of propagation of the incident field. From a mathematical point of view, the condition of regularity is essential in selecting the proper solution in both the numerical and the asymptotic determination of the field distribution inside the cylinder. Moreover, our representation has the advantage that the contributions from the closing contours in the complex-order plane can be analyzed directly using the asymptotic expressions obtained and Jordan's lemma.

For the slab, the situation is less involved. In that case, the asymptotic expressions that are used to analyze the validity of the SEM representation directly provide an interpretation of the excited field in terms of repeatedly reflected and transmitted waves. Such an interpretation also holds for each of the one-dimensional scattering problems following from the angular decomposition of the cylinder problem. The solution of each subproblem represents a wave traveling in the radial direction only with a fixed angular distribution. This interpretation is complementary to the one outlined above, where the solution is decomposed in constituents with an almost fixed radial distribution that propagate in the angular direction.

The organization of this chapter is as follows. In order to gain insight into the type of contributions that we can expect in the SEM approach, we first consider in Section 2.2 the case of a homogeneous, lossless slab. In Section 2.3 we investigate the application of the SEM to an inhomogeneous slab. Most of the material treated in these two sections was discussed previously in Tijhuis and Blok (1980, 1984a,b). Compared with those papers, the asymptotic analysis, in particular, has been improved. For the inhomogeneous case, results for more complicated slab configurations are presented and discussed. In particular, the modes of a symmetric Epstein layer are investigated in Appendix 2.3.A. The discussion of the numerical procedure used to determine the frequency-domain field distributions inside the slab is postponed until Section 2.4, where we investigate the direct evaluation of the time-domain fields by Fourier inversion. In that section, we also formulate some integral equations for the electric-field strength inside the slab, which will be useful in the upcoming chapters. In Section 2.5 both methods are applied to the cylindrical scattering problem. As far as the physical interpretation is concerned, we restrict ourselves for the time being to analyzing the radial behavior. The subject matter of this section was treated earlier in Tijhuis and Van der Weiden (1986). As in the case of the slab, additional numerical results are presented and discussed. The decomposition of the excited field in terms of angularly propagating waves and the associated ordering of the natural modes is considered in Section 2.6, which is an enlarged version of Tijhuis (1986). Finally, the conclusions are stated in Section 2.7.

2.2 Scattering by a homogeneous, lossy dielectric slab: singularity expansion method

2.2.1 Formulation of the problem

In this and the following section, we consider the application of the singularity expansion method (SEM) to the transient scattering of a pulsed electromagnetic wave of finite duration by an isotropic, lossy dielectric slab embedded in vacuo (Figure 2.2.1). In order to obtain insight into the problems connected with the application of the SEM, we consider, in the present section, first the case where the slab is homogeneous. In Section 2.3, we then generalize the analysis to the case of an inhomogeneous slab. Anticipating this generalization, we start our analysis by formulating the problem for the inhomogeneous case. A Cartesian coordinate system is introduced, with z as the coordinate normal to the boundaries of the slab and x and y as coordinates parallel to them. The configuration then consists of three domains $\mathcal{D}_i$ (with i = 1, 2, 3) as indicated in Table 2.2.1, where d denotes the thickness of the slab, and ε_0, μ_0 denote the permittivity and permeability in vacuo, respectively. From $\mathcal{D}_1$, a linearly polarized electromagnetic pulse of finite duration T is normally incident on the slab (the case of oblique incidence is a straightforward extension). Then the problem is one-dimensional and the field intensities are functions of z and t only. We write the electric-field intensity and the magnetic-field intensity of the incident field as

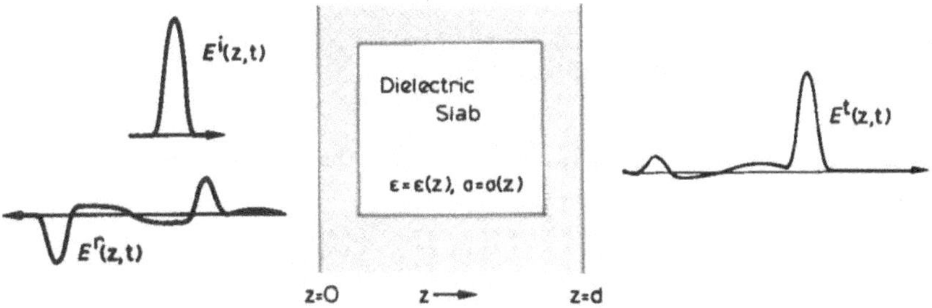

Figure 2.2.1 A pulsed plane wave normally incident on an inhomogeneous, lossy dielectric slab embedded in vacuo.

Table 2.2.1 Subdivision of the configuration into domains.

domain	z-coordinate	permittivity	conductivity	permeability
$\mathcal{D}_1$	$-\infty<z<0$	$\varepsilon(z)=\varepsilon_0$	$\sigma(z)=0$.	$\mu(z)=\mu_0$
$\mathcal{D}_2$	$0<z<d$	$\varepsilon(z)=\varepsilon_2(z)\geq\varepsilon_0$	$\sigma(z)=\sigma_2(z)\geq 0$	$\mu(z)=\mu_0$
$\mathcal{D}_3$	$d<z<\infty$	$\varepsilon(z)=\varepsilon_0$	$\sigma(z)=0$	$\mu(z)=\mu_0$

$$\underline{E}^i = F(t - z/c_0)\underline{i}_y, \quad \underline{H}^i = -Y_0F(t - z/c_0)\underline{i}_x, \tag{2.2.1}$$

where $Y_0 = (\varepsilon_0/\mu_0)^{\frac{1}{2}}$, $c_0 = (\varepsilon_0\mu_0)^{-\frac{1}{2}}$ (c_0 is the speed of electromagnetic waves in vacuo) and where $F(t)$ is an integrable function that vanishes outside the interval $0 < t < T$. The total electromagnetic field is then written as

$$\underline{E} = E(z,t)\underline{i}_y, \quad \underline{H} = H(z,t)\underline{i}_x, \tag{2.2.2}$$

$$\begin{aligned} E(z,t) &= E_1(z,t) = E^i(z,t) + E^r(z,t) && \text{in } \mathcal{D}_1, \\ E(z,t) &= E_2(z,t) && \text{in } \mathcal{D}_2, \\ E(z,t) &= E_3(z,t) = E^t(z,t) && \text{in } \mathcal{D}_3. \end{aligned} \tag{2.2.3}$$

Similar definitions hold for $H(z,t)$. The superscripts r and t denote the reflected and transmitted fields, respectively. Since the incident field reaches the front of the slab at the instant $t = 0$, causality ensures that $\{E^r,H^r\}$, $\{E_2,H_2\}$ and $\{E^t,H^t\}$ vanish in the time interval $-\infty < t < 0$. Therefore, the Laplace transforms of these field quantities are identical to the corresponding one-sided Laplace transforms, e.g.

$$\mathrm{E}^r(z,s) = \int_0^\infty \exp(-st)E^r(z,t)dt, \tag{2.2.4}$$

which are regular for $\mathrm{Re}(s) > 0$. The Laplace-transformed incident field follows from (2.2.1) as

$$\mathrm{E}^i(z,s) = \exp(-sz/c_0)\ \mathrm{F}(s), \tag{2.2.5}$$

where

$$F(s) = \int_0^T \exp(-st)\ F(t)dt,$$

and hence it is regular for all s. In view of subsequent calculations, we normalize all distances with respect to d and all times with respect to the vacuum travel time across the slab, i.e. d/c_0; e.g.

$$\bar{z} = zd, \quad \bar{t} = c_0 t/d,$$

and introduce the dimensionless quantities

$$\bar{s} = sd/c_0, \quad \bar{\sigma}(\bar{z}) = Z_0\sigma(z)d, \quad \bar{\varepsilon}_r(\bar{z}) = \varepsilon_r(z),$$

and the normalized field quantities

$$\bar{E}(\bar{z},\bar{s}) = E(z,s), \quad \bar{H}(\bar{z},\bar{s}) = Z_0 H(z,s),$$

where $Z_0 = (\mu_0/\varepsilon_0)^{\frac{1}{2}}$ and where $\varepsilon_r(z) = \varepsilon(z)/\varepsilon_0$ denotes the relative permittivity. In further discussion, we omit the bars, except when displaying numerical results.

The Laplace-transformed, normalized, source-free electromagnetic field equations are then given by

$$\begin{aligned} \partial_z\, H(z,s) &= s(\varepsilon_r(z) + \sigma(z)/s)E(z,s), \\ \partial_z\, E(z,s) &= s\, H(z,s). \end{aligned} \tag{2.2.6}$$

Elimination of H leads to

$$[\partial_z{}^2 - s^2(\varepsilon_r(z) + \sigma(z)/s)]E(z,s) = 0, \tag{2.2.7}$$

which will be regarded as our fundamental differential equation. With the aid of the one-dimensional Green's function

$$G(z,z';s) = (2s)^{-1} \exp(-s|z - z'|), \tag{2.2.8}$$

the following integral relation, which is equivalent to (2.2.7), is obtained:

$$E(z,s) = E^i(z,s) - s^2\int_{-\infty}^{\infty} C(z',s)G(z,z';s)E(z',s)dz', \qquad (2.2.9)$$

where the *contrast function* C, defined as

$$C(z,s) = \varepsilon_r(z) - 1 + \sigma(z)/s, \qquad (2.2.10)$$

only differs from zero when $0 < z < 1$. When $0 < z < 1$, (2.2.9) constitutes an integral equation of the second kind in $E(z,s)$. Equation (2.2.9) can be written in operator form as

$$L(s)E(z,s) = E^i(z,s), \qquad (2.2.11)$$

in which

$$L(s)f(z) = \int_{-\infty}^{\infty} [\delta(z - z') + s^2 C(z',s)G(z,z';s)]f(z')dz'. \qquad (2.2.12)$$

Once $E(z,s)$ in $0 < z < 1$ has been obtained from either of the equations (2.2.6), (2.2.7) or (2.2.9), it is, through (2.2.9), known for all z and the time-domain response follows from the evaluation of the inversion integral

$$\mathcal{E}(z,t) = (2\pi i)^{-1} \int_{L_\beta} \exp(st)E(z,s)ds, \qquad (2.2.13)$$

where L_β is the Bromwich contour

$$L_\beta = \{s \mid Re(s) = \beta,\ \beta \geq 0\}. \qquad (2.2.14)$$

One way of evaluating (2.2.13) is to continue $E(z,s)$ analytically into the domain $Re(s) < 0$, supplement L_β by circular arcs at infinity in the complex s-plane and apply Cauchy's theorem. In case the contribution of the integral along the supplementary arcs vanishes, only the singularities in the analytic continuation of $E(z,s)$ contribute to the result. The conditions under which this procedure can be followed, will be investigated in subsequent subsections.

2.2.2 Theory of the singularity expansion method

In the remainder of this section, we restrict the analysis to the case of a homogeneous slab, i.e. ε_r and σ have constant values. Then, an analytical closed-form solution is available. Following standard transmission-line theory, we write the electric field in the form:

$$E(z,s) = \begin{cases} F(s)[\exp(-sz) + r(s)\exp(sz)] & \text{in } \mathcal{D}_1, \\ F(s)[a(s)u^+(z;0) + b(s)u^-(z;1)] & \text{in } \mathcal{D}_2, \\ F(s)[t(s)\exp(-s(z-1))] & \text{in } \mathcal{D}_3. \end{cases} \tag{2.2.15}$$

In (2.2.15), the field inside the slab is expressed in terms of the plane-wave solutions

$$u^{\pm}(z;z_0) = \exp[\mp s\, n(s)\, (z - z_0)], \tag{2.2.16}$$

where $n(s) = (\varepsilon_r + \sigma/s)^{\frac{1}{2}}$, with $\mathrm{Re}(n) \geq 0$. They have been normalized such that they represent waves originating from $z = z_0$ and traveling in the direction of increasing (u^+) and decreasing (u^-) z, respectively. Application of the boundary conditions (continuity of E and $\partial_z E$ across $z = 0$ and $z = 1$) yields:

$$\begin{aligned} r(s) &= [-R(s) + T^+(s)R(s)T^-(s)\exp(-2n(s)s)/G(s)], \\ a(s) &= T^+(s)/G(s), \\ b(s) &= T^+(s)R(s)\exp(-n(s)s)/G(s), \quad \text{and} \\ t(s) &= T^+(s)T^-(s)\exp(-n(s)s)/G(s), \quad \text{with} \\ G(s) &= 1 - R^2(s)\exp(-2n(s)s), \\ R(s) &= (n(s)-1)/(n(s)+1), \\ T^+(s) &= 2/(n(s)+1) \text{ and } T^-(s) = 2n(s)/(n(s)+1). \end{aligned} \tag{2.2.17}$$

In (2.2.17), $T^{\pm}$ are the plane-wave transmission factors at an interface and R is the plane-wave reflection factor (see Figure 2.2.2).

Since we have $E(z,s)$ in closed form, we can now evaluate the in-

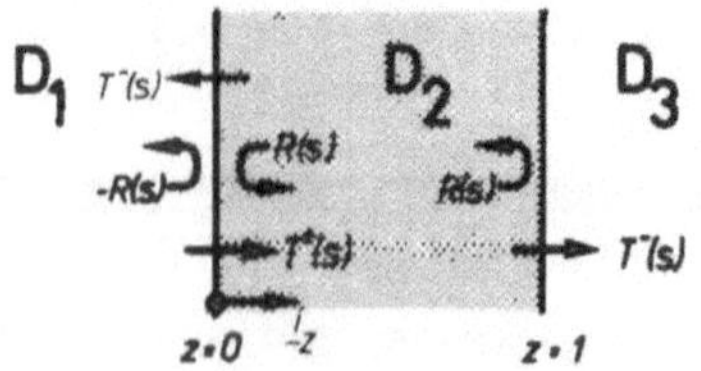

Figure 2.2.2 Illustration of frequency-dependent reflection and transmission factors for a homogeneous, lossy slab embedded in vacuum.

version integral (2.2.13) as indicated in Subsection 2.2.1. In particular, the analytic continuation of E(z,s) into the half-plane Re(s) < 0 is explicit. Henceforth, we shall not notationally distinguish between E(z,s) and its analytic continuation. Firstly, the occurrence of n(s) in (2.2.17) suggests that E(z,s) may have a branch cut coinciding with the branch cut of n(s). The latter is, in accordance with the choice of the square root, given by

$$\mathcal{B}_n = \{s \mid -\sigma/\varepsilon_r < \mathrm{Re}(s) < 0,\ \mathrm{Im}(s) = 0\}. \tag{2.2.18}$$

For a point approaching $\mathcal{B}_n$, we have

$$\lim_{\delta \downarrow 0} n(s + i\delta) = -\lim_{\delta \uparrow 0} n(s + i\delta). \tag{2.2.19}$$

However, the expression for the total field E(z,s) is invariant under a change of sign in n(s) and hence the total field has no branch cut along $\mathcal{B}_n$.

Secondly, we have to consider the contribution from the poles of E(z,s) to the expression for the time-domain field $\mathcal{E}(z,t)$. Since $\mathcal{E}(z,t)$ is a real-valued function, these poles $\{s_m\}$, also referred to as natural frequencies, occur either on the real s-axis or in complex conjugate pairs. From (2.2.17), we see that these natural frequencies are then found as the zeros of the slab denominator G(s):

$$G(s) = 1 - R^2(s)\exp(-2n(s)s) = 0. \tag{2.2.20}$$

From (2.2.4) and (2.2.5), we know that G(s) only has zeros in the half-plane $\mathrm{Re}(s) \le 0$. This also follows from the positiveness of $\mathrm{Re}(n(s)s)$ and $\mathrm{Re}(n(s))$ for $\mathrm{Re}(s) > 0$.

As discussed in Appendix 2.2.A, for special combinations of ε_r and σ, E(z,s) can have a pole of higher order. However, since the occurrence of poles of higher order can always be handled by considering limiting cases pertaining to simple poles, we shall restrict ourselves in what follows to the general case of simple poles only. The residual contribution of the simple pole $s = s_m$ follows by expanding G(s) around $s = s_m$. We find

$$E_{\text{poles}}(z,t) = \sum_{m=-\infty}^{\infty} F(s_m)e_m(z)\exp(s_m t), \tag{2.2.21}$$

with

$$e_m(z) = \begin{cases} 2\,D_m \exp(s_m z)/(n^2(s_m) - 1) & \text{in } \mathcal{D}_1, \\ D_m[\exp(-n(s_m)s_m z)/(n(s_m) + 1) & \\ \quad + \exp(n(s_m)s_m z)/(n(s_m) - 1)]/n(s_m) & \text{in } \mathcal{D}_2, \\ 2\,D_m \exp(-s_m(z-1) - n(s_m)s_m)/(n(s_m) + 1)^2 & \text{in } \mathcal{D}_3, \end{cases}$$

where

$$D_m = \frac{(\varepsilon_r s_m + \sigma)(\varepsilon_r s_m + \sigma - s_m)}{(\varepsilon_r s_m + \sigma/2)(\varepsilon_r s_m + \sigma - s_m) + \sigma}\,.$$

From (2.2.21), we see that to each pole there corresponds a complex natural mode field distribution $e_m(z)$, the latter being a non-vanishing solution of

$$L(s_m)e_m(z) = 0. \tag{2.2.22}$$

The sequence of quantities $\{s_m, e_m(z)\}$ depends on the slab configuration only, while $F(s_m)$ is determined by the incident field. Therefore, once the modal parameters are known, the residual contribution of the poles of E(z,s) can be found from (2.2.21). Finally, we have to consider the contribution from the supplementing circular arcs at infinity. In order to analyze the relevant contribution, we first construct an asymptotic approximation of $E(z,s)$ as $|s|\to\infty$ by expanding $n(s)$ in terms of powers of s^{-1}. In this way, we obtain

$$E(z,s) = E^a(z,s)(1 + \mathcal{O}(s^{-1})), \tag{2.2.23}$$

where $E^a(z,s)$ is given by:

$$E^a(z,s) = \begin{cases} F(s)[\exp(-sz) + r_a(s)\exp(sz)] & \text{in } \mathcal{D}_1, \\ F(s)[a_a(s)u_a^+(z;0) + b_a(s)u_a^-(z;1)] & \text{in } \mathcal{D}_2, \\ F(s)[t_a(s)\exp(-s(z-1))] & \text{in } \mathcal{D}_3, \end{cases} \tag{2.2.24}$$

where

$$\begin{aligned} u_a^{\pm}(z;z_0) &= \exp[\mp(s\varepsilon_r^{1/2} + \sigma/(2\varepsilon_r^{1/2}))(z - z_0)], \\ r_a(s) &= -R_a + T_a^+R_aT_a^-D_a^2 \exp(-2s\tau)/G_a(s), \\ a_a(s) &= T_a^+/G_a(s), \\ b_a(s) &= T_a^+D_aR_a\exp(-s\tau)/G_a(s), \text{ and} \\ t_a(s) &= T_a^+D_aT_a^- \exp(-s\tau)/G_a(s), \text{ with} \\ G_a(s) &= 1 - (R_aD_a)^2 \exp(-2s\tau), \\ \tau &= \varepsilon_r^{1/2}, \\ R_a &= (\varepsilon_r^{1/2} - 1)/(\varepsilon_r^{1/2} + 1), \\ T_a^+ &= 2/(\varepsilon_r^{1/2} + 1), \quad T_a^- = 2\varepsilon_r^{1/2}/(\varepsilon_r^{1/2} + 1), \text{ and} \\ D_a &= \exp(-\sigma/(2\varepsilon_r^{1/2})). \end{aligned} \tag{2.2.25}$$

Note that the asymptotic transmission factors $T_a^{\pm}$ and the asymptotic reflection factor R_a are independent of s and in fact apply to a lossless dielectric slab. In (2.2.25), two new quantities have been introduced, viz. the damping factor D_a for a single wave transit across the slab and the corresponding travel time τ. The poles of $E^a(z,s)$, from now on referred to as the asymptotic poles $\{s_m^a\}$, are given by

$$s_m^a = \ln(R_a)/\varepsilon_r^{\frac{1}{2}} - \sigma/(2\varepsilon_r) + im\pi/\varepsilon_r^{\frac{1}{2}},\ m\in\mathbb{Z}\,. \tag{2.2.26}$$

Asymptotic evaluation of (2.2.20) yields:

$$s_m = s_m^a + O(m^{-1}) \quad \text{as} \quad |m|\to\infty. \tag{2.2.27}$$

Since the incident pulse is given by an integrable function $F(t)$, application of the Riemann-Lebesgue theorem yields the following asymptotic estimate for $F(s)$:

$$\begin{aligned} &F(s) = o(1) && \text{as } |s|\to\infty \text{ in } \mathrm{Re}(s) \geq 0, \\ &F(s)\exp(sT) = o(1) && \text{as } |s|\to\infty \text{ in } \mathrm{Re}(s) \leq 0. \end{aligned} \tag{2.2.28}$$

We now take the radius of the supplementary arcs such that the closing contour in the left half-plane always intersects the line $\{s\,|\,\mathrm{Re}(s) = \ln(R_a)/\varepsilon_r^{\frac{1}{2}} - \sigma/2\varepsilon_r\}$ in the middle between two asymptotic poles. For s on these contours, the asymptotic approximations given above then yield the asymptotic behavior of $\exp(st)E(z,s)$ as indicated in Table 2.2.2. Note that in $\mathcal{D}_1$, we only consider the reflected field. The incident field, which has a different asymptotic behavior, is known in closed form and therefore need not be evaluated. With the results in Table 2.2.2, we are now in the position to analyze the presence of a contribution from the essential singularity of $E(z,s)$ as $|s|\to\infty$. Using Jordan's lemma, we find that the contributions along the closing contours vanish in the time intervals indicated in Table 2.2.3. As the next to last line of Table 2.2.3 shows, there is for each z a termination instant, before

Table 2.2.2 Asymptotic behavior as $|s|\to\infty$ of $\exp(st)E(z,s)$.

behavior in half-plane	reflected field $(-\infty < z < 0)$	field inside slab $(0 < z < 1)$	transmitted field $(1 < z < \infty)$
right	$o(1)\exp[s(t+z)]$	$o(1)\exp[s(t-z\tau)]$	$o(1)\exp[s(t-\tau-(z-1))]$
left	$o(1)\exp[s(t+z-T)]$	$o(1)\exp[s(t+z\tau-T)]$	$o(1)\exp[s(t-T+\tau-(z-1))]$

Table 2.2.3 Time intervals for which the contribution from closing contours at infinity vanishes in the case of a homogeneous slab.

closure in half-plane	reflected field $(-\infty<z<0)$	field inside slab $(0<z<1)$	transmitted field $(1<z<\infty)$
right	$-\infty<t<-z$	$-\infty<t<z\tau$	$-\infty<t<\tau+(z-1)$
left	$T-z<t<\infty$	$T-z\tau<t<\infty$	$T-\tau+(z-1)<t<\infty$

which the contribution of the closing contour vanishes upon closure to the right. Then, the time-domain field $E(z,t)$ is, apart from the incident field in $\mathcal{D}_1$, zero. From the last line of Table 2.2.3, we see that for each z there is also an initial instant, after which the contribution from a closing contour in the left half-plane vanishes. Then the time-domain field $E(z,t)$ is equal to $E_{poles}(z,t)$. Combining both lines in the table, we find that the time-domain field for all instants either vanishes or can be calculated from pole contributions only, provided that the following conditions are satisfied:

$$\begin{aligned} &\text{for the reflected field} &&: \text{never},\\ &\text{for the field inside the slab} &&: \text{if } T/2\tau < z < 1,\\ &\text{for the transmitted field} &&: \text{if } T/2\tau < 1. \end{aligned} \tag{2.2.29}$$

For the case $T/2\tau < 1$, the closing conditions and the asymptotic transmission and damping factors defined in (2.2.25) are illustrated in Figure 2.2.3.

If the conditions (2.2.29) are violated, there is always a time interval of nonzero duration prior to the initial instant for closing to the left, in which the behavior of the function on the closing contour violates the conditions for the applicability of Jordan's lemma. In that interval, a nonvanishing field cannot be represented by modal contributions only because of a nonvanishing contribution from the essential singularity at infinity. In the next subsection, we will discuss the evaluation of the field in these special intervals.

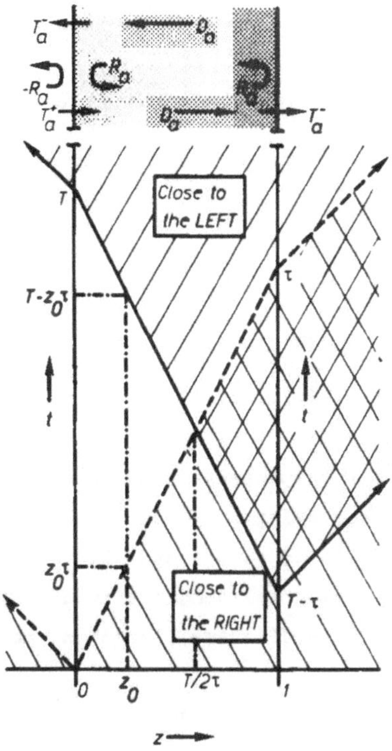

Figure 2.2.3 Upper part: physical interpretation of the reflection, transmission and damping factors occurring in $E^a(z,s)$ (see (2.2.25)). Lower part: space-time illustration of the conditions for vanishing contributions from closing contours. In the unshaded region, the Bromwich contour cannot be closed without a possible contribution along the supplementary arcs.

2.2.3 Decomposition of the integrand

In this subsection, we discuss one modification of the theory presented in Subsection 2.2.2, which allows us to handle the essential singularity of $E(z,s)$ at infinity also in the "forbidden" time intervals. The conditions (2.2.29) suggest that we distinguish between the case $T < 2\tau$ and the case $T > 2\tau$.

Asymptotics for $T < 2\tau$

As we already saw in Subsection 2.2.2, the time-domain field in $\mathcal{D}_3$ can, for $T < 2\tau$, always be determined by closing the Bromwich contour such that the contribution of the closing contour vanishes. In

$\mathcal{D}_2$ and $\mathcal{D}_1$, however, special care is needed. Let us first consider region $\mathcal{D}_2$.

From (2.2.15), we see that there E(z,s) is decomposed as

$$E(z,s) = E^L(z,s,n(s)) + E^R(z,s,n(s)), \tag{2.2.30}$$

with

$$\begin{aligned} E^L(z,s,n(s)) &= F(s)a(s)u^+(z;0), \\ E^R(z,s,n(s)) &= F(s)b(s)u^-(z;1). \end{aligned} \tag{2.2.31}$$

In (2.2.31), E^L and E^R represent waves that travel in the direction of increasing and decreasing z, respectively. Using the asymptotic approximations (2.2.24), (2.2.27) and (2.2.28), we can then analyze the behavior of the essential singularity at infinity for the two terms separately. The results are listed in Table 2.2.4. We find

Table 2.2.4 Asymptotic behavior as $|s|\to\infty$ of $\exp(st)E^L(z,s,n(s))$ and $\exp(st)E^R(z,s,n(s))$ in $\mathcal{D}_2$ $(0<z<1)$.

half-plane	$E^L(z,s,n(s))\exp(st)$	$E^R(z,s,n(s))\exp(st)$
right	$O(1)\exp[s(t-z\tau)]$	$O(1)\exp[s(t-2\tau+z\tau)]$
left	$O(1)\exp[s(t-z\tau-T+2\tau)]$	$O(1)\exp[s(t+z\tau-T)]$

that the integral along one of the closing contours discussed in Subsection 2.2.2 vanishes as indicated in Table 2.2.5. From Table 2.2.5, we observe that Jordan's lemma can now be applied to the terms E^L and E^R separately with a vanishing contribution from the supplementing arc at infinity, since we have assumed $T < 2\tau$. For $0 < z < T/2\tau$ and $z\tau < t < T - z\tau$, where the conditions listed in Table 2.2.3 are violated, we then close the Bromwich contour to the left for E^L and to the right for E^R. In this case, the residual contribution $e^L_m(z)$ follows from expanding E^L around $s = s_m$. Note, however, that E^L contains the branch cut associated with the square-root expression for n(s). The consequence of this will be analyzed further on.

Next, we turn our attention to the domain $\mathcal{D}_1$. The decomposition

Table 2.2.5 Conditions for vanishing closing contours in $\mathcal{D}_2$ $(0<z<1)$ for E^L and E^R.

closing in half-plane	E^L	E^R
right	$-\infty<t<z\tau$	$-\infty<t<2\tau-z\tau$
left	$T-2\tau+z\tau<t<\infty$	$T-z\tau<t<\infty$

procedure applying to z in $\mathcal{D}_2$ can also be utilized to evaluate the field in the "forbidden" time interval for z in $\mathcal{D}_1$. From (2.2.16) and (2.2.17), we have

$$E^r(z,s) = F(s)r(s)\exp(sz), \tag{2.2.32}$$

with

$$\begin{aligned} r(s) &= -R(s) + T^+(s)R(s)T^-(s)\exp(-2n(s)s)/G(s) \\ &= [-R(s) + R(s)\exp(-2n(s)s)]/G(s). \end{aligned} \tag{2.2.33}$$

As in (2.2.30), we now decompose $E^r(z,s)$ into

$$E^r(z,s) = E^L(z,s,n(s)) + E^R(z,s,n(s)), \tag{2.2.34}$$

where

$$\begin{aligned} E^L(z,s,n(s)) &= -F(s)R(s)\exp(sz)/G(s), \\ E^R(z,s,n(s)) &= F(s)R(s)\exp(sz - 2n(s)s)/G(s). \end{aligned} \tag{2.2.35}$$

Carrying out the same asymptotic evaluation procedure as in $\mathcal{D}_2$, we find the asymptotic behavior and the conditions for a vanishing contribution from the closing contours for each of the separate terms defined in (2.2.35). The results are summarized in Table 2.2.6. Again we see that for $-z < t < T - z$, we can close the Bromwich contour to the left for E^L and to the right for E^R since $T < 2\tau$. Here, too, E^L contains the branch cut associated with the square-root expression for $n(s)$.

Table 2.2.6 Asymptotic behavior as $|s|\to\infty$ and conditions for vanishing contributions of closing contours for $\exp(st)E^L(z,s,n(s))$ and $\exp(st)E^R(z,s,n(s))$ in $\mathcal{D}_1$ $(-\infty<z<0)$.

closing in half-plane	Asymptotic behavior E^L	E^R	Closing conditions E^L	E^R
right	$O(1)\exp[s(t+z)]$	$O(1)\exp[s(t+z-2\tau)]$	$-\infty<t<-z$	$-\infty<t<2\tau-z$
left	$O(1)\exp[s(t+z+2\tau-T)]$	$O(1)\exp[s(t+z-T)]$	$T-2\tau-z<t<\infty$	$T-z<t<\infty$

Contributions of branch cut for $T < 2\tau$

In Subsection 2.2.3, we have found that the total field $E(z,s)$ in each domain is continuous across the branch cut $\mathcal{B}_n$ of $n(s)$. This property, however, is lost for the separate terms E^R and E^L as introduced in (2.2.30) and (2.2.34), respectively. Investigating the influence of a change of sign in $n(s)$ on $E(z,s)$, we observe the property

$$E^L(z,s,n(s)) = E^R(z,s,-n(s)), \tag{2.2.36}$$

in both $\mathcal{D}_1$ and $\mathcal{D}_2$, which is in accordance with the continuity of $E(z,s)$ across $\mathcal{B}_n$. However, upon evaluating E^L by closing the contour to the left, the additional contribution from the branch cut must be taken into account. Note that poles may be present on $\mathcal{B}_n$ (see Appendix 2.2.A), which entails an evaluation of the integral along the contour $\mathcal{L}_n$ shown in Figure 2.2.4.

Letting δ_1 and δ_2 subsequently approach zero, we then obtain, besides the residual contributions of the relevant poles, a principal-value integral. For example, in $\mathcal{D}_1$ and for $-z < t < T - z$, we end up with

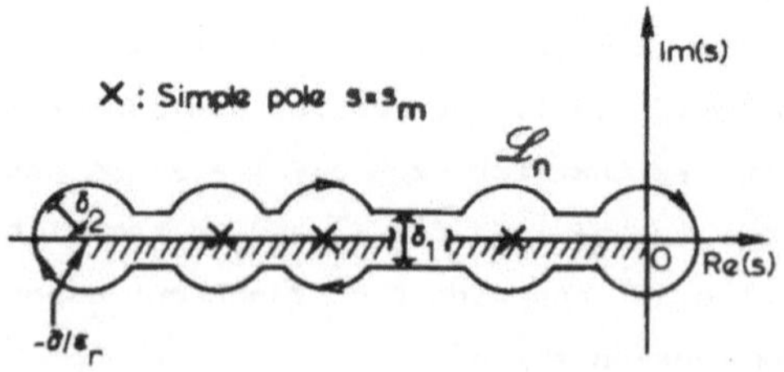

Figure 2.2.4 Closing contour $\mathcal{L}_n$ around the branch cut.

$$E(z,t) = E^i(z,t) + \sum_{m=-\infty}^{\infty} F(s_m)e_m^L(z)\exp(s_m t)$$
$$+ P\int_{-\sigma/\varepsilon_r}^{0} ds \left\{F(s)\exp[s(t+z)] \right. \tag{2.2.37}$$
$$\left. \times \frac{[(\varepsilon_r - 1)s + \sigma]\cos(n_0(s)s)}{[(\varepsilon_r + 1)s + \sigma]\sin(n_0(s)s) + 2n_0(s)s\cos(n_0(s)s)}\right\},$$

where P denotes the principal value of the relevant integral. In (2.2.37), we have $n_0(s) = (-\varepsilon_r - \sigma/s)^{\frac{1}{2}}$, with $\mathrm{Re}(n_0) \geq 0$, and

$$e_m^L(s) = \begin{cases} D_m \exp(s_m z)/(\varepsilon_r - 1 + \sigma/s_m) & \text{for } s_m \in B_n, \\ -D_m R(s_m)\exp(s_m z)/2n(s_m) & \text{otherwise.} \end{cases} \tag{2.2.38}$$

With this, the discussion for the case $T < 2\tau$ has been completed.

Decomposition of the field for incident pulses of duration $T > 2\tau$

For incident pulses of duration $T > 2\tau$, there is, for each z, a nonvanishing time interval where the conditions in Tables 2.2.3, 2.2.5 and 2.2.6 do not provide a way to add a supplementary arc, with vanishing contribution, to the Bromwich contour. However, utilizing the linearity of the problem, we can always write the incident field as the superposition of succeeding pulses of finite duration less than or equal to 2τ, i.e.

$$F(t) = \sum_{n=0}^{N} F_n(t), \tag{2.2.39}$$

where $F_n(t) = F(t)$ in the time interval $2n\tau \leq t < (2n+2)\tau$ and vanishes otherwise. On each $F_n(t)$ separately, we can apply the procedure described above, provided that we take into account the proper time delay $2n\tau$.

Physical interpretation

With the theory developed up to now, we are able to connect the time-domain field with its natural-mode representation, together with possible additional contributions. It should be observed that in those cases, where these additional contributions do not vanish, the representation of the field in terms of natural-mode contributions is incomplete. This is explained as follows. Within the slab, each natural oscillation, being a standing wave, consists of two waves: one traveling in the direction of increasing z and one traveling in the direction of decreasing z. In view of the principle of causality, the waves traveling in the direction of decreasing z are not excited before the front of the incident pulse has reached the rear end of the slab. Upon reflection at the latter interface, waves that travel in the direction of decreasing z with (dimensionless) speed $1/\tau$ are excited. Only after these waves have reached the point of observation, does the full modal expansion start to hold there. The decomposition of $E(z,s)$ in the domains $\mathcal{D}_1$ and $\mathcal{D}_2$ for $T < 2\tau$ discussed earlier in this section can now be understood as a way to distinguish between the reflections at the front and at the back of the slab. Points of observation in the domain $\mathcal{D}_3$ behind the slab can only be reached by waves that have traversed the entire slab. Therefore, for $T < 2\tau$ no further measures are necessary. In the case $T > 2\tau$, the reflected wave traveling in the direction of decreasing z coincides, upon arrival at $z = 0$, with the part of the incident pulse that reaches the front of the slab after $t = 2\tau$. The decomposition of the incident pulse into succeeding pulses of maximum duration 2τ according to (2.2.39) can thus be understood as a way to distinguish between wave constituents having undergone a different number of reflections at the slab interfaces.

2.2.4 Numerical implementation and results

In order to obtain insight into the applicability of the SEM in practice, the theory presented in Subsections 2.2.2 and 2.2.3 has been implemented numerically. We first consider the location of the natural frequencies. From Appendix 2.2.A, we know that priority has to be given to the determination of the number of poles on the negative real axis. To this end, we first compute the critical value σ_m^{crit} with $m = \mathrm{ent}(\sigma/2\pi\varepsilon_r^{\frac{1}{2}})$. For $\sigma \gtrless \sigma_m^{crit}$, we then have $2m\pm1$ poles on the negative real axis. Both σ_m^{crit} and the location of these poles are found by applying a real version of Muller's method (see Traub (1964)). Next, we determine the poles off the real axis by solving (2.2.54) for subsequent values of m with a complex version of Muller's method. Since these equations then always have exactly a single solution in the complex s-plane, convergence of the numerical procedure towards the proper pole is automatically ensured. As an illustration, we show in Figure 2.2.5 the location of the natural frequencies $\{s_m, |m| \le 15\}$ in the complex s-plane for the case where five additional pole pairs occur on the real axis. In Figure 2.2.6, we have plotted the behavior of the root loci of the first ten poles in the upper half-plane for two typical fixed values of $n = \varepsilon_r^{\frac{1}{2}}$ and for σ increasing from zero to the numerically determined critical value σ_m^{crit}. In addition, the critical values are indicated. It is noted that for low permittivity contrast, the real part of s_m may have a maximum at a nonzero value of σ. At this value, the corre-

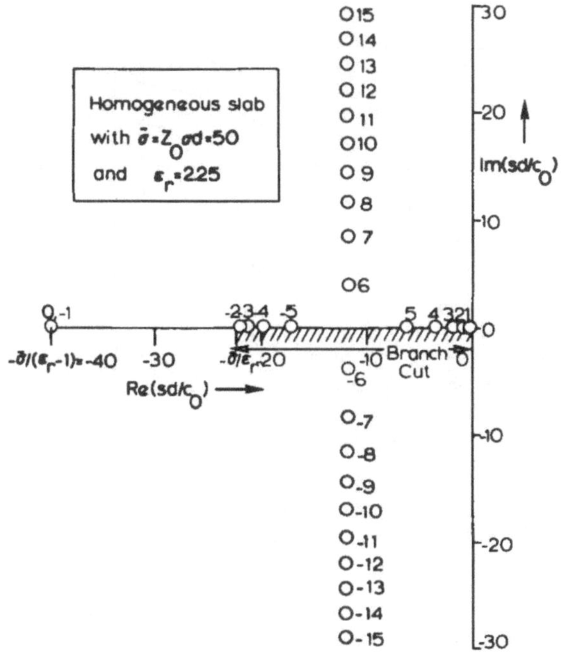

Figure 2.2.5 Location of the natural frequencies with $|m| \le 15$ in the complex s-plane for a homogeneous, lossy dielectric slab with $\varepsilon_r = 2.25$ and $Z_0\sigma d = 50$.

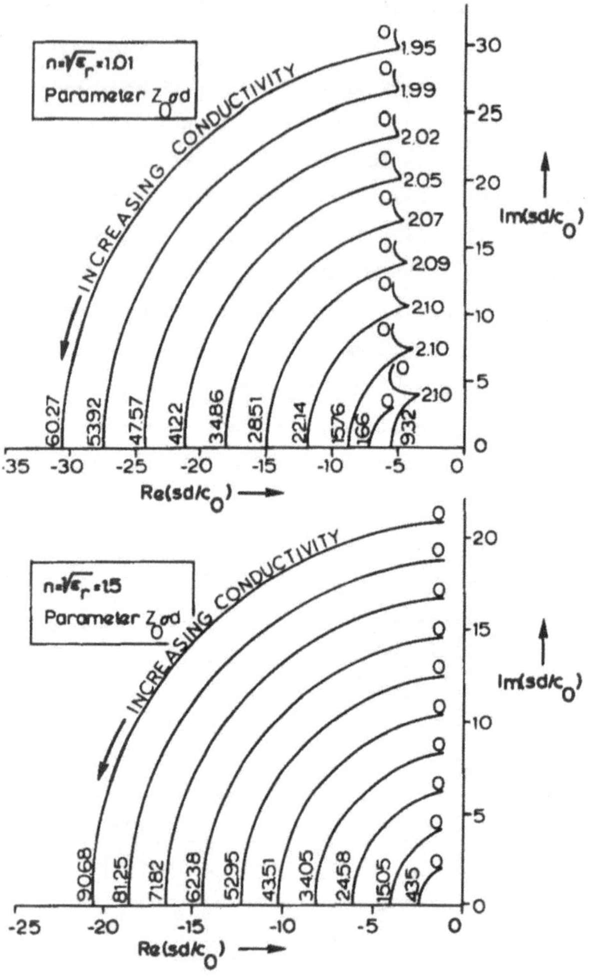

Figure 2.2.6 Root loci of the first ten natural frequencies of a homogeneous, lossy dielectric slab for constant permittivity and varying conductivity.

sponding residual contribution (see (2.2.21)) exhibits less damping than the corresponding mode for a lossless slab. The physical explanation of this phenomenon is that the conductivity contrast causes an increased reflection at the slab boundaries, which, for a certain σ-interval, outweighs the increased attenuation for waves traveling through the slab. For higher permittivity contrast, the reflection at a slab boundary is mainly determined by ε_r and the phenomenon is no longer observed.

Once the natural frequencies $\{s_m\}$ have been determined, they can be substituted in (2.2.21) or (2.2.37) to compute the time-domain field. However, the rate of convergence of the summation of the pole series remains to be investigated. Expanding $e_m(z)$ asymptotically, we obtain

$$e_m(z) = e_m^a(z)(1 + O(m^{-1})) \quad \text{as } |m|\to\infty, \tag{2.2.40}$$

where $e_m^a(z)$ is the field distribution of the natural mode that determines the residual contribution to $E^a(z,s)$ at $s = s_m^a$:

$$e_m^a(z) = \begin{cases} 2\exp[s_m^a z]/(\varepsilon_r - 1) & \text{in } \mathcal{D}_1, \\ (D_a)^z \exp[-s_m^a z\tau]/\varepsilon_r^{\frac{1}{2}}(\varepsilon_r^{\frac{1}{2}} + 1) & \\ \quad + (D_a)^{1-z} \exp[s_m^a(z-1)\tau]/\varepsilon_r^{\frac{1}{2}}(\varepsilon_r^{\frac{1}{2}} - 1) & \text{in } \mathcal{D}_2, \\ 2 D_a \exp[s_m^a(-(z-1)-\tau)]/(\varepsilon_r^{\frac{1}{2}} + 1)^2 & \text{in } \mathcal{D}_3. \end{cases} \tag{2.2.41}$$

Combining (2.2.21), (2.2.25), (2.2.26) and (2.2.40), we see that the higher-order terms of the pole series have asymptotically the same rate of decay in time, with weight factors proportional to $F(s_m)$. Therefore, the rate of the convergence is determined by the asymptotic behavior of $F(s)$ as $|s|\to\infty$. Although convergence of the summation in a principal-value sense was shown in Subsections 2.2.2 and 2.2.3, this convergence is in general too slow for an efficient numerical computation and therefore needs to be improved.

If, as a first step, we limit $F(t)$ for the numerical computation to the class of piecewise continuously differentiable functions, then we can refine (2.2.28) to

$$\begin{aligned} F(s) &= O(s^{-1}) \quad \text{as } |s|\to\infty \text{ in } \mathrm{Re}(s) \ge 0, \\ F(s)\exp(sT) &= O(s^{-1}) \quad \text{as } |s|\to\infty \text{ in } \mathrm{Re}(s) \le 0. \end{aligned} \tag{2.2.42}$$

For fixed z and t, the terms in the pole series are then of order

m^{-1} as $m\to\infty$, which still leads to a numerical summation which is too slow. In principle, a better rate of convergence could be obtained by imposing further restrictions on F. However, this would exclude discontinuous incident pulses. Since we do want to handle discontinuous pulses, for instance in the decomposition (2.2.39), other measures have to be taken to improve the numerical rate of convergence.

To this end, we utilize the asymptotic field $E^a(z,s)$ and rewrite the inversion integral (2.2.13) as

$$\mathcal{E}(z,t) = (2\pi i)^{-1}\int_{L_\beta} \exp(st)[E(z,s) - E^a(z,s)]ds + \mathcal{E}^a(z,t), \quad (2.2.43)$$

where

$$\mathcal{E}^a(z,t) = (2\pi i)^{-1}\int_{L_\beta} \exp(st)E^a(z,s)ds. \quad (2.2.44)$$

From (2.2.24) and (2.2.25) we see that the integral in (2.2.44) can be determined analytically by substitution of the expansion

$$1/G_a(s) = \sum_{k=0}^{\infty}(R_aD_a)^{2k}\exp(-2ks\tau), \quad (2.2.45)$$

which holds for $\mathrm{Re}(s) > \ln(R_aD_a)/\tau$. Interchanging summation and integration, and using (2.2.5), we find

$$\mathcal{E}^a(z,t) = \begin{cases} \mathcal{E}^i(z,t) - R_a\mathcal{F}(t+z) + T_a^+D_a^2R_aT_a^-\mathcal{G}(t+z-2\tau) & \text{in } \mathcal{D}_1, \\ T_a^+(D_a)^z\mathcal{G}(t-z\tau) + T_a^+(D_a)^{2-z}R_a\mathcal{G}(t-(2-z)\tau) & \text{in } \mathcal{D}_2, \\ T_a^+D_aT_a^-\mathcal{G}(t-(z-1)-\tau) & \text{in } \mathcal{D}_3, \end{cases} \quad (2.2.46)$$

with

$$\mathcal{G}(t) = \sum_{k=0}^{\infty}(R_aD_a)^{2k}\mathcal{F}(t - 2k\tau), \quad (2.2.47)$$

which is a representation of $\mathcal{E}^a$ in terms of traveling waves. The integral in (2.2.43) can then be evaluated by following the procedure outlined in Subsections 2.2.2 and 2.2.3. For space-time points where the full modal representation (2.2.21) is valid, we end up with

$$E(z,t) = \sum_{m=-\infty}^{\infty} [F(s_m)e_m(z)\exp(s_m t) - F(s_m^a)e_m^a(z)\exp(s_m^a t)] + E^a(z,t). \quad (2.2.48)$$

Using (2.2.26) and (2.2.40), we see that the terms in the summation in (2.2.48) are of order m^{-2}, which means that we have improved the rate of convergence by one order as compared with (2.2.21).

For the space-time points where only $E^L(z,s,n(s))$ contributes, a similar representation can be given. In that case (see (2.2.37)) also the nonzero contribution of the branch cut has to be computed. In this computation, we have to take special measures in the neighborhood of both the square-root-like singularities at the end points of the interval and, possibly, the first-order pole singularities on the branch cut. To this end, we subdivide the integration interval into subintervals containing at most one singularity such that each possible pole is in the center of a subinterval. The square-root-like singularities are handled by a stretching procedure and the principal-value integral is evaluated by choosing a symmetric integration formula (in fact, a repeated four-point Gaussian rule) on each of the subintervals.

As an illustration, we present in Figure 2.2.7 the computed reflected and transmitted fields caused by a rectangular pulse incident on a homogeneous slab with $\varepsilon_r = 2.25$ for various values of the conductivity.

Appendix 2.2.A Influence of the conductivity on the location of the poles

In this appendix, we investigate the location of the complex natural frequencies of a homogeneous, lossy dielectric slab, the permittivity of which is held constant and whose conductivity varies. From (2.2.20), we know that the natural frequencies $\{s_m\}$ satisfy the transcendental equation $G(s) = 1 - R^2(s)\exp(-2n(s)s) = 0$, or

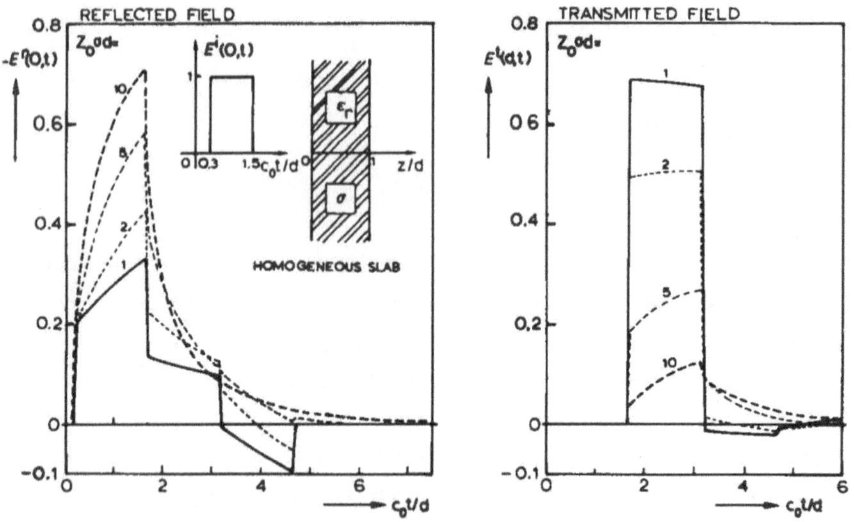

Figure 2.2.7 Reflected and transmitted fields due to a rectangular pulse as indicated in the inset, incident on a lossy, homogeneous dielectric slab with $\varepsilon_r = 2.25$ and $Z_0\sigma d = 1, 2, 5, 10$. Time duration of the incident pulse is $c_0T/d = 1.2$.

$$\exp(2n(s)s) = R^2(s), \tag{2.2.49}$$

where $n(s) = (\varepsilon_r + \sigma/s)^{\frac{1}{2}}$ and $R(s) = (n(s) - 1)/(n(s) + 1)$. From (2.2.18) and (2.2.19), we know that special care is needed near the branch cut of $n(s)$:

$$B_n = \{s| - \sigma/\varepsilon_r < \mathrm{Re}(s) < 0, \mathrm{Im}(s) = 0\}, \tag{2.2.50}$$

on which we have

$$\lim_{\delta\downarrow 0} n(s + i\delta) = -\lim_{\delta\uparrow 0} n(s + i\delta) \tag{2.2.51}$$

and

$$\lim_{\delta\downarrow 0} R(s + i\delta) = \lim_{\delta\uparrow 0} 1/R(s + i\delta). \tag{2.2.52}$$

For zero conductivity ($\sigma = 0$), the solutions of (2.2.49) are known in closed form, viz.

$$s_m = \varepsilon_r^{-\frac{1}{2}} \ln[(\varepsilon_r^{\frac{1}{2}} - 1)/(\varepsilon_r^{\frac{1}{2}} + 1)] + im\pi\varepsilon_r^{-\frac{1}{2}}, \quad m \in \mathbb{Z} . \tag{2.2.53}$$

The logarithm in (2.2.53) should be understood as the real-valued function of its real and positive argument.

For nonzero conductivity, (2.2.49) can no longer be solved in closed form. In fact, the numerical solution of (2.2.49) and a unique ordering of the poles both become quite complicated and can practically not be carried out without some insight into the approximate location of the poles. Therefore, we use (2.2.53) as the starting point of our evaluation. Since both sides of (2.2.49) vary continuously with σ, its solution will do so as well. This allows us to determine the root locus of a specific pole with increasing values of σ, starting from the value $\sigma = 0$.

The special form of (2.2.49) and (2.2.53) then induces a numbering system by a decomposition of (2.2.49) into the equivalent set of equations:

$$n(s)s = \ln[R(s)] + im\pi, \qquad m \in \mathbb{Z}, \tag{2.2.54}$$

where the logarithm is now complex with $-\pi < \arg[R(s)] \leq \pi$. In accordance with this definition, $\ln[R(s)]$ has an additional branch cut along B_R, given by

$$B_R = \{s \mid -\sigma/(\varepsilon_r - 1) < \mathrm{Re}(s) < -\sigma/\varepsilon_r, \ \mathrm{Im}(s) = 0\}, \tag{2.2.55}$$

on which $R(s)$ is real and negative. At $s = -\sigma/(\varepsilon_r - 1)$ we have $R(s) = 0$ and the logarithm is singular. In view of the special behavior of $n(s)$ and $R(s)$ near the negative real s-axis, it is convenient to start the analysis with the poles lying on that axis.

Poles on the real s-axis outside B_n.

In order to avoid difficulties at the branch cut of $n(s)$, we first consider the interval $-\infty < s \leq -\sigma/\varepsilon_r$. Direct substitution shows that

$s = -\sigma/\varepsilon_r$ satisfies (2.2.49). However, this value is not a pole of the expression (2.2.15) for E(z,s) since the numerator also vanishes. For decreasing s, exp(2n(s)s) decreases monotonically towards zero, whereas $R^2(s)$ first decreases to zero at $s = -\sigma/(\varepsilon_r - 1)$ and then again monotonically increases towards $R^2(\infty) = [(\varepsilon_r^{1/2} - 1)/(\varepsilon_r^{1/2} + 1)]^2$. Comparing the decay of both functions at $s = -\sigma/\varepsilon_r$, we then find the following pole distribution on the interval $-\infty < s < -\sigma/\varepsilon_r$ (see Figure 2.2.8a):

(i) For all σ, there is one pole (m = 0) on the interval $-\infty < s < -\sigma/(\varepsilon_r - 1)$.

(ii) For $2\varepsilon_r < \sigma < \infty$, there is one pole on $\mathcal{B}_R$.

(iii) For $0 \le \sigma < 2\varepsilon_r$, there are 0 or 2 poles on $\mathcal{B}_R$. The question of the presence of this pole pair can only be evaluated numerically and will be discussed further on in this appendix.

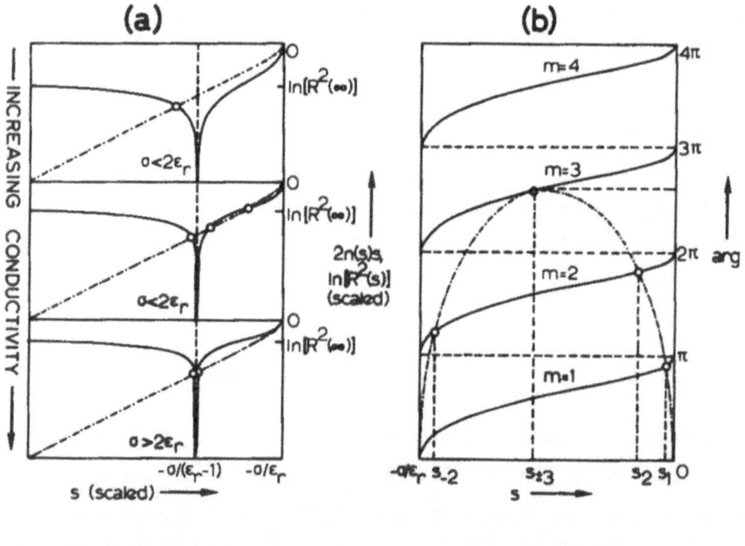

Figure 2.2.8 Graphical illustration of the determination of the natural frequencies located on the real axis. (a): Solution of (2.2.49) in the interval $-\infty < s < -\sigma/\varepsilon_r$. Dash-dot line: 2 n(s)s; solid line: $\ln[R^2(s)]$. The curves were plotted for the special values $\varepsilon_r = 1.44$ and σ = 1.5, 2.2 and 3.3, respectively. (b): Solution of (2.2.56) in the interval $-\sigma/\varepsilon_r < s < 0$ for s_3 and s_{-3} coalescing on the real s-axis. Solid lines: $2\arctan(n_0(s)) + (m-1)\pi$; dash-dot line: $-sn_0(s)$. The curves were plotted for $\varepsilon_r = 2.25$.

Poles on the branch cut.

For s on the branch cut, n(s) is imaginary and hence $|\exp(2n(s)s)|$ as well as $|R^2(s)| = 1$. Consequently (2.2.49) can be replaced by

$$-sn_0(s) = 2 \arctan(n_0(s)) + (m - 1)\pi, \; m = 1,2,3,\dots,\infty, \qquad (2.2.56)$$

where, as in (2.2.37), $n_0(s) = (-\varepsilon_r - \sigma/s)^{\frac{1}{2}}$ with $\mathrm{Re}(n_0) \geq 0$. In (2.2.56) the shift of 1 in m is introduced to obtain agreement with the choice of m in (2.2.54). The solution of (2.2.56) is illustrated graphically in Figure 2.2.8b. Note that the solution at $s = -\sigma/\varepsilon_r$ turns up again. Comparing the behavior of both sides of (2.2.56) near $s = -\sigma/\varepsilon_r$, we find that the condition for (2.2.56) to have at least one solution on $\mathcal{B}_n$ is given by $\sigma > 2\varepsilon_r$. As Figure 2.2.8b shows, the maximum value of $-sn_0(s)$ is $\sigma/2\varepsilon_r^{\frac{1}{2}}$. Therefore, for $\sigma > 2\varepsilon_r$, (2.2.56) has at least $2\,\mathrm{ent}(\sigma/2\pi\varepsilon_r^{\frac{1}{2}}) - 1$ solutions and at most two more. The end points of the subintervals in which these poles are to be found can then be determined analytically by solving the equations

$$-sn_0(s) = n\pi, \quad 1 \leq n \leq \mathrm{ent}(\sigma/2\pi\varepsilon_r^{\frac{1}{2}}). \qquad (2.2.57)$$

Double poles

In Table 2.2.7, we have summarized the number of poles on each subinterval of the real axis resulting from the analysis so far. From Table 2.2.7, we see that for fixed values of σ, the number of poles on the real axis is now known except for the possible presence of one last additional pole pair. From Figure 2.2.8 we see that the question as to the presence of such a pole pair can be resolved by determining the values of σ for which *a double pole* is present on

Table 2.2.7 Pole distribution on the negative real axis for a homogeneous slab with relative permittivity ε_r and conductivity σ: number of poles per subinterval.

σ-interval	$-\infty<s<-\sigma/(\varepsilon_r-1)$	$-\sigma/(\varepsilon_r-1)<s<-\sigma/\varepsilon_r$	$-\sigma/\varepsilon_r<s<0$
$0\leq\sigma<2\varepsilon_r$	1	0 or 2	0
$2\varepsilon_r<\sigma<\infty$	1	1	$2\,\mathrm{ent}(\sigma/2\pi\varepsilon_r^{\frac{1}{2}})\pm1$

the real axis. In that case, s has to satisfy both (2.2.49) and

$$[sd_s n(s) + n(s)]\exp(2n(s)s) = (d_s n(s))R(s)\partial_n R(s). \tag{2.2.58}$$

Evaluation of the derivatives and substitution of (2.2.49) reduces (2.2.58) to the algebraic equation of the second degree

$$2\varepsilon_r(\varepsilon_r - 1)s^2 + \sigma(3\varepsilon_r - 1)s + \sigma(\sigma + 2) = 0, \tag{2.2.59}$$

whose solutions are

$$\begin{aligned} s_\pm = \{&-(3\varepsilon_r - 1)\sigma \pm [(3\varepsilon_r - 1)^2\sigma^2 \\ &- 8\varepsilon_r(\varepsilon_r - 1)(\sigma^2 + 2\sigma)]^{\frac{1}{2}}\}/4\varepsilon_r(\varepsilon_r - 1). \end{aligned} \tag{2.2.60}$$

The critical values $\sigma = \sigma_m^{crit}$, for which a double pole occurs, are found by requiring that (2.2.49) and (2.2.60) be simultaneously satisfied. Since the relevant values of $s_\pm$ only occur on the real axis, only the values of σ for which

$$\sigma > 16\varepsilon_r(\varepsilon_r - 1)/(\varepsilon_r + 1)^2 \tag{2.2.61}$$

are of importance. For the critical value of σ at which the first additional pole pair appears on the real s-axis, we can distinguish between two cases. For $\varepsilon_r < 3$, we have a critical value, σ_1^{crit}, in the interval $16\varepsilon_r(\varepsilon_r - 1)/(\varepsilon_r + 1)^2 < \sigma < 2\varepsilon_r$, which can be determined by substitution of s_- into (2.2.49). If σ is increased beyond that value, the double pole splits up into two simple poles, one of which remains on $\mathcal{B}_R$, while the other traverses into $\mathcal{B}_n$ from $\sigma = 2\varepsilon_r$ onwards. For $\varepsilon_r \geq 3$, the double pole for m = 1 always occurs for $\sigma = 2\varepsilon_r$ at $s = -\sigma/\varepsilon_r$. For increasing σ, this pole splits up into one single pole on $\mathcal{B}_R$ and one on $\mathcal{B}_n$. All other critical values σ_m^{crit}, $m = 2,3,4,\dots,\infty$, can be found by substitution of s_+ in (2.2.56) for the appropriate value of m.

Poles in the cut s-plane

We are now left with the determination of the poles off the real s-axis. There, the original equation (2.2.49) can be replaced by the equivalent sequence (2.2.54). In order to determine the number of solutions of a single equation from this sequence, we rewrite it as

$$g_m(s) = n(s)s - \ln[R(s)] - im\pi = 0. \qquad (2.2.62)$$

Since $g_m(s)$ is analytic in the complex s-plane that has been cut along $\mathcal{B}_n + \mathcal{B}_R$, we can find the total number of zeros N_z of $g_m(s)$ by applying the argument principle (see e.g. Copson (1976)) to the domain in between the contour $\mathcal{L}_{n+R}$ defined in Figure 2.2.9 and a

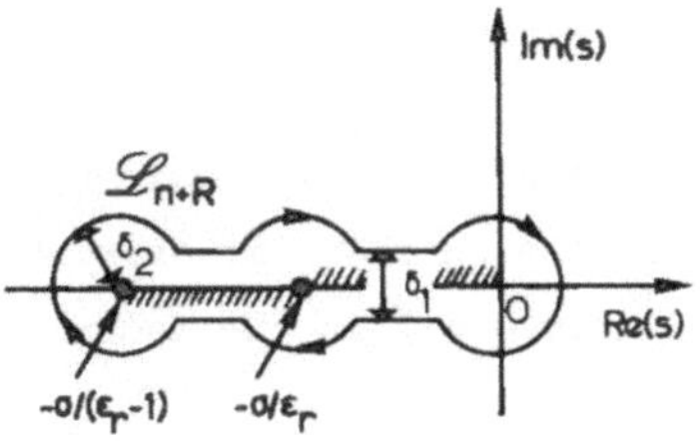

Figure 2.2.9 Inner contour $\mathcal{L}_{n+R}$ in the application of the argument principle for $g_m(s)$.

circular contour $\mathcal{L}_\infty$ at infinity:

$$\begin{aligned} N_z &= (2\pi i)^{-1}\oint_{\mathcal{L}_\infty} d\ln[g_m(s)] + (2\pi i)^{-1}\oint_{\mathcal{L}_{n+R}} d\ln[g_m(s)] \\ &= (2\pi i)^{-1}\oint_{\mathcal{L}_\infty}[g_m'(s)/g_m(s)]ds + (2\pi)^{-1}\oint_{\mathcal{L}_{n+R}} d\arg[g_m(s)]. \end{aligned} \qquad (2.2.63)$$

Expanding $g_m(s)$ asymptotically, we obtain

$$g_m'(s)/g_m(s) = 1/s + O(s^{-2}) \quad \text{as} \quad |s|\to\infty. \qquad (2.2.64)$$

Hence,

$$(2\pi i)^{-1}\oint_{\mathcal{L}_\infty}[g_m'(s)/g_m(s)]ds = 1, \qquad (2.2.65)$$

for all m, ε_r and σ. The integral along L_{n+R} is more difficult to evaluate. However, careful consideration of the location of $g_m(s)$ in the complex g-plane for s on L_{n+R} shows that, for nonzero m, the second integral in (2.2.63) equals minus one if there are at least $2|m|$ poles on $B_n + B_R$ and vanishes otherwise. This implies that $g_m(s)$ has either no zeros or exactly a single zero in the cut s-plane, respectively. In accordance with the results derived above for the number of poles on the interval $-\infty < s < -\sigma/(\varepsilon_r - 1)$, a similar analysis shows that for $m = 0$ there is always a single zero in the cut s-plane.

The results described above also allow us to connect the poles on and off the real axis. We arrive at the following picture. Starting at $\sigma = 0$, where the poles are given by (2.2.53), the pole pair $\{s_m, s_{-m}\}$, with $m = 1,2,3,\ldots,\infty$, gradually travels towards the negative real s-axis, where at $\sigma = \sigma_m^{crit}$, it coalesces into a double pole. For $\sigma > \sigma_m^{crit}$, the double pole again splits up into a pole pair $\{s_m, s_{-m}\}$ that remains on the real axis. We choose $s_m > s_{-m}$ in order to restore the uniqueness in the indexing of the poles.

2.3 Scattering by an inhomogeneous, lossy dielectric slab: singularity expansion method

2.3.1 Formulation of the problem

In this section, we return to the general problem formulated in Subsection 2.2.1, i.e. the application of the SEM to the transient scattering of a pulsed electromagnetic wave of finite duration by an inhomogeneous, lossy dielectric slab. For such a slab, the Laplace-transformed electric field is no longer known in closed form. As a consequence, alternative techniques have to be employed for the determination of the natural frequencies and the corresponding residual amplitudes, for the analysis of the contribution from the closing contours and for the decomposition of the integrand in the space-time region where this contribution does not vanish. In the upcoming subsections, these problems will be addressed.

2.3.2 Residual contributions

In this subsection, we describe the determination of the residual contributions corresponding to the poles $\{s_m\}$. These poles occur either on the negative real axis or in complex conjugate pairs in the left half of the s-plane. To each pole, there corresponds a complex natural-mode field distribution, defined by

$$L(s_m)E_m(z) = 0, \qquad E_m(1) = 1. \tag{2.3.1}$$

As we shall see later, the normalization in (2.3.1) facilitates the numerical computation of the natural frequencies and the corresponding field distributions. As in the case of the homogeneous slab, for special combinations of $\varepsilon_r(z)$ and $\sigma(z)$, $E(z,s)$ can have a pole of higher order. However, since the occurrence of poles of higher order can always be handled by considering limiting cases pertaining to simple poles, we shall restrict ourselves in what follows to the general case of simple poles only. From the asymptotic behavior of $E(z,s)$, which will be discussed in detail in Subsection 2.3.3, it follows that the poles can be ordered according to their increasing modulus. The Mittag-Leffler expansion of $E(z,s)$ can then be written as (see Nussenzveig (1972))

$$E(z,s) = \sum_m e_m(z)/(s - s_m) + E^e(z,s), \tag{2.3.2}$$

where $E^e(z,s)$ is an entire function and $e_m(z) = \text{res } E(z,s)|_{s=s_m}$.

Now, we can determine the residues $e_m(z)$ following a procedure that has been outlined by Marin (1973). First, we rewrite (2.3.2) as

$$E(z,s) = e_m(z)/(s - s_m) + E^e_m(z,s), \tag{2.3.3}$$

where $E^e_m(z)$ is a function which is analytic in a finite domain around $s = s_m$. Substituting (2.3.3) in (2.2.11) and using the analytic properties of $L(s)$, $E^i(z,s)$ and $E^e_m(z,s)$, we then find

$$e_m(z) = \lambda_m \, E_m(z), \tag{2.3.4}$$

in which λ_m is an unknown complex constant. Next, we define the *object product* of two functions f and g as

$$\langle f(z,s_1)|g(z,s_2)\rangle = \int_0^1 f(z,s_1)C(z,s_2)g(z,s_2)dz. \tag{2.3.5}$$

Note that the kernel in this object product is the contrast function $C(z,s)$ for the complex frequency $s = s_2$ that also occurs in the function g. Using the integral representation (2.2.9), we find that the operator $L(s)$ has the following property:

$$\langle f(z,s_1)|L(s_2)g(z,s_2)\rangle = \langle L(s_2)f(z,s_1)|g(z,s_2)\rangle. \tag{2.3.6}$$

Finally, combining (2.2.11), (2.3.1) and (2.3.6), we find

$$\langle [L(s) - L(s_m)]E(z,s)|E_m(z)\rangle = \langle E^i(z,s)|E_m(z)\rangle. \tag{2.3.7}$$

The unknown constant λ_m is determined by substituting in (2.3.7) the formulas (2.3.3) and (2.3.4) and the expansion

$$L(s) = L(s_m) + L'(s_m)(s - s_m) + O[(s - s_m)^2], \tag{2.3.8}$$

where L'(s) is the derivative operator of L(s):

$$L'(s)f(z) = \tfrac{1}{2}\int_0^1 \partial_s[sC(z',s)\exp(-s|z - z'|)]f(z')dz'. \tag{2.3.9}$$

Taking the limit $s \to s_m$ in (2.3.7), we then determine λ_m as

$$\lambda_m = F(s_m)D_m, \tag{2.3.10}$$

where the coupling coefficient D_m is now given by

$$D_m = \frac{\langle \exp(-s_m z)|E_m(z)\rangle}{\langle L'(s_m)E_m(z)|E_m(z)\rangle}. \tag{2.3.11}$$

D_m depends on the slab configuration only and defines the strength of the relevant natural mode. It is noted that because of the normalization of $E_m(z)$, D_m differs from the corresponding coefficient for the homogeneous slab (see (2.2.21)). In the actual numerical computation of D_m from $E_m(z)$ both a single and a double integral have to be computed.

Substitution of (2.3.2), (2.3.4) and (2.3.10) in the Bromwich contour integral finally yields the residual contribution of the poles to the time-domain response as

$$\mathcal{E}_{poles}(z,t) = \sum_m D_m F(s_m)E_m(z)\exp(s_m t). \tag{2.3.12}$$

Therefore, once the modal parameters $\{s_m, E_m(z), D_m\}$ have been computed numerically, the residual contribution of the poles of E(z,s) can be found by substituting $F(s_m)$ in (2.3.12).

2.3.3 Contribution from the closing contours

In the total time-domain response, the entire function $E^e(z,s)$ in (2.3.2) yields in the time domain the contribution of the supplementary arc either in the left or in the right half of the s-plane (see also Baum (1978)). As in the homogeneous case, the role of the entire function can be analyzed by constructing an asymptotic solution for $|s|\to\infty$. We restrict the constitutive parameters $\varepsilon_r(z)$ and $\sigma(z)$ to the class of functions that are continuous in $\mathcal{D}_2$ and whose derivatives are discontinuous at no more than a finite number of points in $\mathcal{D}_2$. From Erdélyi (1956), we know that inside such a slab, there exist two linearly independent solutions $u^{\pm}(z)$ whose first-order WKB approximations $u_a^{\pm}(z;z_0)$ are given by

$$u_a^{\pm}(z;z_0) = (N(z_0)/N(z))^{\frac{1}{2}}\exp\{\mp s\int_{z_0}^{z}[N(z') + \sigma(z')/2N(z')s]dz'\}, \tag{2.3.13}$$

with z and z_0 in $\mathcal{D}_2$ and $N(z) = \varepsilon_r^{\frac{1}{2}}(z)$. $u_a^{+}(z;z_0)$ and $u_a^{-}(z;z_0)$ represent wave-like solutions originating from $z = z_0$ and traveling in the direction of increasing and decreasing values of z, respectively. Since $u^{\pm}(z) = u_a^{\pm}(z;z_0)[1 + \mathcal{O}(s^{-1})]$, these formal solutions will show approximately the same behavior. In analogy with (2.2.15), we write the electric field in the form:

$$E(z,s) = \begin{cases} F(s)[\exp(-sz) + r(s)\exp(sz)] & \text{in } \mathcal{D}_1, \\ F(s)[a(s)u^{+}(z)/u^{+}(0) + b(s)u^{-}(z)/u^{-}(1)] & \text{in } \mathcal{D}_2, \\ F(s)[t(s)\exp(-s(z-1))] & \text{in } \mathcal{D}_3. \end{cases} \tag{2.3.14}$$

Application of the boundary conditions (continuity of E and $\partial_z E$ across $z = 0$ and $z = 1$) results in:

$$\begin{aligned} r(s) &= R_0^{+}(s) + T_0^{+}(s)D^{+}(s)R_1^{+}(s)D^{-}(s)T_0^{-}(s)/G(s), \\ a(s) &= T_0^{+}(s)/G(s), \\ b(s) &= T_0^{+}(s)D^{+}(s)R_1^{+}(s)/G(s), \\ t(s) &= T_0^{+}(s)D^{+}(s)T_1^{+}(s)/G(s), \\ G(s) &= 1 - R_0^{-}(s)D^{+}(s)R_1^{+}(s)D^{-}(s), \end{aligned} \tag{2.3.15}$$

with

$$\begin{aligned}
R_0^+(s) &= [su^+(0) + \partial_z u^+(0)]/[su^+(0) - \partial_z u^+(0)],\\
R_0^-(s) &= [\partial_z u^-(0) - su^-(0)]u^+(0)/[su^+(0) - \partial_z u^+(0)]u^-(0),\\
R_1^+(s) &= -[su^+(1) + \partial_z u^+(1)]u^-(1)/[su^-(1) + \partial_z u^-(1)]u^+(1),\\
T_0^+(s) &= 2su^+(0)/[su^+(0) - \partial_z u^+(0)], \qquad (2.3.16)\\
T_0^-(s) &= [u^+(0)\partial_z u^-(0) - u^-(0)\partial_z u^+(0)]/[su^+(0) - \partial_z u^+(0)]u^-(0),\\
T_1^+(s) &= [u^+(1)\partial_z u^-(1) - u^-(1)\partial_z u^+(1)]/[su^-(1) + \partial_z u^-(1)]u^+(1),\\
D^+(s) &= u^+(1)/u^+(0) \quad \text{and} \quad D^-(s) = u^-(0)/u^-(1).
\end{aligned}$$

In (2.3.16), T is the transmission factor at an interface and R the reflection factor. D combines the damping due to the conductivity and the time-delay phase factor corresponding to a single wave transit of the slab. The superscripts + and - denote propagation in the direction of increasing and decreasing z, respectively, while the subscripts 0 and 1 indicate the z-coordinates of the corresponding terminations of the slab. This physical interpretation is illustrated in the upper part of Figure 2.3.1. Note that the expressions (2.3.14) - (2.3.16) are invariant for a change in the normalization in either of the solutions $u^{\pm}(z)$, and that for each of the pairs $\{R_0^+(s),T_0^+(s)\}$, $\{R_0^-(s),T_0^-(s)\}$ and $\{R_1^+(s),T_1^+(s)\}$, we have the customary relation $1 + R = T$. From (2.3.15), it is observed that the natural frequencies $\{s_m\}$ are the zeros of the formal characteristic equation:

$$G(s) = 1 - R_0^-(s)D^+(s)R_1^+(s)D^-(s) = 0. \qquad (2.3.17)$$

The asymptotic approximation $E^a(z,s)$ to the electric field $E(z,s)$ is now obtained directly by replacing the functions $u^{\pm}(z)$ by their WKB approximations as given in (2.3.13). The derivatives $\partial_z u^{\pm}(z)$ are found in first-order approximation by utilizing the fact that differentiation of a WKB approximation of a given accuracy yields the derivative of the solution to the same order of accuracy (see Erdélyi (1956)). For the derivative of $u_a^{\pm}$ with respect to z, we find:

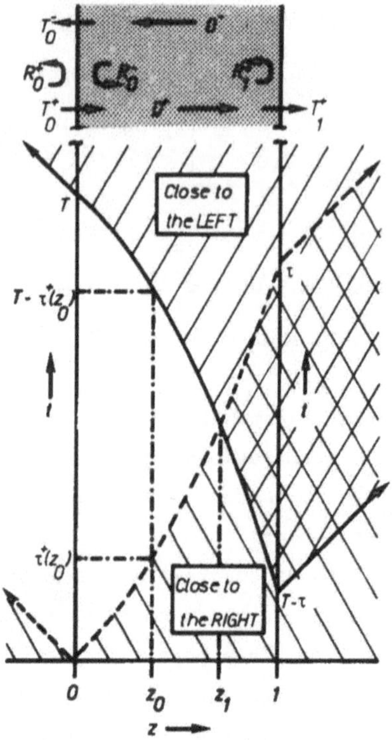

Figure 2.3.1 Upper part: physical interpretation of the reflection, transmission and damping factors occurring in the formal solution (see (2.3.16)). Lower part: space-time illustration of the conditions for vanishing closing-contour contributions for the inhomogeneous slab. In the unshaded region, the Bromwich contour cannot be closed without a possible contribution along the supplementary arcs.

$$\partial_z u_a^{\pm}(z;z_0) = \mp sN(z)u_a^{\pm}(z;z_0)[1 + O(s^{-1})]. \tag{2.3.18}$$

With the aid of (2.3.13) and (2.3.18), the asymptotic solution $E^a(z,s)$ can then, as in (2.2.24), be written in the form:

$$E^a(z,s) = \begin{cases} F(s)[\exp(-sz) + r_a(s)\exp(sz)] & \text{in } \mathcal{D}_1, \\ F(s)[a_a(s)u_a^+(z;0) + b_a(s)u_a^-(z;1)] & \text{in } \mathcal{D}_2, \\ F(s)[t_a(s)\exp(-s(z-1))] & \text{in } \mathcal{D}_3, \end{cases} \tag{2.3.19}$$

with, analogous to (2.2.25),

$$\begin{aligned} r_a(s) &= -R_{0a} + T_{0a}^+ D_a^+ R_{1a} D_a^- T_{0a}^- \exp(-2s\tau)/G_a(s), \\ a_a(s) &= T_{0a}^+/G_a(s), \\ b_a(s) &= T_{0a}^+ D_a^+ R_{1a} \exp(-s\tau)/G_a(s), \quad \text{and} \\ t_a(s) &= T_{0a}^+ D_a^+ T_{1a}^+ \exp(-s\tau)/G_a(s), \quad \text{with} \\ G_a(s) &= 1 - R_{0a} D_a^+ R_{1a} D_a^- \exp(-2s\tau). \end{aligned} \tag{2.3.20}$$

In (2.3.20), we have

$$\begin{aligned} R_{0a} &= (N_0 - 1)/(N_0 + 1), \quad R_{1a} = (N_1 - 1)/(N_1 + 1), \\ T_{0a}^+ &= 2/(N_0 + 1), \quad T_{0a}^- = 2N_0/(N_0 + 1), \quad T_{1a}^+ = 2N_1/(N_1 + 1), \\ D_a^{\pm} &= (N_0/N_1)^{\pm\frac{1}{2}} \exp[-\textstyle\int_0^1 (\sigma(z)/2N(z))dz], \quad \text{and} \\ \tau &= \textstyle\int_0^1 N(z)dz, \quad \text{with} \\ N_0 &= \lim_{z\downarrow 0} N(z) \quad \text{and} \quad N_1 = \lim_{z\uparrow 1} N(z). \end{aligned} \tag{2.3.21}$$

The (asymptotic) plane-wave reflection and transmission factors R and T introduced in (2.3.21) allow the same interpretation as their equivalents defined in (2.3.16). D now represents only the damping due to the conductivity; in (2.3.20), the time-delay phase factors $\exp(-s\tau)$ and $\exp(-2s\tau)$ have been taken out explicitly, with τ representing the one-way travel time. It is noted that only the possible asymmetry of the slab around $z = \frac{1}{2}$ causes (2.3.20) and (2.3.21) to differ in form from (2.2.25). Since the asymptotic behavior is uniform in $0 \le z \le 1$, we have

$$E(z,s) = E^a(z,s)[1 + \mathcal{O}(s^{-1})]. \tag{2.3.22}$$

Now, $E^a(z,s)$ becomes singular when

$$G_a(s) = 1 - R_{0a} D_a^+ R_{1a} D_a^- \exp(-2s\tau) = 0. \tag{2.3.23}$$

The solutions of the asymptotic transcendental equation (2.3.23) are the asymptotic natural frequencies $\{s_m^a\}$, with

$$s_m^a = \ln(R_{0a}D_a^+R_{1a}D_a^-)/2\tau + im\pi/\tau, \quad m\in\mathbb{Z}. \tag{2.3.24}$$

The relation between these asymptotic natural frequencies $\{s_m^a\}$ and their actual counterparts $\{s_m\}$ follows when, in the formal characteristic equation (2.3.17), a first-order Taylor expansion is carried out around $s = s_m^a$. This results in:

$$G(s_m^a) + (s - s_m^a)\partial_s G(s_m^a) + O((s - s_m^a)^2) = 0. \tag{2.3.25}$$

Substitution of the asymptotic approximations (2.3.13) and (2.3.18) in (2.3.17) yields the following estimates for the terms in (2.3.25) containing the characteristic denominator as $m\to\infty$:

$$\begin{aligned} G(s_m^a) &= G_a(s_m^a) + O(m^{-1}) = O(m^{-1}), \\ \partial_s G(s_m^a) &= \partial_s G_a(s_m^a)[1 + O(m^{-1})] = O(1), \end{aligned} \tag{2.3.26}$$

since $G_a(s_m^a) = 0$. With these estimates, (2.3.25) directly reduces to:

$$s_m = s_m^a + O(m^{-1}). \tag{2.3.27}$$

As in the homogeneous case, we can analyze the contribution of the supplementary arcs by choosing their radius such that the closing contour in the left half-plane always intersects the line $\{s|\ \mathrm{Re}(s) = \ln[R_{0a}D_a^+R_{1a}D_a^-]/2\tau\}$ in the middle between two asymptotic poles. For s on these contours, we then find the asymptotic behavior of $\exp(st)E(z,s)$ as $|s|\to\infty$, as indicated in Table 2.3.1. In that table,

$$\tau^+(z) = \int_0^z N(z')dz' \tag{2.3.28}$$

denotes the time it takes for the front of the pulse to travel in

Table 2.3.1 Asymptotic behavior as $|s|\to\infty$ of $\exp(st)E(z,s)$ in the case of an inhomogeneous slab.

closure in half-plane	reflected field $(-\infty<z<0)$	field inside slab $(0<z<1)$	transmitted field $(1<z<\infty)$
right	$O(1)\exp[s(t+z)]$	$O(1)\exp[s(t-\tau^{+}(z))]$	$O(1)\exp[s(t-\tau-(z-1))]$
left	$O(1)\exp[s(t+z-T)]$	$O(1)\exp[s(t+\tau^{+}(z)-T)]$	$O(1)\exp[s(t-T+\tau-(z-1))]$

the direction of increasing z through the slab from 0 to z. Following the same asymptotic evaluation procedure that has led to Table 2.2.3, we now find that the contributions along the closing contours vanish in the time intervals indicated in Table 2.3.2. The conditions in Table 2.3.2 are illustrated graphically in the space-time diagram shown in the lower part of Figure 2.3.1 for $T < 2\tau$. From

Table 2.3.2 Time intervals for which the contribution from closing contours at infinity vanishes in the case of an inhomogeneous, lossy dielectric slab.

closure in half-plane	reflected field $(-\infty<z<0)$	field inside slab $(0<z<1)$	transmitted field $(1<z<\infty)$
right	$-\infty<t<-z$	$-\infty<t<\tau^{+}(z)$	$-\infty<t<\tau+(z-1)$
left	$T-z<t<\infty$	$T-\tau^{+}(z)<t<\infty$	$T-\tau+(z-1)<t<\infty$

Figure 2.3.1, we observe that in that case there is some z_1 in $\mathcal{D}_2$ such that $T - \tau^{+}(z_1) = \tau^{+}(z_1)$. For $z > z_1$, the electric field can always be determined by closing the Bromwich contour such that the integral along the supplementary arc at infinity vanishes. For $z < z_1$, however, there is, as in the homogeneous case, always a time interval in which the possible contribution from the supplementary arcs remains to be investigated. When $T > 2\tau$, such an interval is, as in the homogeneous case, present for all z.

For an inhomogeneous slab, E(z,s) can generally only be determined numerically. Therefore, we can no longer decompose E(z,s) as indicated in (2.2.30) and (2.2.34). In principle, we can resolve this problem by evaluating numerically the integral along either of the closing contours or the original Bromwich contour for points in the unshaded region in Figure 2.3.1. However, for large s the function E(z,s) becomes hard to compute (see Subsection 2.3.4) and, furthermore, shows a rapidly oscillating behavior. Therefore, such a direct numerical integration is hard to carry out. In order to avoid the relevant computation, one can generalize (2.2.39) and take into account only the part of the incident pulse that has emerged into the slab at $t = t'$ by decomposing $F(t)$ as

$$F(t) = F_<(t;t') + F_>(t;t'), \tag{2.3.29}$$

where

$$F_<(t;t') = \begin{cases} F(t) & \text{for} \quad -\infty < t < t', \\ 0 & \text{for} \quad t' < t < \infty. \end{cases}$$

This decomposition is illustrated in Figure 2.3.2. From causality, $F_>(t;t')$ then only causes a reflected field in $\mathcal{D}_1$ for $t > t' - z$ and total fields in $\mathcal{D}_2$ and $\mathcal{D}_3$ for $t > t' + z$. As in (2.2.5), we introduce the, now t'-dependent, Laplace transform of $F_<$ as

$$F_<(s;t') = \begin{cases} 0 & \text{for } -\infty < t' \leq 0, \\ \int_0^{t'} \exp(-st)F(t)dt & \text{for } 0 \leq t' \leq T, \\ \int_0^{T} \exp(-st)F(t)dt = F(s) & \text{for } T \leq t' \leq \infty. \end{cases} \tag{2.3.30}$$

Carrying out the asymptotic analysis for $F_<$, we can then replace T by t' in both Table 2.3.2 and Figure 2.3.1. For z in $\mathcal{D}_3$ and $\tau + (z - 1) \leq t \leq T - \tau + (z - 1)$ as well as for z in $\mathcal{D}_2$ and

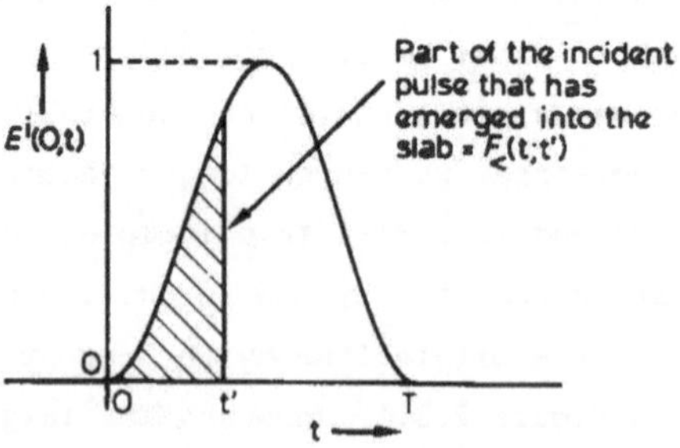

Figure 2.3.2 Truncation of the incident pulse at $t = t'$.

$\tau^+(z) \le t \le T - \tau^+(z)$, the integral along the closing contour in the left half-plane then vanishes if $F_<(s;t)$ plays the role of $F(s)$.

For the reflected field in $\mathcal{D}_1$, the contribution from $F_<(s;t')$ along the closing contour in the left half-plane vanishes for $t' - z < t < \infty$. In order to apply in $\mathcal{D}_1$ a similar decomposition as in $\mathcal{D}_2$ and $\mathcal{D}_3$, this time interval has, for each particular z, to be extended to $t' - z \le t < \infty$, since the total contribution from $F_>(s,t')$ vanishes for $-\infty < t \le t' - z$. Then, we can take into account $F_<(s;t + z)$ only. Now,

$$F_<(s,t + z)r_a(s)\exp(s(t + z)) = O(1) \text{ as } |s|\to\infty \text{ in } \mathrm{Re}(s) \le 0, \tag{2.3.31}$$

which is insufficient for making the integral along the closing contour vanish. We resolve this problem by rewriting the inversion integral in (2.2.13) as

$$\begin{aligned} \mathcal{E}(z,t) &= (2\pi i)^{-1}\int_{L_\beta} \exp(st)E^a(z,s)ds \\ &+ \lim_{M\to\infty} (2\pi i)^{-1}\int_{L_M} \exp(st)[E(z,s) - E^a(z,s)]ds. \end{aligned} \tag{2.3.32}$$

The second integral is written in the form that is later needed for the acceleration of the convergence in its numerical evaluation; in it L_M is the cut-off Bromwich contour, given by

$$L_M = \{s|\mathrm{Re}(s) = \beta,\ \beta \ge 0,\ -\Omega_M < \mathrm{Im}(s) < \Omega_M\}. \tag{2.3.33}$$

In (2.3.33), Ω_M is chosen such that the point $s = (s^a_{M+1} + s^a_M)/2$ lies on the semi-circular arc that closes L_M in the left half-plane. As in Subsection 2.2.4, the first contour integral can be evaluated analytically by expanding $1/G_a(s)$. We find

$$(2\pi i)^{-1}\int_{L_\beta} \exp(st)E^a(z,s)ds = \mathcal{E}^a(z,t), \tag{2.3.34}$$

with

$$E^a(z,t) = \begin{cases} E^i(z,t) - R_{0a}F(t+z) \\ \quad + T^+_{0a}D^+_aR_{1a}D^-_aT^-_{0a}G^a(t+z-2\tau) & \text{in } \mathcal{D}_1, \\ T^+_{0a}D^+_a(z)G^a(t-\tau^+(z)) \\ \quad + T^+_{0a}D^+_aR_{1a}D^-_a(z)G^a(t-\tau-\tau^-(z)) & \text{in } \mathcal{D}_2, \\ T^+_{0a}D^+_aT^+_{1a}G^a(t-(z-1)-\tau) & \text{in } \mathcal{D}_3, \end{cases} \tag{2.3.35}$$

where

$$G^a(t) = \sum_{k=0}^{\infty}(R_{0a}D^+_aR_{1a}D^-_a)^kF(t-2k\tau). \tag{2.3.36}$$

In (2.3.35), we have introduced the z-dependent damping factors

$$\begin{aligned} D^+_a(z) &= [N_0/N(z)]^{\frac{1}{2}}\exp[-\textstyle\int_0^z(\sigma(z')/2N(z'))dz'], \\ D^-_a(z) &= [N_1/N(z)]^{\frac{1}{2}}\exp[-\textstyle\int_z^1(\sigma(z')/2N(z'))dz'], \end{aligned} \tag{2.3.37}$$

and the corresponding travel times

$$\begin{aligned} \tau^+(z) &= \textstyle\int_0^z N(z')dz', \\ \tau^-(z) &= \textstyle\int_z^1 N(z')dz', \end{aligned} \tag{2.3.38}$$

where + and - again refer to the direction of propagation through the slab. As in (2.2.15), we can write the electric field outside the slab in the form:

$$E(z,s) = \begin{cases} F(s)[\exp(-sz) + r(s)\exp(sz)] & \text{in } \mathcal{D}_1, \\ F(s)[t(s)\exp(-s(z-1))] & \text{in } \mathcal{D}_3, \end{cases} \tag{2.3.39}$$

which, with (2.3.22), yields

$$\begin{aligned} F_<(s,t+z)[r(s) - r_a(s)]\exp(s(t+z)) &= \mathcal{O}(s^{-1}) \\ \text{as } |s|\to\infty \text{ in } \mathrm{Re}(s) &\le 0. \end{aligned} \tag{2.3.40}$$

This behavior enables us to close the contour L_M for z in $\mathcal{D}_1$ and $-z < t < T - z$ along a semi-circular arc in the left half of the s-plane such that the contribution of the closing contour vanishes as $M\to\infty$.

For the sake of completeness, we remark that if F(s) is replaced by $F_<(s;t')$ for some value of $t' < T$, this replacement also has to be carried out in the residual contributions. However, since both D_m and $E_m(z)$ depend only on the configuration, this will not cause computational difficulties.

Finally, we now have to consider the poles of $E^a(z,s)$ in the left half of the s-plane. Expanding $E^a(z,s)$ around $s = s_m^a$, we find

$$\operatorname{res} E^a(z,s)\Big|_{s=s_m^a} = e_m^a(z), \tag{2.3.41}$$

with

$$e_m^a(z) = \begin{cases} (2\tau)^{-1}T_{0a}^+D_a^+R_{1a}D_a^-T_{0a}^- \exp[s_m^a(z - 2\tau)] & \text{in } \mathcal{D}_1, \\ (2\tau)^{-1}\{T_{0a}^+D_a^+(z)\exp[-s_m^a\tau^+(z)] & \\ \quad + T_{0a}^+D_a^+R_{1a}D_a^-(z)\exp[-s_m^a(\tau + \tau^-(z))]\} & \text{in } \mathcal{D}_2, \\ (2\tau)^{-1}T_{0a}^+D_a^+T_{1a}^+ \exp[-s_m^a(z - 1 - \tau)] & \text{in } \mathcal{D}_3. \end{cases} \tag{2.3.42}$$

Decomposing the incident field as before and substituting (2.3.12), (2.3.34) and (2.3.41) into (2.3.32), we then arrive at

$$E(z,t) = \lim_{M\to\infty} E_M(z,t), \tag{2.3.43}$$

where $E_M(z,t)$ is defined as

$$\begin{aligned} E_M(z,t) = E^a(z,t) + \textstyle\sum_{m=-M}^{M}\{&D_mF_<(s_m;t')E_m(z)\exp(s_mt) \\ &- F_<(s_m^a;t')e_m^a(z)\exp(s_m^at)\}, \end{aligned} \tag{2.3.44}$$

with $t' = t + z$ in $\mathcal{D}_1$ and $t' = t$ elsewhere. For a given incident field, the expressions (2.3.43) and (2.3.44) yield $E(z,t)$ at all space-time points using modal parameters only.

The necessity of the decomposition (2.3.29) can also be illustrated by considering the residual contribution of the poles of $E^a(z,s)$ alone. Namely, the special form of (2.3.24), (2.3.41) and (2.3.42) allows the summation over m to be carried out analytically. We find

$$\sum_{m=-\infty}^{\infty} e_m^a(z)\exp(s_m^a t) = E_{poles}^a(z,t), \tag{2.3.45}$$

with

$$E_{poles}^a(z,t) = \begin{cases} T_{0a}^+ D_a^+ R_{1a} D_a^- T_{0a}^- G_{poles}^a(t + z - 2\tau) & \text{in } D_1, \\ T_{0a}^+ D_a^+(z) G_{poles}^a(t - \tau^+(z)) & \\ \quad + T_{0a}^+ D_a^+ R_{1a} D_a^-(z) G_{poles}^a(t - \tau - \tau^-(z)) & \text{in } D_2, \\ T_{0a}^+ D_a^+ T_{1a}^+ G_{poles}^a(t - (z - 1) - \tau) & \text{in } D_3, \end{cases} \tag{2.3.46}$$

and

$$G_{poles}^a(t) = \sum_{k=-\infty}^{\infty} (R_{0a} D_a^+ R_{1a} D_a^-)^k F(t - 2k\tau). \tag{2.3.47}$$

Comparing (2.3.46) with (2.3.35), we find that $E_{poles}^a(z,t)$ differs from $E^a(z,t)$ in three aspects. In the first place, the incident field for $z < 0$ travels towards the slab, i.e. in the direction of increasing z. Secondly, the directly reflected field corresponds to a wave inside the slab traveling in the direction of increasing z. Therefore, neither of these fields is included in the waves traveling away from the slab with a standing-wave counterpart inside the slab and they are therefore absent in (2.3.46). Finally, the pole contribution contains noncausal terms, namely the terms with $k < 0$ in $G_{poles}^a(t)$ (see (2.3.47)). Clearly both these differences have to be canceled by the integral along the closing contour in the left half of the complex s-plane.

In conclusion, we remark that the results obtained in this section are, as in the homogeneous case, in agreement with the principle of causality: an incident pulse cannot excite natural modes before it has entered the scattering obstacle.

2.3.4 Numerical implementation

For the inhomogeneous slab, which was discussed in Subsections 2.3.2 and 2.3.3, we have also implemented the SEM theory numerically. The natural frequencies and the corresponding natural modes are found as follows. Outside the slab, we redefine the electric field in analogy with (2.3.39) as

$$E(z,s) = \begin{cases} \exp(-sz)/t(s) + r(s)\exp(sz)/t(s) & \text{in } \mathcal{D}_1, \\ \exp[-s(z-1)] & \text{in } \mathcal{D}_3. \end{cases} \tag{2.3.48}$$

Since the boundary conditions require the continuity of $E(z,s)$ and $H(z,s)$ at both ends of the slab, the unknown coefficients $r(s)/t(s)$ and $1/t(s)$ in (2.3.48) can be computed by direct numerical integration of the system (2.2.6) from $z = 1$ to $z = 0$.

In our case, this integration is performed by a Runge-Kutta-Verner fifth- and sixth-order method (IMSL Library Reference Manual, 1982). The numerical aspects of this procedure will be discussed in more detail for a more general configuration in Section 2.4. The natural frequencies $\{s_m\}$ are then found by searching the zeros of $1/t(s)$ with Muller's method (see Traub (1964)). For the lower-order poles, the convergence to the correct pole is ensured by starting from the results for a homogeneous slab. Gradually increasing the inhomogeneity of the slab, we are then able to trace the root locus of each pole in the complex plane. For higher-order poles, the root-finding procedure is started at the corresponding asymptotic natural frequencies.

Special care is needed when two or more poles are located close together in the complex s-plane. In that case, convergence problems may arise in Muller's method, which can only handle isolated single poles. This problem can be resolved by determining the zeros of $1/t(s)$ as defined in (2.3.48) from integrals of the argument type:

$$(2\pi i)^{-1}\int_{L} s^{\ell} t'(s)/t(s)\,ds = -\sum_m (s_m)^{\ell}. \tag{2.3.49}$$

In (2.3.49), L denotes any closed contour in the complex s-plane, and the summation runs over the zeros inside L. To this end, a modified version of the algorithm described by Singaraju et al. (1976) was employed. Compared with Muller's method, this algorithm has the advantage that it locates *all* the zeros in a given search area. On the other hand, using it to accurately determine the zeros consumes a considerable amount of computation time due to the large number of function values needed in the evaluation of the contour integrals. Hence, the algorithm should only be used to determine the location of the zeros up to 2 or 3 significant digits in cases where Muller's method does not apply. The estimates thus obtained can then be refined by using them as starting values in Muller's method. Although the procedure above need not in general be applied in slab problems, it will be of use in more complicated configurations, for which suitable starting values are not available. Examples will be encountered in Sections 2.5 and 2.6, where we consider a two-dimensional scattering problem.

The natural modes $\{E_m(z)\}$ are automatically obtained in the search process. With their aid, the coupling coefficients $\{D_m\}$ can be determined by computing the relevant integrals (see (2.3.11)). For the single integral, a trapezoidal rule is applied. For the double integral, we repeat this trapezoidal rule, taking the singularity of the operator $L'(s)$ at $z = z'$ (see (2.3.9)) into account in an appropriate way.

A final difficulty arises in the summation of the thus-obtained residual contributions. For almost all space-time points, a limitation of $F(t)$ to the class of piecewise continuously differentiable functions is, as in the homogeneous case, sufficient to ensure a fast numerical convergence of the summation procedure given by (2.3.43) and (2.3.44). For z in D_1 and $-z < t < T - z$, however, where the acceleration of the convergence (2.3.32) is needed to suppress a nonzero contribution of the entire function, the numerical convergence is too slow. From the asymptotic analysis presented in Subsection 2.3.3, we find that for those space-time points, the time-domain field can be written as

$$E(z,t) = E_M(z,t) + \alpha(z,t)/M + O(M^{-2}) \quad \text{as } M\to\infty, \tag{2.3.50}$$

with a nonvanishing coefficient $\alpha(z,t)$. Since the computation time for the determination of the SEM-quantities s_m, $E_m(z)$ and D_m increases with increasing m, this implies that the time-domain field $E(z,t)$ at those space-time points cannot be obtained by direct numerical summation of the residual contributions. In order to circumvent this difficulty, the correction parameter $\alpha(z,t)$ is estimated from a least-squares fit of the known $E_M(z,t)$ as a function of 1/M on an M-interval where all higher-order terms in (2.3.50) are negligible. The true time-domain field is then obtained by extrapolation to 1/M = 0. For a typical scattering problem, this procedure is illustrated in Figure 2.3.3.

2.3.5 Inhomogeneous slab: numerical results

Results obtained

Using the computational scheme described in Subsection 2.3.4, we have obtained numerical results for several permittivity and conductivity profiles and various incident-pulse shapes. In this Subsection we show some representative examples. In Figures 2.3.4-2.3.7,

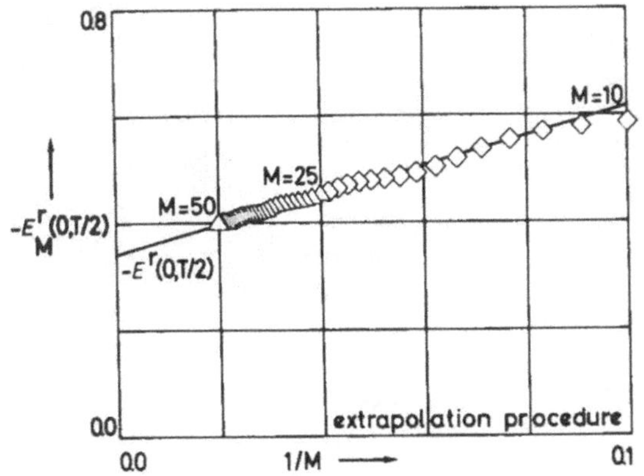

Figure 2.3.3 Illustration of the (second) acceleration of the convergence for the reflected field specified in Figure 2.3.7b. Only the points indicated by a triangle were taken into account in the extrapolation process.

we consider the scattering by a slab with a parabolic permittivity profile and a constant conductivity. Figure 2.3.4a shows the location of the first eleven natural frequencies in the complex s-plane both for the lossless case and for $Z_0\sigma d = 2$. In Figures 2.3.4b and 2.3.4c, the magnitudes of the corresponding coupling coefficients are plotted. From Figure 2.3.4, we observe that the actual modal parameters rapidly approach their asymptotic approximations. The approximations for $\mathrm{Im}(s_m)$ and D_m do not depend on the conductivity profile, which is in agreement with the theory developed in Subsection 2.3.3. The field distributions of some of the natural modes with $\sigma = 0$ are given in Figure 2.3.5. It is noted that both the real and the imaginary parts of $E_m(s)$ have exactly $|m|$ zero crossings in the interval $0 < z/d < 1$. Furthermore, the symmetry of the configuration induces the natural modes to be alternately symmetric and anti-symmetric around $z/d = \frac{1}{2}$. For $Z_0\sigma d = 2$, the field distributions closely resemble the ones shown in Figure 2.3.5. In Figure 2.3.6, we

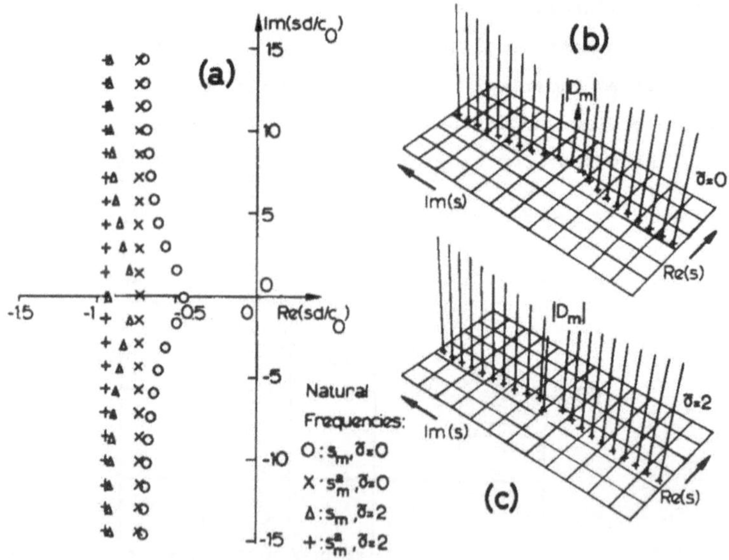

Figure 2.3.4 Configuration-dependent parameters plotted in the complex s-plane for a dielectric slab with a parabolic permittivity profile $\varepsilon_r(z) = 6.25 - 16(z/d - 0.5)^2$ and constant conductivity $Z_0\sigma d = 0$ and 2. (a): Complex natural frequencies s_m with $|m| \le 10$. (b,c): Magnitudes of the corresponding coupling coefficients D_m.

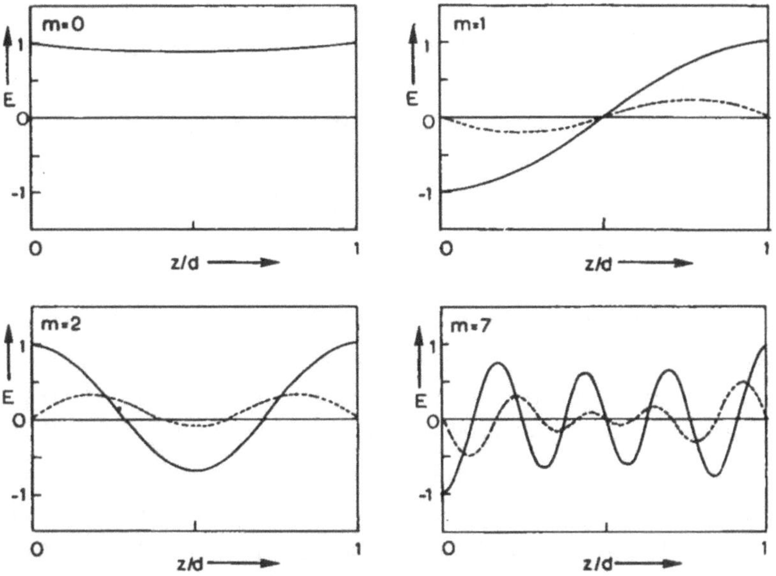

Figure 2.3.5 Field distributions of the natural modes with m = 0, 1, 2 and 7 for the lossless slab specified in Figure 2.3.4 (—— = real part, --- = imaginary part).

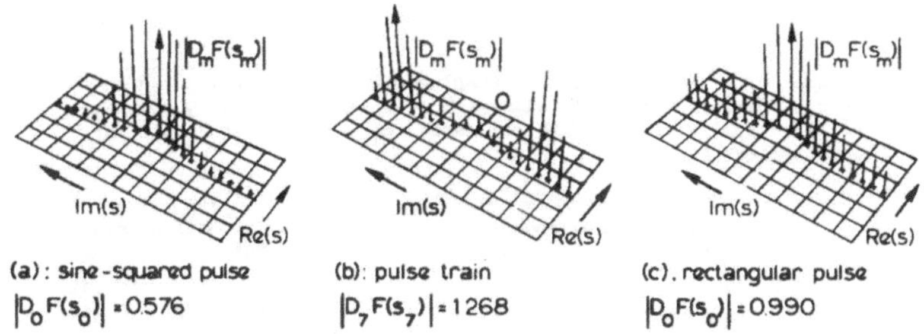

Figure 2.3.6 The magnitudes of the residual amplitudes $D_mF(s_m)$ for the lossless slab specified in Figure 2.3.4 for three different pulse shapes with $c_0T/d = 1.5$. (a) $F(t) = \sin^2(\pi t/T)\mathrm{rect}(t - T/2;T)$; (b) $F(t) = \sin(5\pi t/T)\mathrm{rect}(t - T/2;T)$; (c) $F(t) = \mathrm{rect}(t - 06T;0.8T)$.

have plotted, for the lossless case, the magnitudes of the actual residual amplitudes $D_m F(s_m)$ for three different incident-pulse shapes: a sine-squared pulse, a sinusoidal pulse train and a rectangular pulse. Clearly, the shape of the incident pulse can be employed to select which natural modes are to be excited. The incident, reflected and transmitted fields for the cases specified in Figure 2.3.6 are shown in Figure 2.3.7a,b,c. (The extrapolation procedure presented in Figure 2.3.3 was done for the rectangular pulse.) To illustrate the effect of conductivity on the scattered fields we have repeated the computation of Figure 2.3.7c for $Z_0 \sigma d = 2$. The resulting fields are shown in Figure 2.3.7d. The dependence of the scattered fields on the conductivity is found to be similar to that for the homogeneous slab (see Figure 2.2.7).

In Table 2.3.3 and Figures 2.3.8 and 2.3.9, we consider the scattering of a sine-squared pulse by an asymmetric configuration, namely a lossless dielectric slab with a linear permittivity profile. In Table 2.3.3, the values of the first eleven natural frequencies s_m and the coupling coefficients D_m are compared with their asymptotic approximation. For this configuration, the WKB-approximation is even better than for the case of a parabolic permittivity profile. In Figure 2.3.8, the field distribution of the natural modes with m = 0, 1, 2 and 7 is shown. Now the symmetry in the field distributions is no longer present. This implies (see (2.3.12)) that the residual amplitudes for the reflected field differ from the ones for

Table 2.3.3 The first eleven complex natural frequencies $s_m d/c_0$ for a lossless dielectric slab with a linear permittivity profile $\varepsilon_r(z) = 2.25 + 4z/d$ and the corresponding values of the coupling coefficient D_m.

m	Natural frequencies Actual	Natural frequencies WKB	Coupling coefficients Actual	Coupling coefficients WKB
0	-0.520	-0.602	0.637	0.741
1	-0.552 + i1.591	-0.602 + i1.539	-0.663 + i0.038	-0.741
2	-0.579 + i3.119	-0.602 + i3.077	0.700 - i0.045	0.741
3	-0.589 + i4.648	-0.602 + i4.616	-0.717 + i0.038	-0.741
4	-0.594 + i6.181	-0.602 + i6.155	0.725 - i0.033	0.741
5	-0.597 + i7.715	-0.602 + i7.694	-0.730 + i0.028	-0.741
6	-0.598 + i9.251	-0.602 + i9.232	0.734 - i0.023	0.741
7	-0.599 + i10.787	-0.602 + i10.771	-0.734 + i0.020	-0.741
8	-0.600 + i12.324	-0.602 + i12.310	0.735 - i0.019	0.741
9	-0.600 + i13.861	-0.602 + i13.849	-0.737 + i0.017	-0.741
10	-0.600 + i15.399	-0.602 + i15.387	0.738 - i0.015	0.741

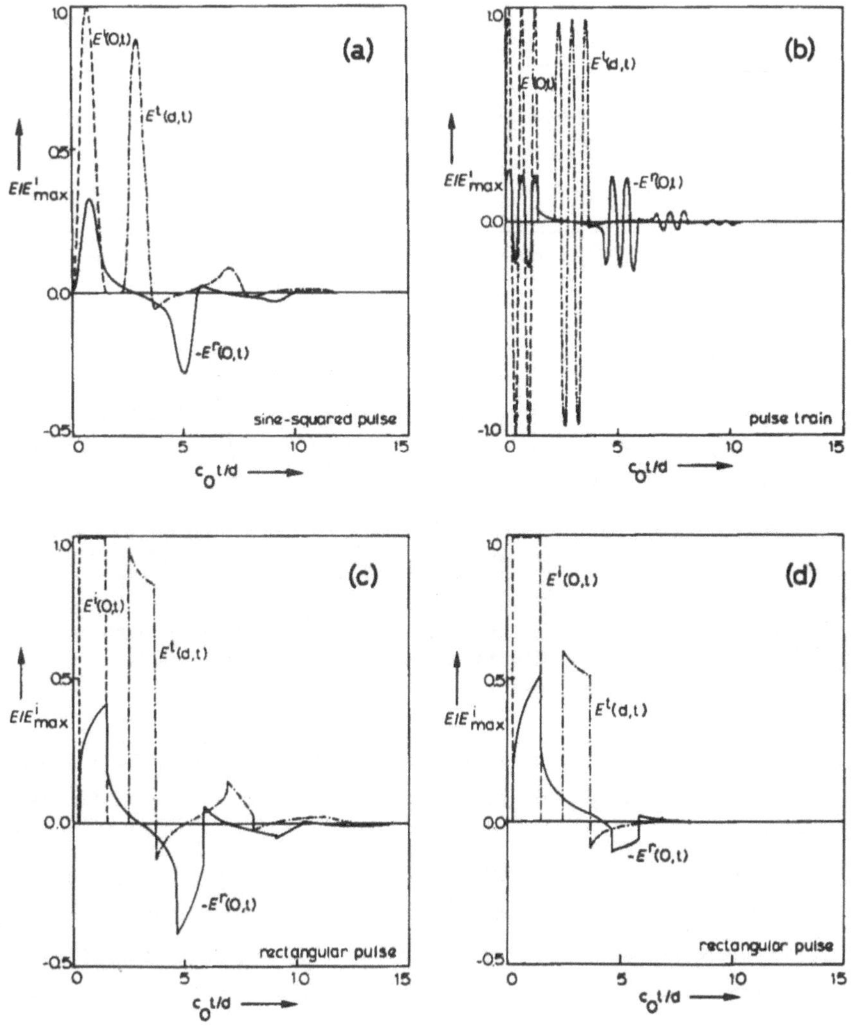

Figure 2.3.7 (a,b,c): Incident, reflected and transmitted fields for the lossless slab specified in Figure 2.3.4 and for the incident pulses specified in Figure 2.3.6. (d): Result of the same computation as in (c) for $Z_0\sigma d = 2$.

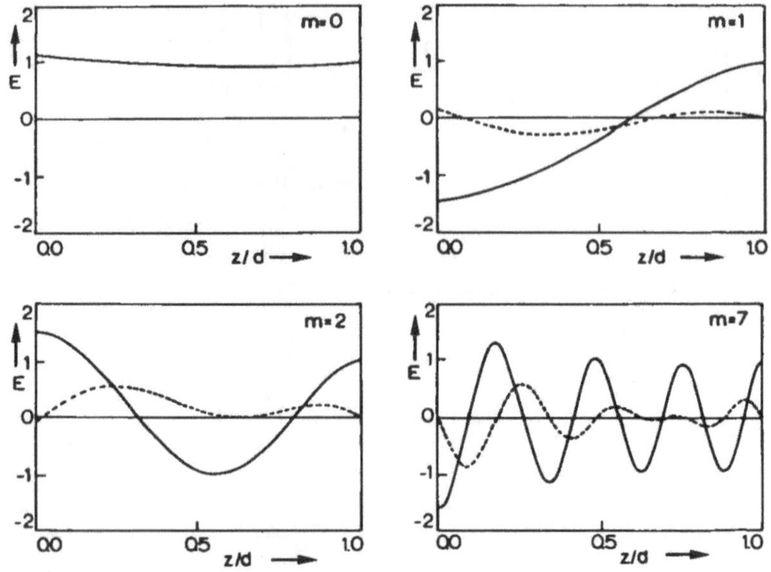

Figure 2.3.8 Field distributions of the natural modes with m = 0, 1, 2, and 7 for the configuration specified in Table 2.3.3 (—— = real part, --- = imaginary part).

the transmitted field; this is illustrated in Figures 2.3.9a and b. In Figure 2.3.9c, the corresponding incident and scattered fields are shown.

Finally, in Figures 2.3.10 - 2.3.12, natural frequencies are shown for some lossless slabs with more exotic permittivity profiles. In view of the symmetry relation $s_{-m} = s_m^*$, only the poles in the upper half of the complex s-plane are indicated. Figure 2.3.10 contains results for two sinusoidal profiles with two minima and two maxima inside the slab, respectively. In Figure 2.3.11, results are presented for two discontinuous, piecewise-constant profiles, and in Figure 2.3.12 for a truncated Epstein profile and for the corresponding actual configuration. It is observed that the natural frequencies only agree up to $|m| = 4$. For $|m| > 4$, the reflections at the small discontinuities in $\varepsilon_r(z)$ at the slab interfaces show their influence on and eventually dominate the location of the poles in the complex s-plane. In view of the small magnitude of the con-

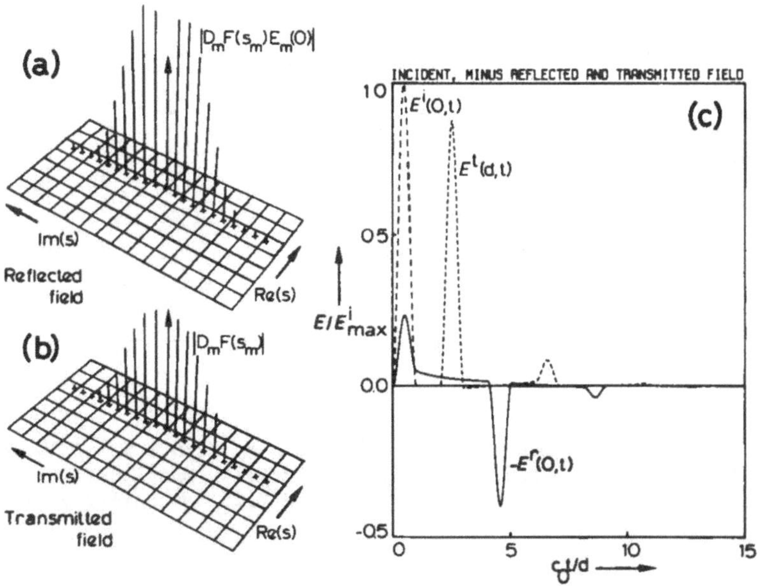

Figure 2.3.9 The magnitudes of the residual amplitudes $D_mF(s_m)E_m(0)$ (reflected field) and $D_mF(s_m)$ (transmitted field) and the corresponding incident, reflected and transmitted fields for the dielectric slab specified in Table 2.3.3 and for the incident pulse $F(t) = \sin^2(\pi t/T)\text{rect}(t - T/2;T)$ with $c_0T/d = 1$.

trast between both profiles (4×10^{-8} at the slab interfaces and decaying exponentially on the outside), it seems fair, therefore, to conclude that, for both configurations, the poles with $|m| > 4$ have little physical significance, if any.

Computational data

In the examples considered in Figures 2.3.4 - 2.3.11, the natural frequencies were computed with an accuracy of 10^{-6}. For those considered in Figures 2.3.4 - 2.3.9, they were computed up to $m = 50$. Taking into account the fact that both the real and the imaginary part of $E_m(z)$ have $|m|$ zero crossings inside the slab, we computed the coupling coefficients with a repeated trapezoidal rule on $51 + 25|m|$ points, i.e. approximately 50 points between two subsequent zero crossings. In the subsequent summation procedure to obtain the reflected and the transmitted field, we encountered two types of convergence. For the computation of the transmitted field

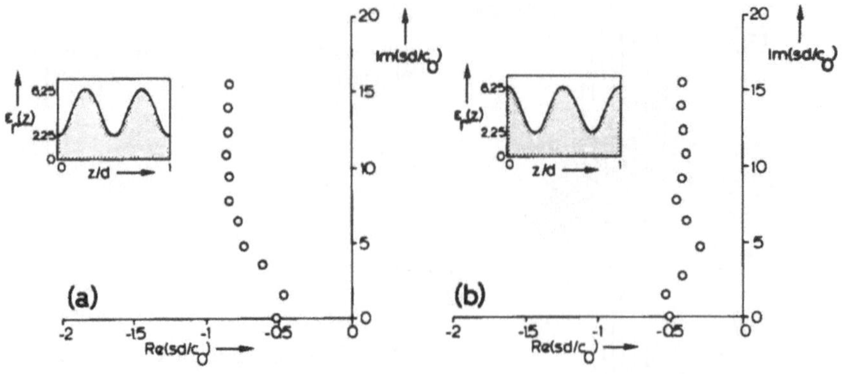

Figure 2.3.10 Complex natural frequencies s_m with $0 \le m \le 10$ for two lossless slabs with sinusoidal permittivity profiles. (a): $\varepsilon_r(z) = 2.25 + 4\sin^2(2\pi z/d)$; (b) $\varepsilon_r(z) = 2.25 + 4\cos^2(2\pi z/d)$.

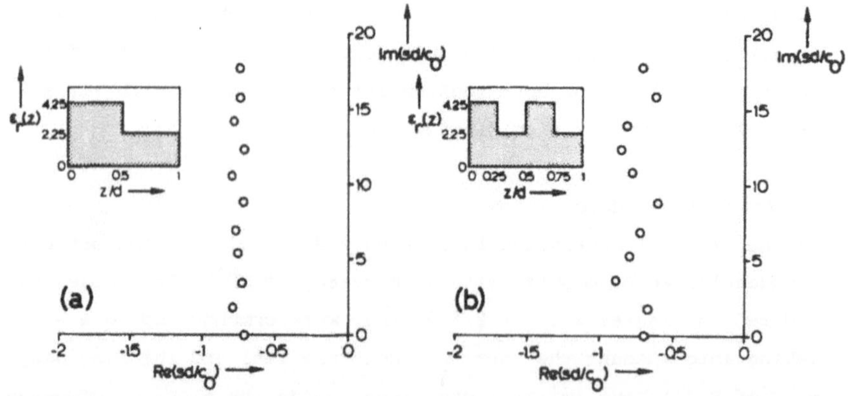

Figure 2.3.11 Complex natural frequencies s_m with $0 \le m \le 10$ for two lossless slabs with discontinuous, piecewise-constant permittivity profiles. (a): $\varepsilon_r(z) = 4.25 - 2U(z/d - 0.5)$; (b): $\varepsilon_r(z) = 4.25 - 2U(z/d - 0.25) + 2U(z/d - 0.5) - 2U(z/d - 0.75)$. U denotes the unit-step function.

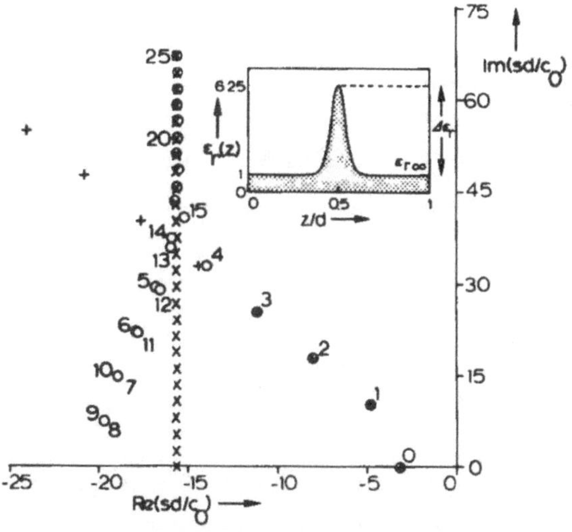

Figure 2.3.12 Complex natural frequencies for the lossless Epstein layer $\varepsilon_r(z) = \varepsilon_{r\infty} + \Delta\varepsilon_r/\cosh^2(z/2D)$ with $\varepsilon_{r\infty} = 1$, $\Delta\varepsilon_r = 5.25$ and $D = d/40$ with $0 \le m \le 7$ and for the truncated and shifted Epstein layer $\varepsilon_r(z) = [\varepsilon_{r\infty} + \Delta\varepsilon_r/\cosh^2((z/d - 0.5)/2D)]$ with $0 \le m \le 25$. +: actual layer; o: truncated layer; x: asymptotic poles for truncated layer. Inset: permittivity profile for truncated layer.

and the reflected field at $t + z/c_0 > T$, about 10 pole pairs were sufficient to obtain convergence in the accelerated series (2.3.44). For the reflected field at $t + z/c_0 < T$, however, use of the extrapolation procedure described in Section 2.3.4 on the partial sums $E_M(z,t)$ with $25 \le M \le 50$ was required.

The computation times for determining the natural frequencies s_m, the corresponding natural modes $E_m(z)$ and the coupling coefficients D_m all increased linearly with increasing $|m|$. For an inhomogeneous, lossy dielectric slab, a computation for $0 \le m \le 10$ took 15 seconds, for $11 \le m \le 30$ 55 seconds and for $31 \le m \le 50$ 2 minutes. The computation of the reflected and transmitted fields, including extrapolation, took about 2 seconds. For the configurations considered in Figures 2.3.10 - 2.3.12, only the natural frequencies indicated were computed. For the Epstein layer of Figure 2.3.12, special care was needed since the poles with $5 \le m \le 12$ occur in pairs that are located close together. As outlined in Subsection 2.3.4, the natural frequencies with $5 \le m \le 14$ were first determined up to four significant digits from argument-type integrals. This computation took about 5 minutes. Subsequently, the values obtained were used as starting values in Muller's method, along with the analytically obtained natural frequencies of the Epstein layer for $0 \le m \le 4$, and the asymptotic natural frequencies s_m^a for $m \ge 15$. Thus the natural frequencies with $0 \le m \le 25$ were improved up to an accuracy of 10^{-8} in about one minute. In both computations, the accuracy in the Runge-Kutta-Verner integration was refined to 10^{-8} to account for the contrast of 4×10^{-8} at the slab boundaries.

Appendix 2.3.A The natural modes of a symmetric Epstein layer

In this appendix, we investigate the natural modes of a horizontally stratified, lossless dielectric medium, whose permittivity is given by the symmetric Epstein profile

$$\varepsilon_r(z) = \varepsilon_{r\infty} + \Delta\varepsilon_r/\cosh^2(z/2D) \tag{2.3.51}$$

(see the inset of Figure 2.3.12). For this configuration, the fundamental differential equation (2.2.7) reduces to

$$[\partial_z^2 + s^2\varepsilon_{r\infty} + s^2\Delta\varepsilon_r/\cosh^2(z/2D)]E(z,s) = 0. \tag{2.3.52}$$

As shown in the literature, the solution to this equation is known in closed form both in terms of hypergeometric functions (see e.g. Brekovskikh (1980)) and in terms of a Barnes contour integral (see Whittaker and Watson (1952)), i.e. a Laplace inversion integral (see De Hoop (1965)). The normalization condition $E(1) = 1$, as introduced in (2.3.1), is now replaced by

$$E(z,s) = \exp(-sN_\infty z)[1 + o(1)] \quad \text{when } z\to\infty, \tag{2.3.53}$$

in which $N_\infty = \varepsilon_{r\infty}^{1/2}$. Like (2.3.1), the condition (2.3.53) selects a solution that, as $z\to\infty$, corresponds to a wave of unit amplitude traveling to the right. The only difference between both conditions is that in (2.3.1) the wave originates from the rear end of the slab, i.e. $z = 1$, and in (2.3.53) from $z = 0$. Following De Hoop, we write the solution of (2.3.52) that meets (2.3.53) as

$$E(z,s) = \frac{[1 + \exp(z/D)]^{b/2}}{2\pi i} \int_{L_p} \exp(pz/D)w(p)dp, \tag{2.3.54}$$

with

$$w(p) = \frac{\Gamma(1+\beta_2 - \beta_1)\Gamma(-p + \alpha_1)\Gamma(-p + \alpha_2)\Gamma(p - \beta_1)}{\Gamma(\alpha_1 - \beta_1)\Gamma(\alpha_2 - \beta_1)\Gamma(1 - p + \beta_2)}, \tag{2.3.55}$$

where $\Gamma(p)$ denotes the gamma function. In (2.3.54) and (2.3.55), we have

$$\begin{aligned} b &= 1 + [1 - 16\Delta\varepsilon_r s^2 D^2]^{1/2}, \\ \alpha_1 &= -N_\infty sD, \\ \alpha_2 &= N_\infty sD, \\ \beta_1 &= -b/2 - N_\infty sD, \\ \beta_2 &= -b/2 + N_\infty sD. \end{aligned} \tag{2.3.56}$$

L_p is the path of integration in the complex p-plane and extends to infinity parallel to the imaginary p-axis; as shown in Figure 2.3.13, the sequences of simple poles $p = \alpha_1 + n$ and $p = \alpha_2 + n$ $(n = 0,1,2,\ldots,\infty)$ lie to the right of L_p, and the sequence $p = \beta_1 - n$ $(n = 0,1,2,\ldots,\infty)$ lies to its left. For $z < 0$, the integral in (2.3.54) can be evaluated by closing L_p to the right in the complex p-plane and applying Cauchy's theorem. With the aid of some standard identities for the gamma functions occurring in the residues (see e.g. De Hoop (1965)), we find

$$\begin{aligned} E(z,s) = &\frac{\Gamma(1 + \beta_2 - \beta_1)\Gamma(-\alpha_1 + \alpha_2)}{\Gamma(\alpha_2 - \beta_1)\,\Gamma(1 - \alpha_1 + \beta_2)} \exp(-sN_\infty z)S_1(z,s) \\ &+ \frac{\Gamma(1 + \beta_2 - \beta_1)\Gamma(\alpha_1 - \alpha_2)}{\Gamma(\alpha_1 - \beta_1)\Gamma(1 - \alpha_2 + \beta_2)} \exp(sN_\infty z)S_2(z,s), \end{aligned} \tag{2.3.57}$$

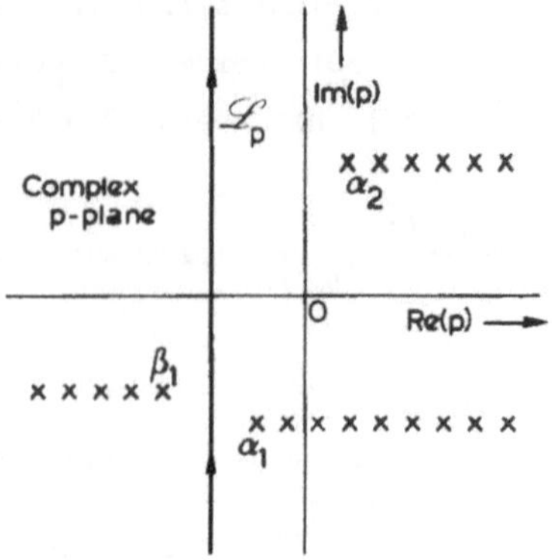

Figure 2.3.13 Barnes contour L_p and simple poles of w(p) as specified in (2.3.55) plotted in the complex p-plane.

where

$$S_1(z,s) = \sum_{n=0}^{\infty} \frac{\exp(nz/D)(-1)^n}{n!} \frac{\Gamma(\alpha_1 + n - \beta_1)\Gamma(\alpha_1 + n - \beta_2)\Gamma(1 + \alpha_1 - \alpha_2)}{\Gamma(\alpha_1 - \beta_1)\Gamma(\alpha_1 - \beta_2)\Gamma(1 + \alpha_1 + n - \alpha_2)},$$

and

$$S_2(z,s) = \sum_{n=0}^{\infty} \frac{\exp(nz/D)(-1)^n}{n!} \frac{\Gamma(\alpha_2 + n - \beta_1)\Gamma(\alpha_2 + n - \beta_2)\Gamma(1 + \alpha_2 - \alpha_1)}{\Gamma(\alpha_2 - \beta_1)\Gamma(\alpha_2 - \beta_2)\Gamma(1 + \alpha_2 + n - \alpha_1)}.$$

As $z \to -\infty$, the terms with n = 0 dominate the sums $S_1(z)$ and $S_2(z)$, and hence we have

$$\lim_{z \to -\infty} S_{1,2}(z,s) = 1. \tag{2.3.58}$$

The expression for E(z,s) given in (2.3.57) represents a natural mode when it corresponds to a wave traveling to the left as $z \to -\infty$ (see also (2.3.53)), i.e. when the first term on the right-hand side of (2.3.57) vanishes. In order to find the natural frequencies $s = s_m$ for which this happens, we first investigate the denominator $\Gamma(\alpha_2 - \beta_1)\Gamma(1 - \alpha_1 + \beta_2)$. This denominator has poles when either

$$\alpha_2 - \beta_1 = b/2 + 2N_\infty sD = -m, \tag{2.3.59a}$$

or

$$1 - \alpha_1 + \beta_2 = 1 - b/2 + 2N_\infty sD = -m, \tag{2.3.59b}$$

where $m = 0,1,2,\ldots,\infty$. Substitution of the definition of b (see (2.3.56)) reduces (2.3.59) to:

$$[2m + 1 + 4N_\infty sD]^2 = 1 - 16\Delta\varepsilon_r s^2 D^2, \tag{2.3.60}$$

which is a quadratic equation for sD. Its solutions are:

$$sD = \{-N_\infty(m + \tfrac{1}{2}) \pm [(m + \tfrac{1}{2})^2 \varepsilon_{r\infty} - m(m + 1)\varepsilon_{r0}]^{\frac{1}{2}}\}/2\varepsilon_{r0}, \tag{2.3.61}$$

with $m = 0,1,2,\ldots,\infty$ and $\varepsilon_{r0} \triangleq \varepsilon_r(0) = \varepsilon_{r\infty} + \Delta\varepsilon_r$. In (2.3.61), the square root is defined according to its principal value, i.e. either $\mathrm{Re}(\ldots) > 0$, or $\mathrm{Re}(\ldots) = 0$ and $\mathrm{Im}(\ldots) \geq 0$.

Next we have to check whether any of the poles of the denominator are canceled by a pole in the numerator $\Gamma(1 + \beta_2 - \beta_1)\Gamma(-\alpha_1 + \alpha_2)$. It turns out that only the combination of $m = 0$ and the plus sign in (2.3.61) needs to be excluded. In that case, we have $s = 0$, and hence also $-\alpha_1 + \alpha_2 = 0$. The uniqueness in the indexing of the natural frequencies is restored by defining for $m = 0$

$$s_0 = -N_\infty/2\varepsilon_{r0}D, \tag{2.3.62}$$

and for $m > 0$ s_m and s_{-m} as the solutions with the corresponding signs in (2.3.61). Substitution of (2.3.61) and (2.3.62) in (2.3.59) shows that:

(i) For all $\varepsilon_{r\infty}$ and $\Delta\varepsilon_r$, s_0 satisfies (2.3.59b).

(ii) For $m > 0$ and $\Delta\varepsilon_r < \varepsilon_{r\infty}/(2m + 1)^2$, s_{-m} satisfies (2.3.59a) and s_m satisfies (2.3.59b).

(iii) For $m > 0$ and $\Delta\varepsilon_r > \varepsilon_{r\infty}/(2m + 1)^2$ both s_{-m} and s_m are solutions to (2.3.59b).

Finally, we consider the asymptotic behavior of the natural-mode field distributions as $z \to -\infty$. Substitution of either (2.3.59a) or (2.3.59b) in (2.3.57) directly results in:

$$E_m(z) = E(z,s_m) = (-1)^m \exp(sN_\omega z)[1 + o(1)] \quad \text{when } z\to-\infty, \qquad (2.3.63)$$

which is in agreement with the observation that, for a symmetric configuration, these distributions should be alternately symmetric and anti-symmetric.

2.4 Scattering by an inhomogeneous, lossy dielectric slab in between two homogeneous, lossless half-spaces: Fourier inversion

2.4.1 Formulation of the problem

The transient fields discussed in the previous two sections can also be obtained by a direct numerical evaluation of the inversion integral in (2.2.13). In the present section, we consider this procedure. In view of a later application, we consider a configuration that is slightly more general than the one discussed until now, namely an inhomogeneous dielectric slab as specified in Subsection 2.2.1, situated in between two homogeneous, lossless dielectric half-spaces (Figure 2.4.1). The constitutive parameters for this configuration have been specified in Table 2.4.1. It is remarked that the choice of a lossless surrounding medium is not essential. The only motivation for this choice is that it leads to a closed-

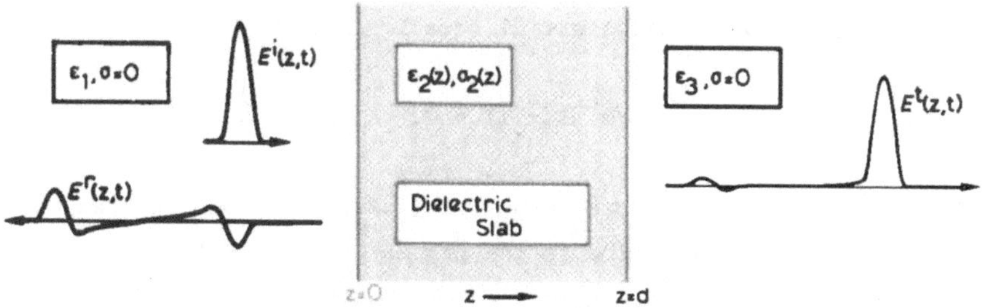

Figure 2.4.1 A pulsed plane wave normally incident on an inhomogeneous, lossy dielectric slab situated in between two homogeneous, lossless dielectric halfspaces.

Table 2.4.1 Subdivision of the configuration into domains

domain	z-coordinate	permittivity	conductivity	permeability
$\mathcal{D}_1$	$-\infty<z<0$	$\varepsilon(z)=\varepsilon_1\geq\varepsilon_0$	$\sigma(z)=0$	$\mu(z)=\mu_0$
$\mathcal{D}_2$	$0<z<d$	$\varepsilon(z)=\varepsilon_2(z)\geq\varepsilon_0$	$\sigma(z)=\sigma_2(z)\geq 0$	$\mu(z)=\mu_0$
$\mathcal{D}_3$	$d<z<\infty$	$\varepsilon(z)=\varepsilon_3\geq\varepsilon_0$	$\sigma(z)=0$	$\mu(z)=\mu_0$

form time-domain representation for the fields outside the slab. For example, for a linearly polarized electromagnetic pulse of finite duration T incident from $\mathcal{D}_1$, the electric-field intensity and the magnetic-field intensity can, for $\sigma(z) = 0$ in $\mathcal{D}_1$, be written as

$$\underline{E}^i = F(t - z/c_1)\underline{i}_y, \quad \underline{H}^i = -Y_1 F(t - z/c_1)\underline{i}_x, \tag{2.4.1}$$

where $Y_1 = (\varepsilon_1/\mu_0)^{\frac{1}{2}}$, $c_1 = (\varepsilon_1\mu_0)^{-\frac{1}{2}}$ (c_1 is the wave speed in $\mathcal{D}_1$). Going through the same Laplace-transform and normalization procedures as in Subsection 2.2.1, we arrive at the same differential equations for the transformed and normalized field quantities, i.e. the system of first-order differential equations (2.2.6):

$$\begin{aligned} \partial_z H(z,s) &= s(\varepsilon_r(z) + \sigma(z)/s)E(z,s), \\ \partial_z E(z,s) &= sH(z,s), \end{aligned} \tag{2.4.2}$$

and the second-order differential equation (2.2.7):

$$[\partial_z^2 - s^2(\varepsilon_r(z) + \sigma(z)/s)]E(z,s) = 0. \tag{2.4.3}$$

The analogon of the integral equation (2.2.9) and some related integral equations pertaining to the present configuration will be discussed further on in Subsection 2.4.4. Here, we will first concentrate on solving the differential equations. From (2.4.3), it follows that, as in the case of the slab embedded in vacuum (see (2.3.14)), we can express the electromagnetic field outside the slab in terms of plane-wave solutions:

$$\begin{aligned} E(z,s) &= \begin{cases} F(s)[\exp(-sN_1 z) + r(s)\exp(sN_1 z)] & \text{in } \mathcal{D}_1, \\ F(s)[t(s)\exp(-sN_3(z-1))] & \text{in } \mathcal{D}_3, \end{cases} \\ H(z,s) &= \begin{cases} F(s)N_1[-\exp(-sN_1 z) + r(s)\exp(sN_1 z)] & \text{in } \mathcal{D}_1, \\ F(s)N_3[-t(s)\exp(-sN_3(z-1))] & \text{in } \mathcal{D}_3, \end{cases} \end{aligned} \tag{2.4.4}$$

where $N_1 = \varepsilon_{1r}^{\frac{1}{2}}$, $N_3 = \varepsilon_{3r}^{\frac{1}{2}}$, with $\varepsilon_{1r} = \varepsilon_1/\varepsilon_0$ and $\varepsilon_{3r} = \varepsilon_3/\varepsilon_0$, and where r(s) and t(s) denote the reflection and transmission coefficients of the slab, respectively. For a general inhomogeneous slab, the solution in $\mathcal{D}_2$ is not known in closed form and has to be determined numerically. One solution procedure will be discussed in Subsection 2.4.2.

Once E(z,s) is solved from either of the equations (2.4.2) or (2.4.3), the corresponding time-domain field follows from the evaluation of the inversion integral (2.2.13):

$$\mathcal{E}(z,t) = (2\pi i)^{-1}\int_{L_\beta} \exp(st)E(z,s)ds. \tag{2.4.5}$$

The customary frequency-domain response is obtained by taking out the spectrum of the incident pulse:

$$E(z,s) = F(s)\hat{E}(z,s) \tag{2.4.6}$$

and replacing s by $i\omega$ with ω being a real-valued angular frequency. Apart from the complex time factor $\exp(i\omega t)$, $\hat{E}(z,i\omega)$ represents the field excited by a monochromatic plane wave with unit amplitude incident from $\mathcal{D}_1$. By taking $\beta = 0$ in the Bromwich contour in (2.4.5) (see (2.2.14)), we obtain

$$\mathcal{E}(z,t) = (2\pi)^{-1}\int_{-\infty}^{\infty}\exp(i\omega t)F(i\omega)\hat{E}(z,i\omega)d\omega, \tag{2.4.7}$$

which expresses the time-domain field in terms of this plane-wave response. In the upcoming subsections, we will investigate the numerical evaluation of the expression (2.4.7).

2.4.2 Numerical evaluation of the plane-wave response

Let us assume that the incident-pulse shape $\mathcal{F}(t)$ and its Laplace transform F(s) are known in closed form. Then the only factor in the integrand in (2.4.7) that needs to be determined numerically is the plane-wave response $\hat{E}(z,s)$. Since, in the application of the SEM, $\hat{E}(z,s)$ is also required for complex s (see Subsection 2.3.4), we formulate the solution scheme for a general, complex s. The form of

$\hat{E}(z,s)$ outside the slab is obtained directly by setting $F(s)$ to 1 in (2.4.4).

As remarked above, the distribution of $\{\hat{E}(z,s),\hat{H}(z,s)\}$ inside the slab is generally not known in closed form and has to be computed numerically. One way to perform this operation is to solve the initial-value problem of the system (2.4.2) by a direct numerical integration. Since such a procedure requires an initial value for the field intensities $\hat{E}$ and $\hat{H}$ at a given starting point, we use the linearity of the problem to introduce the normalized solution $\{\hat{e}(z,s),\hat{h}(z,s)\} = \{\hat{E}(z,s),\hat{H}(z,s)\}/t(s)$, which is known in $\mathcal{D}_3$. Application of the boundary conditions (continuity of $\hat{e}$ and $\hat{h}$ across discontinuities in $\varepsilon_r(z)$ and $\sigma(z)$) at $z = 1$ then yields the starting values

$$\hat{e}(1,s) = 1, \quad \hat{h}(1,s) = -N_3. \tag{2.4.8}$$

This allows $\hat{e}(z,s)$ and $\hat{h}(z,s)$ inside the slab to be determined by the direct numerical integration of (2.4.2). Special care is needed in the choice of the integration procedure. From (2.4.2) and (2.4.3) it is observed that the variation of $\hat{e}(z,s)$ and $\hat{h}(z,s)$ with z speeds up when the magnitude of one or more of the parameters s, $\varepsilon_r(z)$ and $\sigma(z)$ increases. Since we want the computational scheme to be generally applicable and yet efficient in the case of less rapid variations in $\hat{e}$ and $\hat{h}$, the procedure should adjust the step size in the integration to the local variation of the solution. We employed a fifth- and sixth-order Runge-Kutta-Verner procedure that is available in the IMSL Fortran Library (1982). This procedure attempts to keep the global error in the solution proportional to a specified tolerance.

Equating the solution obtained at $z = 0$ with the normalized version of the expressions given in (2.4.4), we then have

$$\begin{aligned} \hat{e}(0,s) &= 1/t(s) + r(s)/t(s), \\ \hat{h}(0,s)/N_1 &= -1/t(s) + r(s)/t(s). \end{aligned} \tag{2.4.9}$$

The values of r(s) and t(s) follow from solving the pair of linear equations (2.4.9) for 1/t(s) and r(s)/t(s). As mentioned in Subsection 2.3.4, this procedure is also suitable for the determination of the natural frequencies $\{s_m\}$, which are found by searching the zeros of 1/t(s) in the complex s-plane.

It should be mentioned that the Runge-Kutta-Verner method is not applicable locally when $\varepsilon_r(z)$, $\sigma(z)$ or one of their first few derivatives is discontinuous at some $z = z_0$ with $0 < z_0 < 1$. This problem can be solved by terminating the Runge-Kutta-Verner integration at $z = z_0$, and using $\hat{e}(z_0,s)$ and $\hat{h}(z_0,s)$ as the initial values for a subsequent integration in the region $z < z_0$. For the case of a single discontinuity in the medium parameters, the entire computational procedure as described above has been visualized in Figure 2.4.2.

The robustness of the procedure is illustrated in Tables 2.4.2 - 2.4.4. In Table 2.4.2, we have listed, for a lossless homogeneous slab embedded in vacuum and excited by a monochromatic plane wave

Table 2.4.2 Computational data from the numerical solution of the scattering by a homogeneous lossless slab with $\varepsilon_{2r} = 4$ in vacuum for increasing angular frequency ω. The tolerance in the Runge-Kutta-Verner algorithm was 10^{-6}.

Frequency ω	Absolute error in $r(i\omega)$	Absolute error in $t(i\omega)$	Computation time	Number of subroutine calls
1	1.5×10^{-8}	2.3×10^{-8}	0.02s	63
10	1.7×10^{-7}	2.4×10^{-7}	0.10s	532
100	1.8×10^{-6}	2.8×10^{-6}	0.90s	4993
1000	1.7×10^{-5}	2.6×10^{-5}	8.77s	49509

with angular frequency ω, the numerical effects of increasing ω. For a number of frequencies we have given the deviations of the numerically obtained $r(i\omega)$ and $t(i\omega)$ from the corresponding exact values. Since these coefficients are obtained from $\hat{e}(0,i\omega)$ and $\hat{h}(0,i\omega)$, which are the least accurate field values obtained in the numerical integration of (2.4.2), their errors are a good indication of the computational accuracy achieved. In addition, we have listed the computation times consumed and the number of calls to the subroutine that evaluates the right-hand side of (2.4.2) for given z, $\hat{e}$ and $\hat{h}$. Both these quantities are representative of the number of steps

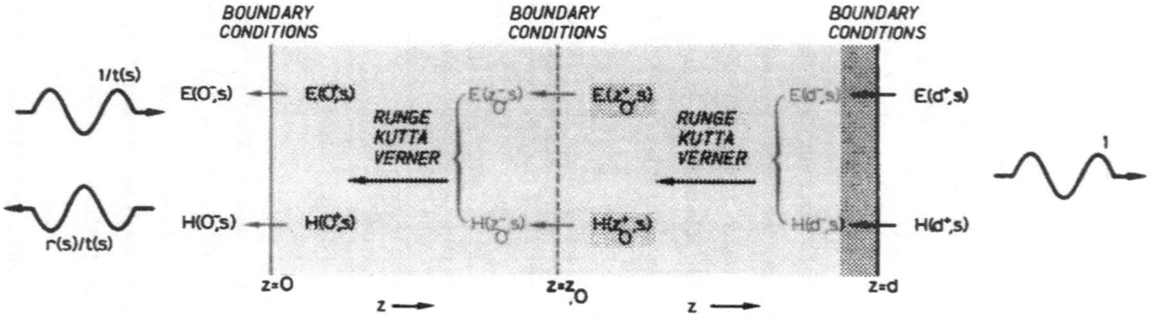

Figure 2.4.2 Visualization of the numerical procedure for the determination of $\{E(z,s),H(z,s)\}$. Inside the slab, the medium parameters are assumed to be continuous, except for a single discontinuity at $z = z_0$.

taken in the numerical integration. It is observed that all data listed in the table increase linearly with increasing frequency. This can be explained as follows. A Runge-Kutta-Verner integration with a specified tolerance over a single oscillation in $\hat{e}(z,i\omega)$ and $\hat{h}(z,i\omega)$, i.e. a single wavelength, requires a fixed number of steps and causes a fixed growth in the global error in the solution. Now, (2.4.3) shows that the number of oscillations in the interval $0 \le z \le 1$ is of $O(|\omega|)$. Consequently, the number of steps and the errors in $\hat{e}(0,i\omega)$ and $\hat{h}(0,i\omega)$ are also of $O(|\omega|)$. If both $r(i\omega)$ and $t(i\omega)$ are of $O(1)$, it then follows from (2.4.9) that their errors also increase linearly with $|\omega|$. In spite of the growing magnitudes of these errors, the results presented in Table 2.4.2 justify the conclusion that up to $|\omega| = 100$, the field distributions are computed with a precision better than the specified tolerance, and up to $|\omega| = 1000$ with an error of at most a factor of 10 larger.

In Tables 2.4.3 and 2.4.4, the numerical effects of enhancing the

Table 2.4.3 Computational data from the numerical solution of the scattering by a homogeneous lossless slab in vacuum for fixed angular frequency $\omega = 10$ and increasing permittivity ε_{2r}. The tolerance in the Runge-Kutta-Verner algorithm was 10^{-6}.

Relative permittivity ε_{r2}	Absolute error in $r(i\omega)$	Absolute error in $t(i\omega)$	Computation time	Number of subroutine calls
1	0	1.9×10^{-7}	0.05s	269
10	6.8×10^{-7}	8.1×10^{-7}	0.15s	827
100	1.1×10^{-6}	1.1×10^{-6}	0.45s	2546
1000	3.9×10^{-7}	2.7×10^{-7}	1.54s	8812

Table 2.4.4 Computational data from the numerical solution of the scattering by a homogeneous, lossy dielectric slab in vacuum for fixed permittivity $\varepsilon_{2r} = 1$, fixed angular frequency $\omega = 10$, and increasing conductivity σ_2. The tolerance in the Runge-Kutta-Verner algorithm was 10^{-6}.

Conductivity σ_2	Absolute error in $r(i\omega)$	Relative error in $t(i\omega)$	Computation time	Number of subroutine calls
1	2.9×10^{-9}	1.6×10^{-7}	0.05s	277
10	4.3×10^{-12}	9.4×10^{-8}	0.08s	437
100	6.2×10^{-17}	1.2×10^{-7}	0.27s	1470
1000	6.2×10^{-17}	3.4×10^{-7}	0.78s	4293

constitutive parameters ε_{2r} and σ_2 are illustrated. Note that in Table 2.4.4, the error in $t(i\omega)$ is specified in relative rather than absolute magnitude. This was done because $t(i\omega)$ decays exponentially with increasing conductivity σ_2. It turns out that the accuracy of $r(i\omega)$ and $t(i\omega)$ is not affected by raising the constitutive coefficients. The only effect is an increase in the number of operations and, hence, in the computation time required. The computation of Table 2.4.4 was also carried out for $\varepsilon_{2r} = 100$ with similar results.

2.4.3 Numerical evaluation of the time-domain response

Once the frequency-domain response is known, the corresponding transient response follows directly from the inversion integral (2.4.7). The most straightforward procedure for the evaluation of this integral seems to be the Fast-Fourier-Transform algorithm. Since this algorithm requires the truncation of the integrand at some $|\omega| = \omega_{max}$, we have to investigate the asymptotic behavior of that function as $|s| = |\omega| \to \infty$. To this end, we utilize the same first-order WKB approximations $u_a^{\pm}(z;z_0)$ that were used in Subsection 2.3.3 (see (2.3.13)). Replacing ε_r by ε_{2r} and σ by σ_2, and matching the resulting expressions with the ones given in (2.4.4), we find $\hat{E}(z,s) = \hat{E}^a(z,s)(1 + \mathcal{O}(s^{-1}))$, with

$$\hat{E}^a(z,s) = \begin{cases} \exp(-sN_1 z) + r_a(s)\exp(sN_1 z) & \text{in } \mathcal{D}_1, \\ a_a(s)u_a^+(z;0) + b_a(s)u_a^-(z;1) & \text{in } \mathcal{D}_2, \\ t_a(s)\exp(-sN_3(z-1)) & \text{in } \mathcal{D}_3, \end{cases} \tag{2.4.10}$$

with, analogous to (2.3.20),

$$\begin{aligned} r_a(s) &= -R_{21}^a + T_{12}^a D_a^+ R_{23}^a D_a^- T_{21}^a \exp(-2s\tau)/G_a(s), \\ a_a(s) &= T_{12}^a/G_a(s), \\ b_a(s) &= T_{12}^a D_a^+ R_{23}^a \exp(-s\tau)/G_a(s), \quad \text{and} \\ t_a(s) &= T_{12}^a D_a^+ T_{23}^a \exp(-s\tau)/G_a(s), \quad \text{with} \\ G_a(s) &= 1 - R_{21}^a D_a^+ R_{23}^a D_a^- \exp(-2s\tau). \end{aligned} \tag{2.4.11}$$

In (2.4.11), the damping factors $D_a^{\pm}$ and the travel time τ are the same as in (2.3.21), while the asymptotic plane-wave reflection and transmission coefficients are defined as

$$\begin{aligned} R_{21}^a &= (N_2(0) - N_1)/(N_2(0) + N_1), \\ T_{12}^a &= 2N_1/(N_1 + N_2(0)), \\ T_{21}^a &= 2N_2(0)/(N_2(0) + N_1), \\ R_{23}^a &= (N_2(1) - N_3)/(N_2(1) + N_3), \\ T_{23}^a &= 2N_2(1)/(N_2(1) + N_3). \end{aligned} \tag{2.4.12}$$

The constants R_{21}^a and R_{23}^a denote the reflection coefficients for a plane wave incident from $\mathcal{D}_2$ on the interface with $\mathcal{D}_1$ and $\mathcal{D}_3$, respectively. The constants T_{12}^a, T_{21}^a, and T_{23}^a are the similarly defined transmission coefficients. The physical interpretation of the coefficients defined in (2.4.12) is illustrated in Figure 2.4.3. From (2.4.10) - (2.4.12), it appears that, for real-valued frequencies, we have for all z:

$$\hat{E}^a(z, i\omega) = \mathcal{O}(1) \quad \text{as } |\omega| \to \infty. \tag{2.4.13}$$

Hence, the rate of convergence of the integral in (2.4.7) is mainly dependent on $F(i\omega)$, i.e. on the shape of the incident pulse. This is essentially the same result that was obtained in Subsections 2.2.4 and 2.3.4 for the rate of convergence of the SEM representation. When $F(t)$ is a continuous, differentiable function we can refine

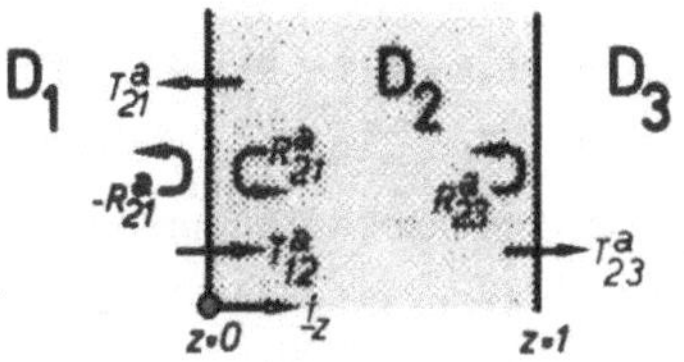

Figure 2.4.3 Physical interpretation of the asymptotic reflection and transmission coefficients defined in (2.4.12).

(2.2.28) and (2.2.42) to:

$$F(i\omega) = O(\omega^{-2}) \quad \text{as } |\omega|\to\infty \quad \text{with } \mathrm{Im}(\omega) = 0, \tag{2.4.14}$$

which, in combination with the oscillatory behavior of the integrand in (2.4.7), suffices for an efficient application of the FFT algorithm. When the incident pulse is only piecewise continuously differentiable, however, we are left with (2.2.42), which is insufficient for such a computation. In that case, the asymptotic solution (2.4.10) must be used to accelerate the rate of convergence by one order of ω^{-1}. This results in

$$\begin{aligned} E(z,t) = {} & (2\pi)^{-1}\int_{-\infty}^{\infty}\exp(i\omega t)F(i\omega) \\ & [\hat{E}(z,i\omega) - \hat{E}^{a}(z,i\omega)]d\omega + E^{a}(z,t), \end{aligned} \tag{2.4.15}$$

with

$$E^{a}(z,t) = \begin{cases} E^{i}(z,t) - R^{a}_{21}F(t + N_1 z) \\ \quad + T^{a}_{12}D^{+}_{a}R^{a}_{23}D^{-}_{a}T^{a}_{21}G^{a}(t + N_1 z - 2\tau) & \text{in } \mathcal{D}_1, \\ T^{a}_{12}D^{+}_{a}(z)G^{a}(t - \tau^{+}(z)) \\ \quad + T^{a}_{12}D^{+}_{a}R^{a}_{23}D^{-}_{a}(z)G^{a}(t - \tau - \tau^{-}(z)) & \text{in } \mathcal{D}_2, \\ T^{a}_{12}D^{+}_{a}T^{a}_{23}G^{a}(t - N_3(z - 1) - \tau) & \text{in } \mathcal{D}_3, \end{cases} \tag{2.4.16}$$

where

$$G^{a}(t) = \sum_{k=0}^{\infty}(R^{a}_{21}D^{+}_{a}R^{a}_{23}D^{-}_{a})^{k}F(t - 2k\tau). \tag{2.4.17}$$

In (2.4.16), the z-dependent damping factors $D^{\pm}_{a}(z)$ and the corresponding travel times $\tau^{\pm}(z)$ are the same as the ones defined in (2.3.37) and (2.3.38). The acceleration procedure summarized in (2.4.15) - (2.4.17) is the generalization for the present configuration of the procedures formulated in (2.2.46) - (2.2.48) and in (2.3.32) - (2.3.36). Note that this procedure is only applicable when $u^{\pm}_{a}(z;z_0)$ is a valid asymptotic approximation, i.e. when $\varepsilon_{2r}(z)$

and $\sigma_2(z)$ are continuous in $\mathcal{D}_2$ with derivatives that are discontinuous at no more than a finite number of points. In the numerical evaluation of the inversion formulas (2.4.7) and (2.4.15), the unwanted periodicity effects of the FFT algorithm (aliasing) were avoided by choosing the frequency-domain sampling such that the resulting time-domain signal extends over twice the interval required. Moreover, a doubling of the frequency interval was simulated by adding to the sampled Fourier transform an equal number of zeros and doubling the order of the FFT.

Some representative results that were obtained with the aid of the computational schemes outlined in the previous and the present subsections are shown in Figures 2.4.4 - 2.4.7. In Figure 2.4.4, the magnitudes of the Fourier-transformed reflected, transmitted and incident fields, i.e. $E^r(0,i\omega) = F(i\omega)r(i\omega)$, $E^t(d,i\omega) = F(i\omega)t(i\omega)$ and $E^i(0,i\omega) = F(i\omega)$, have been plotted as a function of ω for the lossless configuration specified in Figure 2.3.4 excited by the rectangular pulse specified in Figure 2.3.6c. For the reflected and transmitted field, the corresponding magnitudes of the WKB approximations given in (2.4.10) were also indicated. It is observed that these approximations remain accurate even down to $|\omega d/c_0| \approx 2$. This was also observed in a similar comparison of the real and imaginary parts of $r_a(i\omega)$ and $t_a(i\omega)$ with the corresponding exact values. As can be seen by comparing Figures 2.4.4a, b with Figure 2.4.4c, the zeros that show up in the spectra plotted in Figures 2.4.4a and 2.4.4b are due to the factor of $F(i\omega)$ and, hence, to the shape of the incident pulse. In fact, these zeros are located exactly at $\omega d/c_0 = 5m\pi/2T$, with $m = \pm1, \pm2, \ldots, \pm\infty$. The local behavior of the spectra in between these zeros can be identified as being associated with the natural frequencies $s = s_m$, as given in Figure 2.3.4 (see also Fiorito et al. (1981)). This identification is observed in more detail from Figures 2.4.5 and 2.4.6, which contain complex-frequency data. Figure 2.4.5 illustrates the behavior of the reflection and transmission coefficients $r(s)$ and $t(s)$ as functions of s for the same slab as considered in Figures 2.3.4 and 2.4.4. The sharp peaks in the magnitudes displayed mark the location of the natural frequencies (see also Figures 2.3.4 and 2.3.6c). Note that these natu-

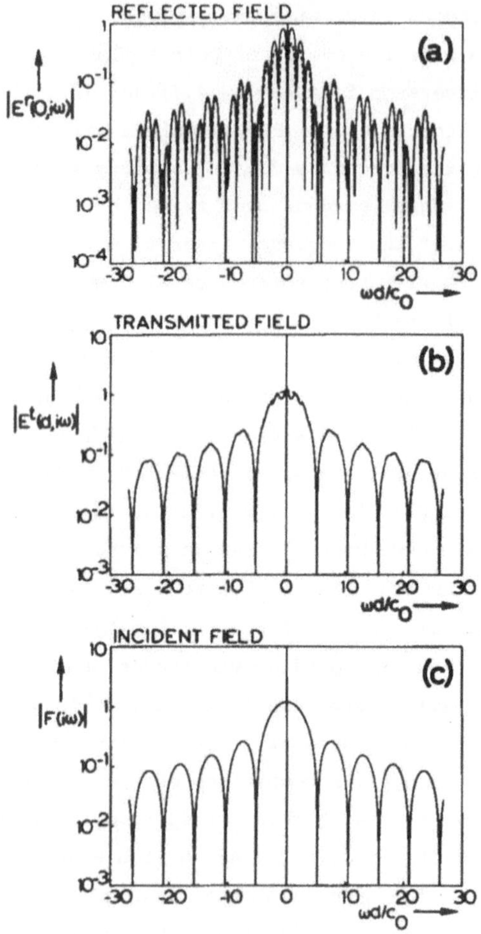

Figure 2.4.4 The magnitudes of the Fourier-transformed fields $E^r(0,i\omega)$, $E^t(d,i\omega)$ and $E^i(0,i\omega) = F(i\omega)$ for the configuration specified in Figure 2.3.4 excited by the rectangular pulse specified in Figure 2.3.6c, plotted as a function of ω. In Figures a, b, the dashed lines denote the WKB approximation $F(s)\hat{E}^a(z,s)$.

ral frequencies give rise to a minimum (an absorbtion peak) in $|r(i\omega)|$ and a maximum (a resonance) in $|t(i\omega)|$. The influence of the incident pulse is elucidated in Figure 2.4.6, where the magnitudes of the Laplace-transformed reflected and transmitted fields for the scattering problem considered in Figures 2.3.4, 2.3.6c,

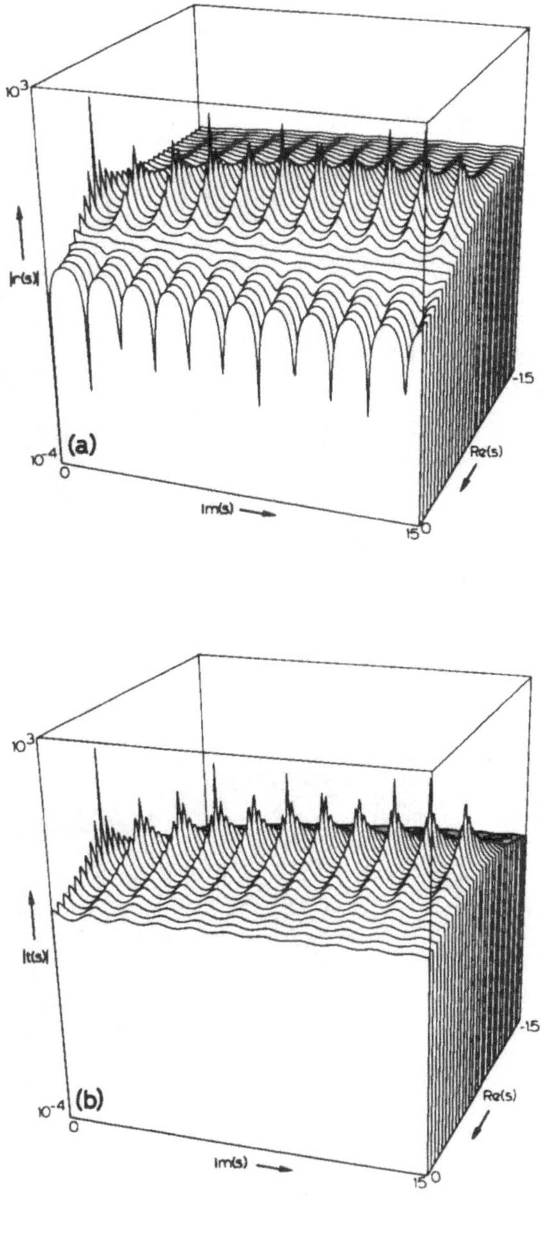

Figure 2.4.5 The magnitudes (logarithmic scale) of the reflection and transmission coefficients r(s) and t(s) for the slab specified in Figure 2.3.4 as a function of the complex frequency s.

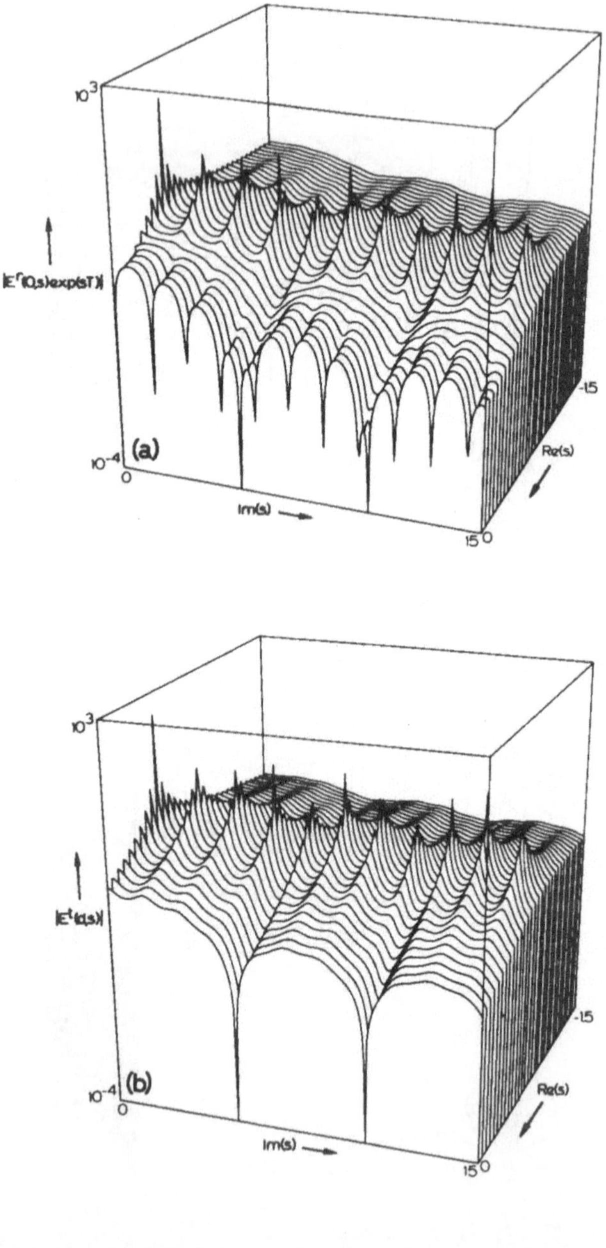

Figure 2.4.6 The magnitudes (logarithmic scale) of the Laplace-transformed quantities $E^r(0,s)\exp(sT)$ and $E^t(d,s)$ for the slab specified in Figure 2.3.4 and the rectangular pulse specified in Figure 2.3.6c as a function of the complex frequency s.

2.3.7c, 2.4.4, and 2.4.5 are plotted as a function of s. In Figure 2.4.6a, the reflected field $E^r(0,s) = F(s)r(s)$ was multiplied by the entire function exp(sT) to account for the exponential increase in F(s) with decreasing Re(s). Note that the cross-sections in the front planes are identical to the corresponding parts of the curves shown in Figures 2.4.4a, b.

The transient signals corresponding to the spectra shown in Figures 2.4.4 and 2.4.6 are presented in Figure 2.4.7a. These signals were determined by numerically evaluating the accelerated inversion formula (2.4.15). The numerical integration was performed by applying a fast Fourier transformation of order 1024 to an array containing 512 zeros and 512 frequency-domain field values covering a frequency interval of twice the range indicated in Figure 2.4.4. The required numerical solution of 257 initial-value problems as outlined in Subsection 2.4.2 took 69 seconds. Within the expected accuracy, the results plotted in Figure 2.4.7a agree well with the SEM results plotted in Figure 2.3.7c.

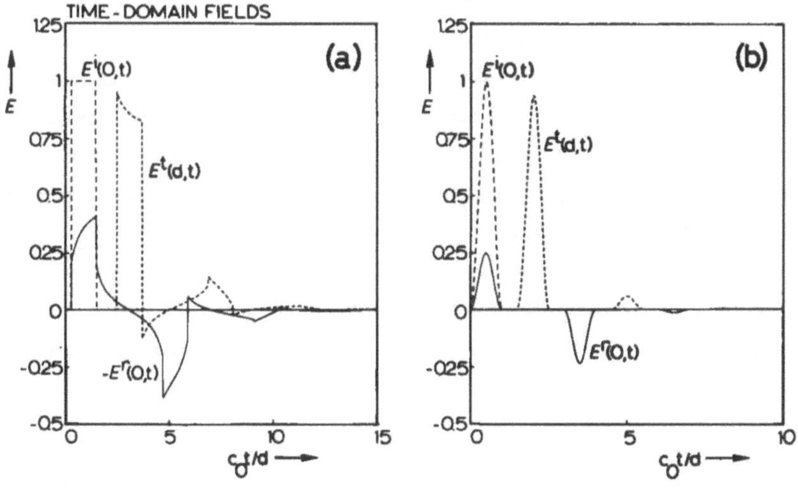

Figure 2.4.7 (a): Transient signals corresponding to the spectra shown in Figures 2.4.4 and 2.4.6 (see also Figure 2.3.7c). (b): Same signals for a lossless three-layer configuration with $\varepsilon_{1r} = \varepsilon_{3r} = 6.25$ and $\varepsilon_{2r} = 2.25$, excited by an incident pulse given by $E^i(0,t) = F(t) = \sin^2(\pi t/T)\mathrm{rect}(t - T/2;T)$ with $c_0T/d = 1$.

In Figure 2.4.7b, we present the reflected and transmitted field for a configuration where the slab has a lower permittivity than the surrounding half-spaces namely a homogeneous, lossless slab with permittivity $\varepsilon_{2r} = 2.25$ embedded in a homogeneous, lossless dielectric with $\varepsilon_{1r} = \varepsilon_{3r} = 6.25$ and excited by a sine-squared pulse. In this case, the transient signals could be computed by a direct numerical evaluation of the inversion integral (2.4.7) with a fast Fourier transformation of order 512. The computation time for the frequency-domain solution was 19 seconds. The reduction in computation time compared with the previous case is due to the lower order of the transform: the Runge-Kutta-Verner integration is carried out for less frequencies covering a smaller frequency range. The signals shown in Figure 2.4.7b and the corresponding closed-form solutions in terms of repeatedly reflected and transmitted waves will be encountered again as reference solutions in Section 3.3.

2.4.4 Integral relations for the frequency-domain field

For the sake of completeness, we will now formulate some integral relations for the frequency-domain field $E(z,s)$. As will be argued further on in this subsection, it is numerically unfavorable to solve $E(z,s)$ from the integral equations that result from them. However, they will be of considerable use, both in solving the time-domain direct-scattering problem (see Section 3.3), and in understanding and solving the inverse-scattering problem (Chapters 5 and 6). The starting point of the analysis is the selection of a *reference medium* with constitutive parameters $\{\bar{\varepsilon}_r(z),\bar{\sigma}(z)\}$. As in (2.2.10), we express the contrast between the actual media $\{\varepsilon_r(z),\sigma(z)\}$ and the reference medium in terms of the *contrast function*

$$C(z,s) \stackrel{\Delta}{=} \varepsilon_r(z) - \bar{\varepsilon}_r(z) + [\sigma(z) - \bar{\sigma}(z)]/s. \qquad (2.4.18)$$

The Green's function of the reference configuration is defined as being the solution of

$$[\partial_z^2 - s^2\bar{\varepsilon}_r(z) - s\bar{\sigma}(z)]G(z,z';s) = -\delta(z - z'). \tag{2.4.19}$$

Combining (2.4.18) and (2.4.19) with the second-order differential equation (2.4.3), which describes the actual scattering problem, directly results in the integral relation

$$\hat{E}(z,s) = \bar{E}(z,s) - s^2\int_{-\infty}^{\infty} C(z',s)G(z,z';s)\hat{E}(z',s)dz', \tag{2.4.20}$$

when $z\in\mathbb{R}$, in which $\bar{E}(z,s)$ represents the field that a plane wave with unit amplitude excites in the reference medium. Equation (2.4.20) is a generalization of (2.2.9), where the reference medium was a vacuum, i.e. $\bar{\varepsilon}_r(z) = 1$, $\bar{\sigma}(z) = 0$ for all z. The relation (2.4.20) becomes an integral equation for $\hat{E}(z,s)$ when z is restricted to the region where the contrast function differs from zero.

Next, we will discuss some possible choices for the reference medium and their consequences for the relation (2.4.20). The most obvious choice for the reference medium outside the slab is to take it identical to the actual one, i.e. $\bar{\sigma}(z) = 0$ and $\bar{\varepsilon}_r(z) = \varepsilon_{1r}, \varepsilon_{3r}$, in $\mathcal{D}_1$ and $\mathcal{D}_3$. In that case, the integration in (2.4.20) ranges over $\mathcal{D}_2$ only, and we avoid all difficulties due to a possible indefinite integral. Moreover, we end up with an integral equation for the field in $\mathcal{D}_2$ only. The choice of $\bar{\varepsilon}_{2r}(z)$ and $\bar{\sigma}_2(z)$ seems less straightforward. Let us first consider the general case of an inhomogeneous slab. For that case, $G(z,z';s)$ is not known in closed form. Formally, we can obtain it as follows. Let $\bar{e}_\pm(z,s)$ be solutions of (2.4.3) in the reference medium, normalized such that

$$\bar{e}_+(z,s) = \begin{cases} \exp(-sN_1z)/\bar{t}^+(s) \\ \quad + [\bar{r}^+(s)/\bar{t}^+(s)]\exp(sN_1z) & \text{in } \mathcal{D}_1, \\ \exp(-sN_3(z-1)) & \text{in } \mathcal{D}_3, \end{cases}$$

$$\bar{e}_-(z,s) = \begin{cases} \exp(sN_1z) & \text{in } \mathcal{D}_1, \\ \exp(sN_3(z-1))/\bar{t}^-(s) \\ \quad + [\bar{r}^-(s)/\bar{t}^-(s)]\exp(-sN_3(z-1)) & \text{in } \mathcal{D}_3. \end{cases} \tag{2.4.21}$$

In (2.4.21) the subscripts and superscripts + and - refer to the direction of propagation of the incident field. Note that $\bar{e}_+(z,s)$ is identical to the normalized solution $\hat{e}(z,s)$ introduced in Subsection 2.4.2. Furthermore, the solutions $\bar{e}_\pm(z,s)$ are essentially the same as the *Jost solutions* used in quantum scattering theory (see Chadan and Sabatier (1977)). From (2.4.19) and the radiation condition, it follows immediately that

$$G(z,z';s) = \bar{e}_+(z_>,s)\bar{e}_-(z_<,s)/W(\bar{e}_-,\bar{e}_+), \tag{2.4.22}$$

where $z_> = \max\{z,z'\}$, $z_< = \min\{z,z'\}$, and where $W(\bar{e}_-,\bar{e}_+)$ denotes the Wronskian

$$\begin{aligned} W(\bar{e}_-,\bar{e}_+) &= \bar{e}_+(z',s)\partial_{z'}\bar{e}_-(z',s) - \bar{e}_-(z',s)\partial_{z'}\bar{e}_+(z',s) \\ &= 2sN_3/\bar{t}^-(s) = 2sN_1/\bar{t}^+(s). \end{aligned} \tag{2.4.23}$$

The relation (2.4.20) takes a special form when we locate the point of observation on the slab interface illuminated by the incident pulse. We then arrive at

$$\begin{aligned} \hat{E}_+(0,s) &= \bar{E}_+(0,s) - \frac{s}{2N_1}\int_0^1 C(z,s)\bar{E}_+(z,s)\hat{E}_+(z,s)dz, \\ \hat{E}_-(1,s) &= \bar{E}_-(1,s) - \frac{s}{2N_3}\int_0^1 C(z,s)\bar{E}_-(z,s)\hat{E}_-(z,s)dz, \end{aligned} \tag{2.4.24}$$

where $\hat{E}_+(z,s)$ denotes the plane-wave solution $\hat{E}(z,s)$ as discussed up to now, $\hat{E}_-(z,s)$ the field excited by $\hat{E}_-^i(z,s) = \exp(sN_3(z-1))$, and $\bar{E}_\pm(z,s)$ the corresponding solutions in the reference configuration. The relations (2.4.24) are in accordance with the principle of reciprocity and will be used in Section 5.2 to analyze the inverse-scattering problem.

For each of the relations listed in (2.4.20) and (2.4.24), the evaluation of the kernel function requires the numerical solution of one or two direct-scattering problems according to the procedure outlined in Subsection 2.4.2. Moreover, the corresponding time-domain equations can only be computed numerically, e.g. by using an

FFT algorithm. A choice of the reference medium for which the kernels are known in closed form and for which the conversion to the time domain goes by inspection is a three-layer medium, in which $\bar{\varepsilon}_{2r}(z) = \bar{\varepsilon}_{2r}$ = constant, $\bar{\sigma}_2(z) = 0$, and the outer media are identical to the actual ones. Limiting ourselves again to incidence from the left, we find the reference solution

$$\bar{E}(z,s) = \begin{cases} \exp(-sN_1 z) + \bar{r}(s)\exp(sN_1 z) & \text{in } \mathcal{D}_1, \\ \bar{a}(s)\exp(-s\bar{N}_2 z) + \bar{b}(s)\exp(-s\bar{N}_2(1 - z)) & \text{in } \mathcal{D}_2, \\ \bar{t}(s)\exp(-sN_3(z - 1)) & \text{in } \mathcal{D}_3, \end{cases} \tag{2.4.25}$$

where the subscripts and superscripts + have been dropped, and where

$$\begin{aligned} \bar{r}(s) &= -R_{21} + T_{12}R_{23}T_{21}\exp(-2s\bar{N}_2)/\bar{G}(s), \\ \bar{a}(s) &= T_{12}/\bar{G}(s), \\ \bar{b}(s) &= T_{12}R_{23}\exp(-s\bar{N}_2)/\bar{G}(s), \quad \text{and} \\ \bar{t}(s) &= T_{12}T_{23}\exp(-s\bar{N}_2)/\bar{G}(s), \quad \text{with} \\ \bar{G}(s) &= 1 - R_{21}R_{23}\exp(-2s\bar{N}_2). \end{aligned} \tag{2.4.26}$$

Similarly, we have the Green's function

$$G(z,z';s) = \begin{cases} T_{21}n_1(z';s)\exp(sN_1 z)/d(s) & \text{in } \mathcal{D}_1, \\ \exp(-s\bar{N}_2|z - z'|)/2s\bar{N}_2 & \\ \quad + R_{21}n_1(z';s)\exp(-s\bar{N}_2 z)/d(s) & \\ \quad + R_{23}n_2(z';s)\exp(-s\bar{N}_2(1 - z))/d(s) & \text{in } \mathcal{D}_2, \\ T_{23}n_2(z';s)\exp(-sN_3(z - 1))/d(s) & \text{in } \mathcal{D}_3, \end{cases} \tag{2.4.27}$$

where it has been assumed that $z'\epsilon\mathcal{D}_2$, and where the numerator functions $n_{1,2}(z',s)$ and the denominator $d(s)$ are given by

$$
\begin{aligned}
n_1(z';s) &= \exp[-s\bar{N}_2 z'] + R_{23}\exp[-s\bar{N}_2 - s\bar{N}_2(1-z')], \\
n_2(z';s) &= \exp[-s\bar{N}_2(1-z')] + R_{21}\exp[-s\bar{N}_2 - s\bar{N}_2 z'], \\
d(s) &= 2s\bar{N}_2\bar{G}(s).
\end{aligned}
\tag{2.4.28}
$$

In (2.4.25) - (2.4.28), we have implicitly introduced the reflection and transmission coefficients

$$
\begin{aligned}
R_{21} &= (\bar{N}_2 - N_1)/(\bar{N}_2 + N_1), \\
R_{23} &= (\bar{N}_2 - N_3)/(\bar{N}_2 + N_3), \\
T_{12} &= 2N_1/(\bar{N}_2 + N_1), \\
T_{21} &= 2\bar{N}_2/(\bar{N}_2 + N_1), \quad \text{and} \\
T_{23} &= 2\bar{N}_2/(\bar{N}_2 + N_3),
\end{aligned}
\tag{2.4.29}
$$

whose physical interpretation corresponds with that of their asymptotic counterparts defined in (2.4.12). The expressions given in (2.4.25) - (2.4.28) simplify considerably when $\bar{\varepsilon}_{2r}$ is chosen equal to the relative permittivity of one of the dielectric half-spaces outside the slab. For example, when $\bar{N}_2 = N_1$, we have $R_{21} = 0$ and, hence, all terms containing this factor vanish from (2.4.25) - (2.4.28). For $z \le 1$, the integral relation (2.4.20) is then elaborated into

$$
\begin{aligned}
\hat{E}(z,s) = \hat{E}^i(z,s) &- s^2\int_0^1 C(z',s)G_1(z,z';s)\hat{E}(z',s)dz' \\
&+ R_{13}\exp(sN_1(z-1))\{\hat{E}^i(1,s) \\
&- \frac{s}{2N_1}\int_0^1 C(z',s)\exp(sN_1(z'-1))\hat{E}(z',s)dz'\},
\end{aligned}
\tag{2.4.30}
$$

where $R_{13} = (N_1 - N_3)/(N_1 + N_3)$, and where

$$
G_1(z,z';s) = (2sN_1)^{-1}\exp(-sN_1|z-z'|)
\tag{2.4.31}
$$

denotes the Green's function of a homogeneous reference medium with $\bar{\varepsilon}_r(z) = \varepsilon_{1r}$ and $\bar{\sigma}(z) = 0$ everywhere.

Obviously, the Green's function (2.4.31) can also be substituted directly into (2.4.20) along with the reference field

$$\bar{E}(z,s) = \hat{E}^i(z,s) = \exp(-sN_1z). \tag{2.4.32}$$

From (2.4.4), (2.4.6) and (2.4.31) it follows that the resulting integral over z' only has significance when $\text{Re}(s) \geq 0$. In that case, the integral over $\mathcal{D}_3$ can be carried out in closed form and we arrive at the integral relation

$$\begin{aligned}\hat{E}(z,s) = \hat{E}^i(z,s) - s^2\int_0^1 C(z',s)G_1(z,z';s)\hat{E}(z',s)dz' \\ + t(s)\exp(sN_1(z-1))(N_1 - N_3)/2N_1,\end{aligned} \tag{2.4.33}$$

which holds for $z \leq 1$. The integral equations (2.4.33) and (2.4.30) can be continued into the left half of the complex s-plane since the relevant integrals are well defined for all complex s. The remaining question is whether (2.4.33) can also be obtained directly from (2.4.30). Taking $z = 1$ in (2.4.30) results in

$$\begin{aligned}\hat{E}^i(1,s) - \frac{s}{2N_1}\int_0^1 C(z',s)\exp(sN_1(z'-1))\hat{E}(z',s)dz' = \\ [(N_1 + N_3)/2N_1]t(s),\end{aligned} \tag{2.4.34}$$

whose substitution immediately reduces (2.4.30) to (2.4.33). From a computational point of view, it seems preferable to discretize both the integrals in (2.4.30) because of the preferential role played by $t(s) = \hat{E}(1,s)$ in (2.4.34). Finally, it should be remarked that the relations obtained by substituting (2.4.25) - (2.4.28) in (2.4.20) are also valid in the entire s-plane, except at the zeros of $\bar{G}(s)$ as defined in (2.4.26), i.e. at the natural frequencies of the reference configuration.

In principle, the direct-scattering problem could be solved by applying the method of moments to one of the integral equations that result from (2.4.20) and (2.4.30) if $0 \leq z \leq 1$. However, because of the oscillatory behavior of E(z,s) as a function of z as described in Subsection 2.4.2, the number of points in the discretization of the integrals on the right-hand sides should be of $\mathcal{O}(|\text{Im}(s)|)$ in order to preserve the accuracy of the discretized integral. In the first place, such a refinement of the discretization may have an adverse effect on the condition of the system matrix of the resulting algebraic equation. In the second place, the computation time consumed in solving this matrix equation will be of $O(|\mathrm{Im}(s)|^3)$, while the Runge-Kutta-Verner method only requires a time of $O(|\mathrm{Im}(s)|)$. Therefore, the latter method seems preferable.

2.5 Scattering by a lossy, radially inhomogeneous dielectric circular cylinder: singularity expansion method and Fourier inversion

2.5.1 Formulation of the problem

In this section, we consider the two-dimensional analogue of the one-dimensional problem discussed up to now, i.e. the scattering of electromagnetic waves by a radially inhomogeneous, lossy dielectric cylinder (Figure 2.5.1) with radius a, embedded in vacuo. A cylindrical coordinate system is introduced, with the z-axis coinciding with the symmetry axis and ρ and ϕ as polar coordinates in the plane perpendicular to it. As indicated, the configuration consists of two domains $\mathcal{D}_1$ and $\mathcal{D}_2$, with $\mathcal{D}_1$ the exterior and $\mathcal{D}_2$ the interior of the cylinder. ε_0 and μ_0 denote the permittivity and permeability in vacuo, respectively. In $\mathcal{D}_2$, we have $\varepsilon = \varepsilon_r(\rho)\varepsilon_0 \geq \varepsilon_0$ and $\sigma(\rho) \geq 0$.

From $\mathcal{D}_1$, an electrically polarized electromagnetic pulse of finite duration is normally incident on the cylinder from the direction $\phi = \pi$. Then the problem is two-dimensional and the field intensities are functions of ρ, ϕ and t only. We write the electric-field intensity and the magnetic-field intensity of the incident field as

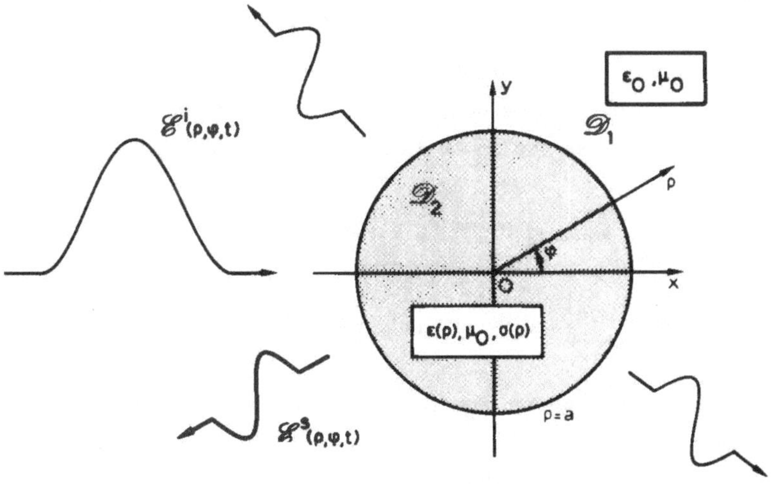

Figure 2.5.1 Scattering of an E-polarized, pulsed electromagnetic plane wave by a radially inhomogeneous, lossy dielectric cylinder.

$$\begin{aligned} E^i &= F(t - (\rho\cos\phi + a)/c_0)\underline{i}_z, \\ H^i &= -Y_0\, F(t - (\rho\cos\phi + a)/c_0)(\sin\phi\underline{i}_\rho + \cos\phi\underline{i}_\phi), \end{aligned} \tag{2.5.1}$$

where $Y_0 = (\varepsilon_0/\mu_0)^{\frac{1}{2}}$, $c_0 = (\varepsilon_0\mu_0)^{-\frac{1}{2}}$ (c_0 is the velocity of electromagnetic waves in vacuo) and where $F(t)$ is a piecewise-continuous function of a bounded variation in t that vanishes outside the interval $0 < t < T$. The total electromagnetic field is then written as

$$\begin{aligned} \underline{E} &= E(\rho,\phi,t)\underline{i}_z, \\ \underline{H} &= H(\rho,\phi,t)\underline{i}_\phi - (\rho\mu_0)^{-1}\int_{-\infty}^{t}\partial_\phi E(\rho,\phi,t')dt'\ \underline{i}_\rho. \end{aligned} \tag{2.5.2}$$

In (2.5.2), we have restricted the unknowns to those field quantities that are continuous at any discontinuity in $\varepsilon(\rho)$ and $\sigma(\rho)$. The scattered field is defined by

$$\begin{aligned} E^s(\rho,\phi,t) &= E(\rho,\phi,t) - E^i(\rho,\phi,t), \\ H^s(\rho,\phi,t) &= H(\rho,\phi,t) - H^i(\rho,\phi,t). \end{aligned} \tag{2.5.3}$$

Since $F(t)$ is piecewise continuous and of bounded variation in t, it follows from Maxwell's equations that, for ρ and t fixed, $E(\rho,\phi,t)$ and $H(\rho,\phi,t)$ also exhibit these properties as a function of ϕ. Hence, these quantities can, as far as their dependence on ϕ is concerned, be represented in terms of a Fourier series, e.g.

$$E(\rho,\phi,t) = \sum_{m=-\infty}^{\infty} \exp(im\phi)E_m(\rho,t),$$

with (2.5.4)

$$E_m(\rho,t) = \frac{1}{2\pi}\int_{-\pi}^{\pi} \exp(-im\phi)E(\rho,\phi,t)d\phi,$$

which converges in mean square to the relevant quantity (see Titchmarsh (1950) or Berg and McGregor (1966)). Furthermore, it follows from the energy balance that, for the scattered field in $\mathcal{D}_1$ and for the total field in $\mathcal{D}_2$, we have as well mean-square convergence on the entire relevant spatial domain, uniformly in t (see

Appendix 2.5.A). This implies that the Fourier series (2.5.4), when truncated at $|m| = M$, approximates the relevant field quantities with the same global accuracy in the entire space-time domain. This is in contradiction with the usual frequency-domain result (see e.g. Franz (1957)), where for an increasing frequency, M has to be increased to reach the same accuracy. Consequently, we are left with the calculation of the corresponding Fourier coefficients for $|m| \le M$, where M is determined by the global accuracy desired.

Since the incident field reaches the boundary of the cylinder at $t = 0$, causality ensures that both $\{\mathcal{E}^s,\mathcal{H}^s\}$ in $\mathcal{D}_1$ and $\{\mathcal{E},\mathcal{H}\}$ in $\mathcal{D}_2$ vanish in the time interval $-\infty < t < 0$. From the definition of the Fourier coefficients $\{\mathcal{E}_m(\rho,t),\mathcal{H}_m(\rho,t)\}$ (see (2.5.4)), it follows that the Laplace transforms of these field quantities are identical to the corresponding one-sided Laplace transforms, e.g.

$$E_m(\rho,s) = \int_0^\infty \exp(-st)\mathcal{E}_m(\rho,t)dt, \qquad (2.5.5)$$

which are regular for $Re(s) > 0$. The corresponding coefficient for the incident field follows from applying an angular Fourier transformation and the shift rule plus a one-sided temporal Laplace transformation to (2.5.1) as

$$E_m^i(\rho,s) = (-1)^m \exp(-sa/c_0)I_m(s\rho/c_0)F(s), \qquad (2.5.6)$$

in which

$$F(s) = \int_0^T \exp(-st)\mathcal{F}(t)dt,$$

and where $I_m(s)$ denotes the modified Bessel function of the first kind of order m. Hence, $E_m^i(\rho,s)$ is regular for all s. In view of subsequent calculations, we carry out a normalization similar to the one applied in the one-dimensional case (see p. 37). For the configuration at hand, we normalize all distances with respect to the radius a and all times with respect to the vacuum travel time from the boundary of the cylinder to its center, i.e. a/c_0; e.g.

$$\bar{\rho} = \rho/a, \qquad \bar{t} = c_0 t/a.$$

In addition, we introduce the dimensionless quantities

$$\bar{s} = sa/c_0, \qquad \bar{\sigma}(\bar{\rho}) = Z_0\sigma(\rho)a, \qquad \bar{\varepsilon}_r(\bar{\rho}) = \varepsilon_r(\rho),$$

and the normalized field quantities

$$\bar{E}_m(\bar{\rho},\bar{s}) = E_m(\rho,s), \qquad \bar{H}_m(\bar{\rho},\bar{s}) = Z_0 H_m(\rho,s),$$

where $Z_0 = (\mu_0/\varepsilon_0)^{\frac{1}{2}}$. Henceforth, we will omit the bars. The transformed, normalized source-free electromagnetic field equations are then given by

$$[\partial_\rho + \rho^{-1}]H_m(\rho,s) = s[\varepsilon_r(\rho) + \sigma(\rho)/s + m^2/s^2\rho^2]E_m(\rho,s), \tag{2.5.7a}$$

$$\partial_\rho E_m(\rho,s) = sH_m(\rho,s). \tag{2.5.7b}$$

Elimination of H_m leads to the second-order differential equation

$$[\partial_\rho^2 + \rho^{-1}\partial_\rho - (m^2/\rho^2 + s^2)]E_m(\rho,s) = s^2\, C(\rho,s)E_m(\rho,s), \tag{2.5.8}$$

where the *contrast function* C, defined as

$$C(\rho,s) = \varepsilon_r(\rho) - 1 + \sigma(\rho)/s, \tag{2.5.9}$$

only differs from zero when $0 \le \rho < 1$. With the aid of the Green's function for the operator working on E_m in the left-hand side of (2.5.8):

$$G_m(\rho,\rho';s) = \rho' K_m(s\rho_>)I_m(s\rho_<), \tag{2.5.10}$$

where $\rho_> = \max(\rho,\rho')$ and $\rho_< = \min(\rho,\rho')$, the following integral relation is obtained, which is equivalent to (2.5.8):

$$E_m(\rho,s) = E_m^i(\rho,s) - s^2\int_0^\infty C(\rho',s)G_m(\rho,\rho';s)E_m(\rho',s)d\rho'. \quad (2.5.11)$$

When $0 \le \rho < 1$, (2.5.11) constitutes an integral equation of the second kind in $E_m(\rho,s)$. Equation (2.5.11) can be written in operator form as

$$L_m(s)E_m(\rho,s) = E_m^i(\rho,s), \quad (2.5.12)$$

in which

$$L_m(s)f(\rho) = \int_0^\infty[\delta(\rho - \rho') + s^2C(\rho',s)G_m(\rho,\rho';s)]f(\rho')d\rho'. \quad (2.5.13)$$

Once $E_m(\rho,s)$ in $0 < \rho < 1$ has been obtained from equation (2.5.7), (2.5.8) or (2.5.11), it is, through (2.5.11), known for all ρ. The space-time domain response follows from the inversion formula:

$$E(\rho,\phi,t) = (2\pi i)^{-1}\lim_{M\to\infty}\left\{\sum_{m=-M}^{M}\exp(im\phi)\int_{L_\beta}\exp(st)E_m(\rho,s)ds\right\}, \quad (2.5.14)$$

where L_β is the same Bromwich contour as in (2.2.14):

$$L_\beta = \{s|\mathrm{Re}(s) = \beta,\ \beta \ge 0\}. \quad (2.5.15)$$

One way of evaluating (2.5.14) is for each angular order m, to continue $E_m(\rho,s)$ analytically into the domain $\mathrm{Re}(s) < 0$, supplement L_β by circular arcs at infinity in the complex s-plane and apply Cauchy's theorem. In case the integrals along the supplementary arcs vanish, only the singularities in the analytic continuation of $E_m(\rho,s)$ contribute to the result. The form of these contributions and the conditions under which this procedure can be followed will be investigated in subsequent subsections.

2.5.2 Contributions from the singularities in the complex s-plane

In this subsection, we consider the contributions from the singularities in the left half of the complex s-plane and the form of their contributions. To this end, we first analyze the singularities that occur.

Inventory of singularities and Mittag-Leffler representation

From (2.5.6) - (2.5.8), it follows that in $\mathcal{D}_1$, the total electromagnetic field takes the form:

$$\begin{aligned} E_m(\rho,s) &= (-1)^m \exp(-s)F(s)[I_m(s\rho) + b_m(s)K_m(s\rho)], \\ H_m(\rho,s) &= (-1)^m \exp(-s)F(s)[I'_m(s\rho) + b_m(s)K'_m(s\rho)], \end{aligned} \qquad (2.5.16)$$

with $b_m(s)$ an as yet unknown parameter. In (2.5.16) and in what follows, the prime denotes differentiation with respect to the argument. Since E_m must remain bounded near $\rho = 0$, we can write the solution inside the cylinder $(0 < \rho < 1)$ as

$$\begin{aligned} E_m(\rho,s) &= (-1)^m \exp(-s)F(s)c_m(s)e_m(\rho,s), \\ H_m(\rho,s) &= (-1)^m \exp(-s)F(s)c_m(s)h_m(\rho,s). \end{aligned} \qquad (2.5.17)$$

In (2.5.17), $e_m(\rho,s)$ denotes the solution of (2.5.8), normalized by

$$\lim_{\rho \downarrow 0} \rho^{-|m|} e_m(\rho,s) = 1, \qquad (2.5.18)$$

$h_m(\rho,s)$ the corresponding magnetic-field strength as defined by (2.5.7b), and $c_m(s)$ an as yet unknown constant. Note that only $|m|$ and m^2 occur in the equations (2.5.8), (2.5.18) and (2.5.7b), which define the pair of solutions $\{e_m,h_m\}$. Therefore, this pair is invariant under a change of sign in m. For a homogeneous cylinder, with $\varepsilon(\rho)$ and $\sigma(\rho)$ constant for $0 < \rho < 1$, we have, for $m \geq 0$:

$$\begin{aligned} e_m^{hom}(\rho,s) &= \frac{m!\,2^m}{s^m\,n(s)^m} I_m(sn(s)\rho), \\ h_m^{hom}(\rho,s) &= \frac{m!\,2^m}{s^m\,n(s)^{m-1}} I'_m(sn(s)\rho), \end{aligned} \qquad (2.5.19)$$

with $n(s) = (\varepsilon_r + \sigma/s)^{\frac{1}{2}}$, $\mathrm{Re}(n(s)) \geq 0$. The corresponding solution for m negative is obtained by using the invariance property discussed above. For a general, inhomogeneous cylinder, the field distributions e_m and h_m have to be computed numerically, e.g. by integrating the two first-order differential equations (2.5.7) from $\rho = 0$ to $\rho = 1$. This procedure will be discussed in detail in Subsection 2.5.4. Here, we will assume these distributions to be known. Application of the boundary conditions at $\rho = 1$ (continuity of E_m and H_m across the interface) yields:

$$\begin{aligned} b_m(s) &= [I_m'(s)e_m(1,s) - I_m(s)h_m(1,s)]/G_m(s), \\ c_m(s) &= 1/sG_m(s), \end{aligned} \tag{2.5.20}$$

with

$$G_m(s) = h_m(1,s)K_m(s) - e_m(1,s)K_m'(s)$$

being the characteristic cylinder denominator. Since the field distributions e_m and h_m defined by (2.5.8), (2.5.18) and (2.5.7b) are entire functions of s, we observe from (2.5.20) that the singularities in the complex s-plane comprise the poles $\{s_{mn}\}$, which are found as zeros of the cylinder denominator $G_m(s)$:

$$G_m(s_{mn}) = G_m(s^*_{mn}) = 0, \quad n = 0, 1, 2, \ldots, \infty, \tag{2.5.21}$$

as well as a branch point at $s = 0$ due to the presence of the modified Bessel function K_m. The corresponding branch cut is chosen along the negative real s-axis:

$$\{s_{m\nu} | s_{m\nu} = -\nu, \quad \mathrm{Im}(\nu) = 0, \quad 0 < \mathrm{Re}(\nu) < \infty\}. \tag{2.5.22}$$

By applying Cauchy's integral formula to a contour circumscribing the whole complex s-plane excepting the singularities and the branch cut, we obtain the generalized Mittag-Leffler representation:

$$E_m(\rho,s) = \sum_{n=0}^{\infty} \left\{ \frac{r_{mn}(\rho)}{s - s_{mn}} + \frac{r^*_{mn}(\rho)}{s - s^*_{mn}} \right\} + \int_0^{\infty} \frac{r_{m\nu}(\rho)}{s - s_{m\nu}} d\nu + E_m^e(\rho,s). \tag{2.5.23}$$

In (2.5.23), $E_m^e(\rho,s)$ is an entire function and $r_{mn}(\rho)$ denotes the residual contribution from the pole at $s = s_{mn}$. $r_{m\nu}(\rho)$ is proportional to the jump across the branch cut:

$$r_{m\nu} \stackrel{\Delta}{=} (i/2\pi)[E_m(\rho,-\nu + i0) - E_m(\rho,-\nu - i0)], \tag{2.5.24}$$

and can be interpreted as the residual contribution from the point $s = -\nu$ on the branch cut.

It should be pointed out that the representation (2.5.23) is not unique; a different choice of the branch cut leads to a different set of poles. For example, when the branch cut is chosen along one of the imaginary s-axes, a different set of poles is found in the upper or lower half-plane in which the cut is located. The special form (2.5.23) was selected by requiring that $E_m(\rho,s)$ satisfy the Schwarz reflection principle in the entire complex s-plane. Accordingly, $E_m^s(\rho,s)$ shows the proper causal behavior as $\rho\to\infty$ for all complex s. In the remainder of this subsection, we will study the form of the residual contributions $r_{mn}(\rho)$ and $r_{m\nu}(\rho)$ in (2.5.23).

Residual contributions from the poles

The residual contributions from the poles can be determined with the aid of the same procedure that was used in Subsection 2.3.2 to determine the residual contributions for an inhomogeneous, lossy dielectric slab. Substituting (2.5.23) in (2.5.12) and using the analytic properties of $L_m(s)$, $E_m^i(\rho,s)$ and $E_m^e(\rho,s)$ in the vicinity of a pole, we obtain

$$r_{mn}(\rho) = \lambda_{mn} E_{mn}(\rho), \tag{2.5.25}$$

where λ_{mn} is an as yet unknown complex constant and where

$$E_{mn}(\rho) = \begin{cases} e_m(\rho,s_{mn}), & 0 \le \rho \le 1, \\ K_m(s_{mn}\rho)e_m(1,s_{mn})/K_m(s_{mn}), & 1 \le \rho < \infty, \end{cases} \tag{2.5.26}$$

denotes the natural-mode field distribution. Next, we define the *object product* of two functions f and g as

$$<f(\rho,s_1)|g(\rho,s_2)> = \int_0^1 f(\rho,s_1)C(\rho,s_2)g(\rho,s_2)\rho d\rho. \qquad (2.5.27)$$

Note that the only difference between this object product and the one defined in (2.3.5) is the occurrence of the factor ρ in the integrand of (2.5.27). This factor can be understood by interpreting the integral on the right-hand side of (2.5.27) as a two-dimensional integral, in which the integral over the angular variable has been evaluated with the aid of the orthogonality relations for the functions $\exp(im\phi)$. The factor ρ is then directly identified as the Jacobian of the transformation from Cartesian to cylindrical coordinates. From the integral representation (2.5.13), we find that the operator $L_m(s)$ has the following property:

$$<f(\rho,s_1)|L_m(s_2)g(\rho,s_2)> = <L_m(s_2)f(\rho,s_1)|g(\rho,s_2)>. \qquad (2.5.28)$$

Using this property, we can almost literally repeat the analysis given in Subsection 2.3.2 to obtain λ_{mn}. We end up with:

$$\lambda_{mn} = F(s_{mn})D_{mn}, \qquad (2.5.29)$$

where the coupling coefficient D_{mn} is now given by

$$D_{mn} = (-1)^m \exp(-s_{mn}) \frac{<I_m(s_{mn}\rho)|E_{mn}(\rho)>}{<L'_m(s_{mn})E_{mn}(\rho)|E_{mn}(\rho)>} . \qquad (2.5.30)$$

In (2.5.30), the prime denotes partial differentiation with respect to s.

Contribution from the branch cut

The residual contribution from the branch cut follows directly by substituting (2.5.16), (2.5.17) and (2.5.20) in (2.5.24). For real-valued positive ν, we have

$$K_m(-\nu \pm i0) = (-1)^m K_m(\nu) \mp i\pi I_m(\nu) \qquad (2.5.31)$$

(see Abramowitz and Stegun (1972), formula 9.6.31), while $e_m(\rho,s)$, $h_m(\rho,s)$ and $I_m(s\rho)$ are continuous at $s = -\nu$. Utilizing these properties in the formulas mentioned above yields

$$r_{m\nu}(\rho) = F(-\nu)D_{m\nu}\, E_{m\nu}(\rho), \tag{2.5.32}$$

where

$$E_{m\nu}(\rho) = \begin{cases} e_m(\rho,-\nu) & 0 \le \rho \le 1, \\ \dfrac{[\beta_m(\nu)K_m(\nu\rho) - \alpha_m(\nu)I_m(\nu\rho)]e_m(1,-\nu)}{\beta_m(\nu)K_m(\nu) - \alpha_m(\nu)I_m(\nu)} & 1 \le \rho < \infty, \end{cases} \tag{2.5.33}$$

is the natural-mode field distribution, and

$$D_{m\nu} = \frac{(-1)^m \exp(\nu)\beta_m(\nu)}{-\nu[\alpha_m^2(\nu) + \pi^2\beta_m^2(\nu)]} \tag{2.5.34}$$

the coupling coefficient. In (2.5.33) and (2.5.34), we have

$$\begin{aligned} \alpha_m(\nu) &= h_m(1,-\nu)K_m(\nu) + e_m(1,-\nu)K'_m(\nu), \quad \text{and} \\ \beta_m(\nu) &= h_m(1,-\nu)I_m(\nu) + e_m(1,-\nu)I'_m(\nu). \end{aligned} \tag{2.5.35}$$

With (2.5.23), (2.5.25) - (2.5.30) and (2.5.32) - (2.5.35), the local form of $E_m(\rho,s)$ in the vicinity of its singularities is known except for an entire function. The contribution from the singularities to the space-time domain field can therefore be obtained directly by evaluating the inversion formula (2.5.14). We arrive at:

$$\begin{aligned} &\mathcal{E}_{\text{singularities}}(\rho,\phi,t) = \\ &\lim_{M\to\infty} \sum_{m=-M}^{M} \exp(im\phi)\Big\{\sum_{n=0}^{\infty}[D_{mn}F(s_{mn})E_{mn}(\rho)\exp(s_{mn}t)] \\ &+ \sum_{n=0}^{\infty}[D^*_{mn}F(s^*_{mn})E^*_{mn}(\rho)\exp(s^*_{mn}t)] \\ &+ \int_0^{\infty}[D_{m\nu}F(-\nu)E_{m\nu}(\rho)\exp(-\nu t)]d\nu\Big\}. \end{aligned} \tag{2.5.36}$$

In (2.5.36), the term between braces is real-valued and invariant under a change of sign in m. Hence, the complex exponential $\exp(im\phi)$ can be replaced by $\cos(m\phi)$. Then, the invariance property also holds for the total term in the resulting angular Fourier series. Consequently, the summation over m can be reduced to one over nonnegative m by including a factor of 2 in the terms with m positive.

2.5.3 Contribution from the closing contours

In the total time-domain response, the entire function $E_m^e(\rho,s)$ in (2.5.23) yields, in the time domain, the contribution of the supplementary arc in either the left or the right half of the s-plane (see also Baum (1978)). As for the case of the dielectric slab, the role of the entire function can be analyzed by constructing an asymptotic solution for $|s|\to\infty$. We will first construct this solution.

WKB asymptotics

To this end we follow the same formal procedure as in Subsection 2.3.3, and write the field in the cylinder as a combination of two linearly independent solutions $u_m^{\pm}(\rho)$. Next, we restrict the constitutive parameters $\varepsilon_r(\rho)$ and $\sigma(\rho)$ to the class of functions that are continuous in $\mathcal{D}_2$ and whose derivatives are discontinuous at no more than a finite number of points in $\mathcal{D}_2$ and we take $|s| >> |m|$. From Erdélyi (1956), we then know that $u_m^{\pm}(\rho) = u_a^{\pm}(\rho;\rho_0)[1 + \mathcal{O}(s^{-1})]$, where $u_a^{\pm}(\rho;\rho_0)$ are the first-order WKB-solutions:

$$u_a^{\pm}(\rho;\rho_0) = [N(\rho_0)/\rho N(\rho)]^{\frac{1}{2}} \exp\left\{\mp s\int_{\rho_0}^{\rho} [N(\rho') + \sigma(\rho')/2N(\rho')s]d\rho'\right\}, \tag{2.5.37}$$

uniformly in each domain $0 < \rho_0 \le \rho \le 1$. In (2.5.37), we have $N(\rho) = \varepsilon_r^{\frac{1}{2}}(\rho)$. $u_a^+(\rho;\rho_0)$ and $u_a^-(\rho;\rho_0)$ represent wave-like solutions originating from $\rho = \rho_0$ and traveling in the direction of increasing and decreasing values of ρ, respectively. Now, we have to find the linear combination of u_a^+ and u_a^- that corresponds to the solution defined by the condition (2.5.18). However, (2.5.37) does not cover the behavior near $\rho = 0$ since, in its derivation, the term m^2/ρ^2 in the second-order differential equation was assumed to be small compared to the terms containing factors of s. To overcome this problem, we consider the local behavior of $e_m(\rho,s)$ near $\rho = 0$. For $\varepsilon_r(\rho)$ and $\sigma(\rho)$ differentiable at $\rho = 0$ with a bounded derivative, we have

$$e_m(\rho,s) = e_m^{hom}(\rho,s)[1 + O(\rho)] \tag{2.5.38}$$

as $\rho\downarrow 0$. In (2.5.38), e_m^{hom} denotes the solution for a homogeneous cylinder as defined in (2.5.19) with $n(s)$ replaced by the corresponding value at the origin, i.e. $n_0(s) = [\varepsilon_r(0) + \sigma(0)/s]^{\frac{1}{2}}$, $\mathrm{Re}(n_0) \geq 0$. The approximation (2.5.38) holds uniformly in s and, hence, also as $|s|\to\infty$. Therefore, we can take this limit in (2.5.19) for a small, nonvanishing value of ρ and match the resulting first-order approximation to the WKB-solutions $u_a^{\pm}(\rho;\rho_0)$. We arrive at $e_m(\rho,s) = e_m^a(\rho,s)[1 + O(s^{-1})]$, with

$$e_m^a(\rho,s) = \frac{m!\ 2^m}{[N(0)s]^{m+\frac{1}{2}}(2\pi)^{\frac{1}{2}}} \times \left\{u_a^-(\rho;0) + \exp[\pm(m+\tfrac{1}{2})\pi i]u_a^+(\rho;0)\right\}, \tag{2.5.39}$$

for $\mathrm{Im}(s) \gtrless 0$. Taking the derivative of e_m^a with respect to ρ, we find with the aid of (2.5.7b), $h_m(\rho,s) = h_m^a(\rho,s)[1 + O(s^{-1})]$, with

$$h_m^a(\rho,s) = \frac{m!\ 2^m N(\rho)}{[N(0)s]^{m+\frac{1}{2}}(2\pi)^{\frac{1}{2}}} \times \left\{u_a^-(\rho;0) - \exp[\pm(m+\tfrac{1}{2})\pi i]u_a^+(\rho;0)\right\}, \tag{2.5.40}$$

for $\mathrm{Im}(s) \gtrless 0$. The asymptotic approximations $\{E_m^a(\rho,s),H_m^a(\rho,s)\}$ to the electromagnetic fields $\{E_m(\rho,s),H_m(\rho,s)\}$ are then obtained by direct substitution of e_m^a and h_m^a and the first-order asymptotic approximations to $I_m(s\rho)$ and $K_m(s\rho)$ in (2.5.16), (2.5.17) and (2.5.20). For example, for the total electric-field strength in $\mathcal{D}_2$, we obtain:

$$E_m^a(\rho,s) = \left(\frac{2}{\pi s}\right)^{\frac{1}{2}} \frac{F(s)\{u_a^-(\rho;0) + \exp[\pm(m+\frac{1}{2})\pi i]u_a^+(\rho;0)\}}{(-1)^m[N(1)+1]u_a^-(1;0) \mp i[N(1)-1]u_a^+(1;0)}. \tag{2.5.41}$$

Similar expressions are obtained for the scattered as well as for the total field in $\mathcal{D}_1$. The poles of $E_m^a(\rho,s)$ in the upper half of the complex s-plane are given by

$$s_{mn}^a = \{\ln R_1 - \int_0^1 \sigma(\rho)/N(\rho)d\rho + i[2n + 1 - \tfrac{1}{2}(-1)^m]\pi\}/2\tau, \qquad (2.5.42)$$

with $n = 0, 1, 2, \ldots, \infty$. In (2.5.42) $\tau \triangleq \int_0^1 N(\rho)d\rho$ denotes the travel time it takes for the front of the incident pulse to traverse the cylinder from the interface to the center and $R_1 = [N(1) - 1]/[N(1) + 1]$ denotes the asymptotic plane-wave reflection coefficient at the interface. The poles in the lower half of the complex s-plane are obtained from (2.5.42) by complex conjugation. If we substitute e_m^a, h_m^a and the first-order asymptotic approximations to K_m and I_m in (2.5.21) and, subsequently, carry out a first-order Taylor expansion around an asymptotic zero $s = s_{mn}^a$, it follows that

$$s_{mn} = s_{mn}^a + \mathcal{O}(n^{-1}) \text{ as } n\to\infty. \qquad (2.5.43)$$

Obviously, the magnitude of the correction term of $\mathcal{O}(n^{-1})$ in (2.5.43) may still depend on m.

Closing conditions

In Subsection 2.5.1, $F(t)$ was restricted to the class of square integrable, piecewise-continuous functions that vanish outside the domain $0 < t < T$. Hence, application of the Riemann-Lebesgue theorem yields the following asymptotic estimate for $F(s)$:

$$\begin{aligned} F(s) &= o(1) \quad \text{as } |s|\to\infty \text{ in } \mathrm{Re}(s) \le 0, \\ F(s)\exp(sT) &= o(1) \quad \text{as } |s|\to\infty \text{ in } \mathrm{Re}(s) \ge 0. \end{aligned} \qquad (2.5.44)$$

As for the case of the dielectric slab (see Sections 2.2 and 2.3), we now take the radius of the supplementary arc such that the closing contour in the left half-plane always intersects the line $\{s|\mathrm{Re}(s) = \mathrm{Re}(s_{m0}^a)\}$ in the middle between two asymptotic poles. Note that this procedure results in two different sets of closing contours for m even and m odd. For s on these contours, the asymptotic approximations given above then yield the asymptotic behavior of $\exp(st)E_m(\rho,s)$ as indicated in Table 2.5.1. In this table,

Table 2.5.1 Asymptotic behavior as $|s|\to\infty$ of $\exp(st)E_m(\rho,s)$.

behavior in half-plane	total field inside cylinder $(0<\rho<1)$	scattered field outside cylinder $(1<\rho<\infty)$	total field outside cylinder $(1<\rho<\infty)$
right	$O(s^{-\frac{1}{2}})\exp[s(t-\tau(\rho))]$	$O(s^{-\frac{1}{2}})\exp[s(t-\tau(\rho))]$	$O(s^{-\frac{1}{2}})\exp[s(t+\tau(\rho))]$
left	$O(s^{-\frac{1}{2}})\exp[s(t-T+\tau(\rho))]$	$O(s^{-\frac{1}{2}})\exp[s(t-T-\tau(\rho)-2)]$	$O(s^{-\frac{1}{2}})\exp[s(t-T-\tau(\rho))]$

$$\tau(\rho) \triangleq \begin{cases} \int_\rho^1 N(\rho')d\rho', & 0 < \rho < 1, \\ (\rho - 1), & 1 < \rho < \infty, \end{cases} \tag{2.5.45}$$

denotes the time it takes for a wave to travel in the radial direction from $\rho = 1$ to the point of observation. Using Jordan's lemma, we find that the contributions along the closing contours vanish as indicated in Table 2.5.2.

Table 2.5.2 Time intervals for which the contribution from the closing contours at infinity vanishes.

closure in half-plane	total field inside cylinder $(0<\rho<1)$	scattered field outside cylinder $(1<\rho<\infty)$	total field outside cylinder $(1<\rho<\infty)$
right	$-\infty<t<\tau(\rho)$	$-\infty<t<\tau(\rho)$	$-\infty<t<-\tau(\rho)$
left	$T-\tau(\rho)<t<\infty$	$T+\tau(\rho)+2<t<\infty$	$T+\tau(\rho)<t<\infty$

The conditions in Table 2.5.2 are illustrated graphically in the space-time diagram shown in Figure 2.5.2 for $T < 2\tau$. From Figure 2.5.2, we observe that in that case, there is some ρ_1 in $\mathcal{D}_2$ such that $T - \tau(\rho_1) = \tau(\rho_1)$. For $0 < \rho < \rho_1$, the electric field can always be determined by closing the Bromwich contour such that the integral along the supplementary arc vanishes. For $\rho > \rho_1$, however, there is always a time interval in which the possible contribution from the supplementary arcs remains to be investigated. When $T > 2\tau$, such an interval is present for all ρ. From the last line of Table 2.5.2 and from Figure 2.5.2, we also see that for ρ in $\mathcal{D}_1$, there is a time interval of length 2 where the contour can be closed toward

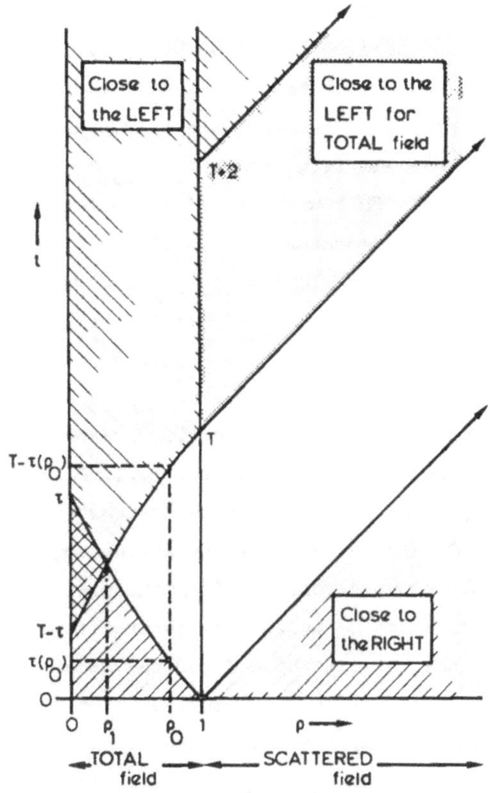

Figure 2.5.2 Space-time illustration of the conditions for vanishing closing contours. In the unshaded region, the Bromwich contour cannot be closed without a possible contribution along the supplementary arcs.

the left for the total field, while it cannot be closed for the scattered field. Physically, this corresponds to the phenomenon that for these space-time points, the part of $E^i_m(\rho,t)$ that represents a wave traveling away from the cylinder is at least partially canceled by the scattered field. In that space-time region, special care is needed in the numerical evaluation of the angular Fourier series (2.5.4), because this series only converges globally for the scattered field.

From Figure 2.5.2 and Table 2.5.2, it is observed that the closing conditions do not depend on the azimuthal angle of observation. This is caused by the fact that, in (2.5.4), we have decomposed the problem into a series of one-dimensional subproblems, each of which describes a wave that travels in the radial direction only and whose angular field distribution does not vary with ρ and t. The true causal nature of the field (e.g. the turn-on time of the field at a given space point) only follows when all the azimuthal constituents $\{E_m(\rho,t)\}$ are added up according to (2.5.4). This is the price we have to pay for obtaining a globally converging representation and, hence, for being able to analyze the role of the entire function as outlined above.

If we want to employ the SEM representation as has been discussed up to now for space-time points in the "forbidden" region, i.e. for points in the unshaded region in Figure 2.5.2, we have to evaluate the integral along either of the closing contours numerically. Alternately, we can determine the field by evaluating the integral along the original Bromwich contour. In principle, the calculation of a contour integral can be avoided by generalizing the pulse-truncation method that was introduced for the case of the dielectric slab (see Subsection 2.3.3) to the configuration presently under discussion. In this method, one takes into account at each instant only the part of the incident pulse that has reached the scattering obstacle. However, it follows from the asymptotic analysis presented in the beginning of this subsection that the error in terminating the summations over n in the resulting modal series over the type (2.5.36) at $n = N$ is of $O(N^{-\frac{1}{2}})$, which is too slow for an efficient numerical computation. A similar problem arose for the case of the slab, where it was handled by accelerating the convergence with the aid of the first-order WKB approximation. In the present case, the occurrence of the factor of $s^{-\frac{1}{2}}$ in (2.5.41) and related expressions prevents such an acceleration. In the first place, it prevents the closed-form evaluation of the Laplace inversion integral leading from $E_m^a(\rho,s)$ to the corresponding time-domain coefficient $E_m^a(\rho,t)$. In the second place, the factor gives rise to a contribution from a branch cut along the negative real s-axis; upon closure to the left, the corresponding integral has to be evaluated numerically. For the configuration at hand, we will therefore not elaborate the pulse-truncation method any further.

The closing conditions of Table 2.5.2 also allow the nature of the entire function $E_m^e(\rho,s)$ in (2.5.23) to be identified in terms of the definitions proposed by Heyman and Felsen (1985a,b). These authors classify an entire function as removable if it describes a wave constituent (e.g. a wavefront) which could also be incorporated into the SEM representation (2.5.36). Any entire function that cannot be treated in this manner is called intrinsic. By letting T become sufficiently small, it follows directly from Table 2.5.2 that $E_m^e(\rho,s)$ basically contains two intrinsic components: the transformed incident field, which is regular for all s (see (2.5.6)), and a function associated with the scattered field that is excited when this field first hits the cylinder. The scattering center of the latter wave constituent is the point $\rho = 1$, $\phi = \pi$ (point 2 in the inset of Figure 2.5.6). For both fields, the presence of an intrinsic entire function could be expected since they are not included in the waves traveling away from the cylinder with a nonvanishing counterpart inside it and are therefore absent in (2.5.36). Nevertheless, we will see from the results presented in Subsection 2.5.5 that, at each point of observation, the transient signal shows up wavefronts after the field constituents described above have passed. Consequently, the entire functions associated with those wavefronts in the approach of Heyman and Felsen are readily identified as being removable.

2.5.4 Numerical implementation

In order to obtain insight into the applicability of the SEM in practice, the theory presented in Subsections 2.5.2 and 2.5.3 has been implemented numerically. In view of the difficulties mentioned at the end of Subsection 2.5.3, the SEM computation was restricted to those space-time points where the full modal representation holds; i.e. where the contribution from the closing contour on the left vanishes (see Table 2.5.2).

We first consider the calculation of the standard solutions $\{e_m(\rho,s), h_m(\rho,s)\}$ defined by (2.5.8), (2.5.18) and (2.5.7b). These definitions are invariant under a change of sign in m. Therefore, we only consider the case $m \geq 0$. Since the normalization condition (2.5.18) provides a starting value at $\rho = 0^+$, it seems logical to determine e_m and h_m by numerically integrating the system of first-order differential equations (2.5.7) from $\rho = 0^+$ to $\rho = 1$. However, the coefficients in (2.5.7) are singular at $\rho = 0$. Furthermore, for $m > 1$, both e_m and h_m vanish at $\rho \downarrow 0$. We circumvent these difficulties by introducing the vector $\underline{f}_m = (f_m^e, f_m^h)$, with

$$f_m^e(\rho,s) = \rho^{-m} e_m(\rho,s), \quad f_m^h(\rho,s) = \rho^{-m+1} h_m(\rho,s) \tag{2.5.46}$$

(see also Dil and Blok (1973)). In terms of this vector, (2.5.7) reduces to the matrix equation:

$$\partial_\rho \underline{f}_m(\rho) = \rho^{-1} \mathbf{A}_m(\rho) \underline{f}_m(\rho), \tag{2.5.47}$$

with

$$\mathbf{A}_m(\rho) = \begin{bmatrix} -m & s \\ \rho^2 s[\varepsilon_r(\rho) + \sigma(\rho)/s] + m^2/s & -m \end{bmatrix} \quad \text{for } m > 0, \tag{2.5.48}$$

and

$$\mathbf{A}_0(\rho) = \begin{bmatrix} 0 & s\rho^2 \\ s[\varepsilon_r(\rho) + \sigma(\rho)/s] & -2 \end{bmatrix}. \tag{2.5.49}$$

The singular coefficients are avoided by substituting in (2.5.47) the first-order Taylor expansions at $\rho = 0$:

$$\begin{aligned} \underline{f}_m(\rho) &= \underline{f}_m(0) + \rho \underline{f}_m'(0) + O(\rho^2), \\ \mathbf{A}_m(\rho) &= \mathbf{A}_m(0) + \rho \mathbf{A}_m'(0) + O(\rho^2), \end{aligned} \tag{2.5.50}$$

where the primes denote differentiation with respect to ρ. Comparing the terms containing factors ρ^{-1} and ρ^0, respectively, we find

$$\underline{f}_m(0) = \begin{bmatrix} 1 \\ m/s \end{bmatrix}, \qquad \underline{f}'_m(0) = \underline{0} \quad \text{for } m > 0, \tag{2.5.51}$$

and

$$\begin{aligned} \underline{f}_0(0) &= \begin{bmatrix} 1 \\ [s\varepsilon_r(0) + \sigma(0)]/2 \end{bmatrix}, \\ \underline{f}'_0(0) &= \begin{bmatrix} 0 \\ [s\varepsilon'_r(0) + \sigma'(0)]/3 \end{bmatrix}. \end{aligned} \tag{2.5.52}$$

Next, we compute, with the aid of (2.5.50) - (2.5.52), a starting vector for $\underline{f}_m(\rho_0)$ for a small, positive value of ρ_0. The system of equations (2.5.47) is then integrated numerically from $\rho = \rho_0$ to $\rho = 1$ by the same Runge-Kutta-Verner fifth- and sixth-order method that was used in the one-dimensional case (IMSL (1982)).

The natural frequencies are found by substituting the computed values of $e_m(1,s)$ and $h_m(1,s)$ in $G_m(s)$ and searching for its zeros. Since we have no a priori information available on the approximate location of the poles, we now have to carry out the search process by the two-step procedure outlined in Subsection 2.3.4. In the first step of this procedure, the approximate location of the zeros is determined from integrals of the argument type:

$$(2\pi i)^{-1} \oint_L s^\ell \, G'_m(s)/G_m(s) ds = \sum_n (s_{mn})^\ell. \tag{2.5.53}$$

In (2.5.53), L denotes any closed contour in the complex s-plane that does not intersect the negative s-axis, and the summation runs over the zeros inside L. To this aim, we employ a modified version of the algorithm described in Singaraju, Giri and Baum (1976). In the second step, the estimates obtained are improved by Muller's method (see Traub (1964)).

The natural-mode field distributions $E_{mn}(\rho)$ defined in (2.5.26) are automatically obtained in the search process. Therefore, the coupling coefficient D_{mn} can also be determined by computing the relevant integrals (see (2.5.30)). For the single integral, a trapezoidal rule is applied. For the double integral, we repeat this trapezoidal rule, taking the singularity of the operator $L'_m(s)$ at $\rho = \rho'$ (see (2.5.10) and (2.5.13)) into account in an appropriate way. Once the coupling coefficients are known, the first two terms on the right-hand side of (2.5.36) are evaluated by a straightforward truncated summation.

Finally, we have to compute the contribution from the integral along the branch cut (2.5.22). Substitution of the numerically determined field values $e_m(\rho,s)$ and $h_m(\rho,s)$ in (2.5.33) - (2.5.35) directly yields the integrand in the last term of (2.5.36). In principle, the resulting integral can then be evaluated by a single Gauss-Laguerre quadrature formula with a sufficient number of points. However, because of the factor of $\exp(-\nu t)$ in (2.5.36), the abcissae and the weight factors of such a quadrature rule depend on the time t. This implies that the relevant values of e_m and h_m have to be recomputed for each instant, i.e. that the Runge-Kutta-Verner procedure has to be carried out for all abcissae at all instants. The greater part of this problem is avoided by evaluating the integral in two parts. First we split off the dominant part by computing the integral over a finite, time-invariant subinterval along which the integrand differs significantly from zero. This integral is determined with a repeated four-point Gauss-Legendre quadrature rule with fixed, time-invariant abcissae and weight factors. Hence, the differential equation (2.5.47) only needs to be solved once for this set of abcissae. Next, we compute the small remaining part of the integral by a single time-dependent Gauss-Laguerre quadrature rule. Thus, only the evaluation of this remaining integral requires a few field values to be recomputed at each instant.

2.5.5 Numerical results and their interpretation

Using the computational scheme described in Subsection 2.5.4, we have obtained numerical results for several cylinder configurations. In the first part of this subsection, we show some representative examples. In the second part, we discuss their physical interpretation.

Results

Computations were carried out for the following configurations:

$$
\begin{aligned}
&\text{i)} && \varepsilon_r(\rho) = 2.25, && \sigma(\rho) = 0 \text{ and } 2,\\
&\text{ii)} && \varepsilon_r(\rho) = 6.25, && \sigma(\rho) = 0 \text{ and } 2,\\
&\text{iii)} && \varepsilon_r(\rho) = 2.25 + 4\rho, && \sigma(\rho) = 2,\\
&\text{iv)} && \varepsilon_r(\rho) = 6.25 - 4\rho^2, && \sigma(\rho) = 2,\\
&\text{v)} && \varepsilon_r(\rho) = 2.25 + 4\sin^2(\tfrac{3}{2}\pi\rho), && \sigma(\rho) = 2.
\end{aligned}
\tag{2.5.54}
$$

For case i), both the closed-form solution (2.5.19) and the solution computed by numerical integration of (2.5.47) were used. In all cases, the Runge-Kutta-Verner integration and Muller's zero-finding procedure were carried out with an accuracy of 10^{-8}. Per angular order m, the poles were taken into account up to $n = 10$. The Fourier series (2.5.4) was truncated at $|m| = 5$ as well as at $|m| = 10$. It turns out that the tolerance in the numerical integration of (2.5.47) must be reduced for increasing m and s. This is probably caused by the factor ρ^{-1} on the right-hand side. Taking into account the fact that both the real and the imaginary part of $E_{mn}(\rho)$ have approximately $n - m/2$ zero crossings inside the cylinder as $n\to\infty$, we computed the coupling coefficients with a repeated trapezoidal rule on $50 + 25\,n$ subintervals, i.e. at least 50 points between two subsequent zero crossings. The branch-cut integral was computed with a 4-point Gauss-Legendre rule repeated 16 or 25 times and a 4- or 6-point Gauss-Laguerre rule, respectively. These results were compared with those of a direct numerical evaluation of the integral over the Bromwich contour (see (2.5.14)) for $\beta = 0$ with an FFT algorithm. In the evaluation of the integrand, the summation over m was truncated at the same value that was used in the pole series. For $|\mathrm{Im}(s)| > 12$, the integrand was replaced by its asymptotic approximation, as described in the first part of Subsection 2.5.3.

Since the results for configurations i) - iv) listed above show great similarity, we only present results for the parabolic and the sinusoidal permittivity profile, i.e. cases iv) and v). In Figures 2.5.3 - 2.5.7, we consider the quadratic profile. In Figure 2.5.3a, the natural frequencies are shown. In Figure 2.5.3b, we have enlarged the portion of the complex plane near the imaginary axis. In both these figures, two types of orderings are observed. As expected, the frequencies can be grouped into families according to their angular order m. There is apparently also an additional ordering. From the results, it is observed that the frequencies can also be grouped in families along which n - ent(m/2) = p, with p a constant, integer number. For $p < 0$, their locations stretch into the negative half of the complex s-plane; for $p \geq 0$ they remain near the imaginary s-axis. In Figure 2.5.4, the first few natural-mode electric-field distributions have been plotted in absolute values for $0 < \rho < 1$ and $m = 7$. Note that for modes with $p < 0$, the magnitude of the field increases exponentially with increasing ρ (see Figure 2.5.4a, b, c), while for modes with $p \geq 0$, p minima are observed in the interval (see Figure 2.5.4d-i). This observation is confirmed in Figure

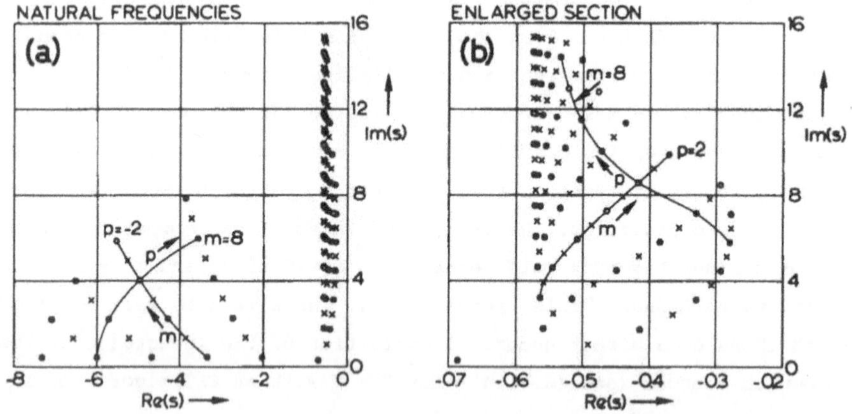

Figure 2.5.3 Natural frequencies s_{mn} with $0 \leq m \leq 10$, $0 \leq n \leq 10$, for a dielectric cylinder with a parabolic permittivity profile $\varepsilon_r(\rho) = 6.25 - 4\rho^2$ and constant conductivity $\sigma(\rho) = 2$. o: m even; x: m odd. The solid lines connect families of poles with m = 8 and with p = ±2, respectively. In Figure (b), the real s-axis was stretched to bring out the ordering in the poles with $p \geq 0$.

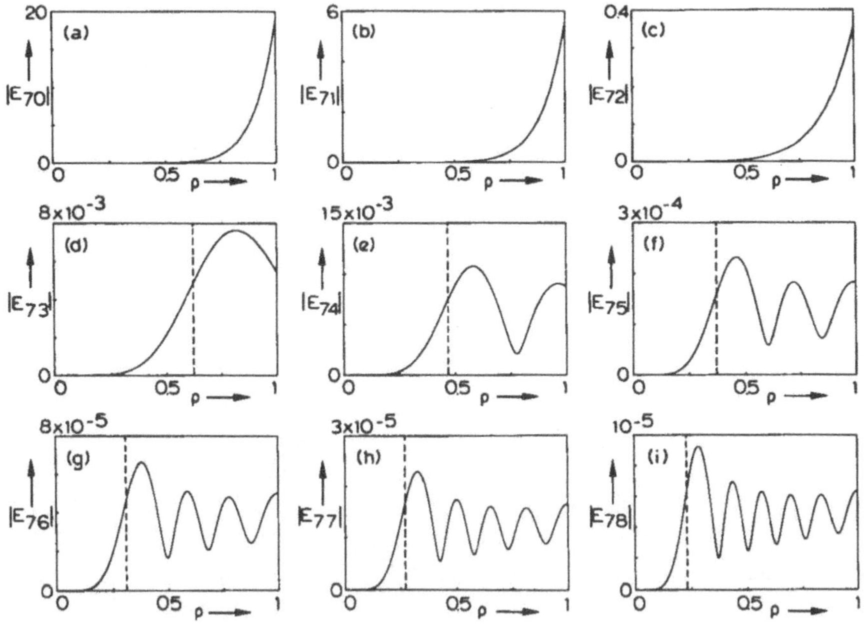

Figure 2.5.4 The magnitude of the natural-mode field distributions $E_{mn}(\rho)$ with $m = 7$, $0 \leq n \leq 8$, for the configuration specified in Figure 2.5.3 plotted for $0 < \rho < 1$. The dashed lines indicate the real parts of the transition points found from (2.6.59).

2.5.5, where similar distributions have been plotted for $p = 3$ and $0 \leq m \leq 8$. In Figure 2.5.6, the electric field at the center of the cylinder is given as a function of time. Because $E_m(0,s) = 0$ for $m \neq 0$, only the term with $m = 0$ needs to be taken into account. Figure 2.5.6 shows the result of the FFT-algorithm, the total contributions from the singularities in the complex s-plane and the contributions from the poles and the branch cuts separately. Fields obtained by a similar computation for points on the boundary are shown in Figure 2.5.7. In this case, the summation over m was truncated at $|m| = 5$. The only effect of this truncation is the presence of some minor oscillations at the beginning of the time interval. In Figure 2.5.7b and 2.5.7c, these oscillations are even observed at instants where the incident plane wave has not yet reached the point of observation. The effect is no longer observed if the summation

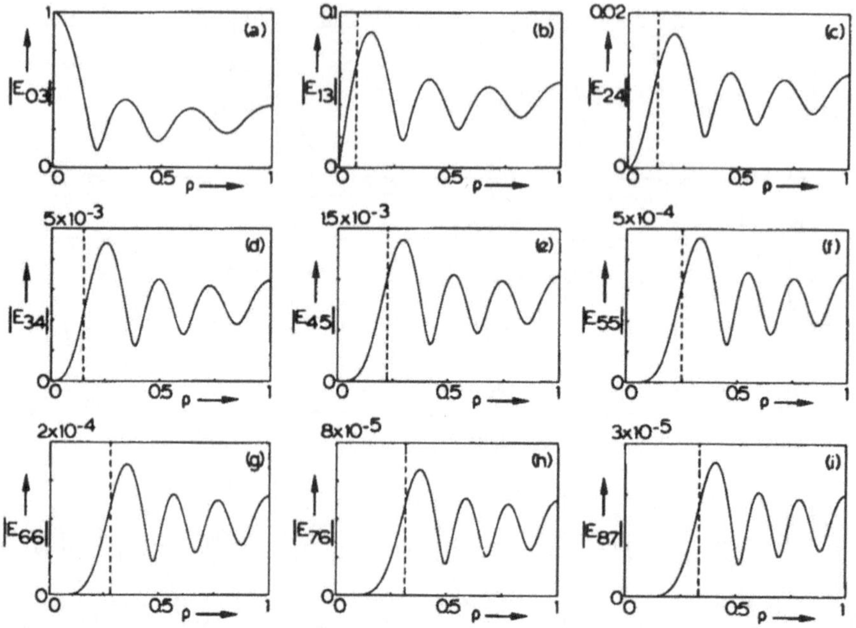

Figure 2.5.5 The magnitude of the natural-mode field distributions with $0 \leq m \leq 8$, $n = 3 + \text{ent}(m/2)$ $(p = 3)$, for the configuration specified in Figure 2.5.3 plotted for $0 < \rho < 1$. The dashed lines mark the real parts of the transition points found from (2.6.59).

over m is truncated at $|m| = 10$. The absence of such oscillations turns out to be a good indication that a sufficient number of angular orders has been taken into account. Wavefronts similar to the ones observed in the transient signals plotted in Figures 2.5.6 and 2.5.7 show up in all the configurations specified in (2.5.54). Therefore, it seems fair to conclude that these wavefronts are associated with reflections at or propagation along the cylinder boundary. A comparison of the arrival times of the various wavefronts at the points of observation confirms this conclusion.

For the present case, the computation time needed to obtain the first eleven natural frequencies for a given angular order varied from 30 seconds for $m = 2$ to 2 minutes and 35 seconds for $m = 10$. Most of this increase can be accounted for by the fact that, for larger angular orders, a larger region of the complex s-plane needs

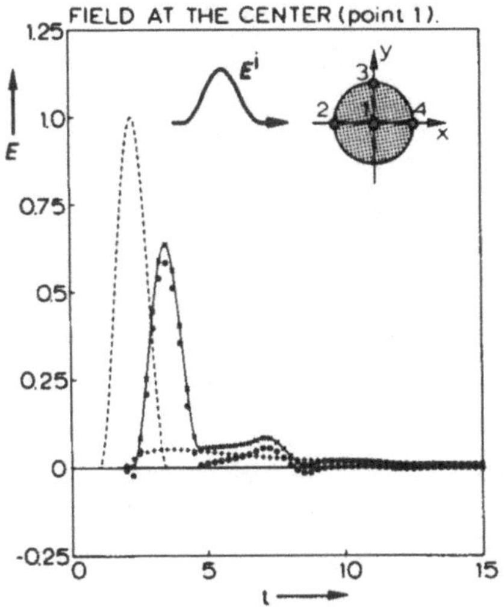

Figure 2.5.6 The electric field at the center of the cylinder specified in Figure 2.5.3 for the incident pulse shape $F(t) = \sin^2(\pi t/T)\text{rect}(t - T/2;T)$ with T = 2.5. Dashed line: incident field; solid line: result of FFT; x: total SEM result; o: total pole contribution; +: branch-cut contribution. Inset: the scattering configuration.

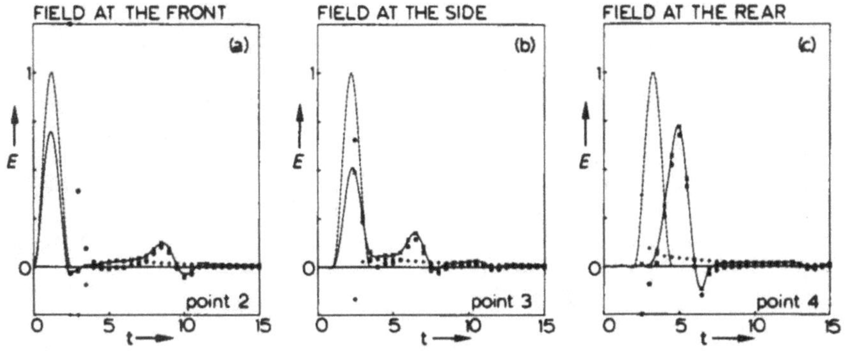

Figure 2.5.7 Time-domain electric fields at the points 2, 3 and 4 shown in Figure 2.5.6 for the configuration specified in Figures 2.5.3 and 2.5.6. The summations were truncated at $|m| = 5$ and $n = 10$. Symbols as given in Figure 2.5.6.

to be investigated. Moreover, it takes more time to evaluate the field values $e_m(1,s)$ and $h_m(1,s)$, which are needed in the calculation of $G_m(s)$. The subsequent computation of the residual contributions from the poles up to m = 5 - including the determination of the coupling coefficient D_{mn} - took 50 seconds. The calculation of the branch-cut integral over the subinterval $-12 < s < 0$ with a 4-point Gauss-Legendre quadrature rule repeated 25 times consumed 1 minute. The evaluation of the remaining integral with a 6-point Gauss-Laguerre formula lasted 2 minutes and 35 seconds. Within the precision of the other contributions, this remaining integral turned out to be negligible. Hence, its computation could have been omitted.

In Figures 2.5.8 - 2.5.10, we consider case v), i.e. the sine-squared profile. In Figure 2.5.8, the natural frequencies are shown. The solid lines connect families $\{s_{mn}\}$ with $p = n - \mathrm{ent}(m/2)$ being constant. We observe that the poles with $p < 0$ are distributed in a

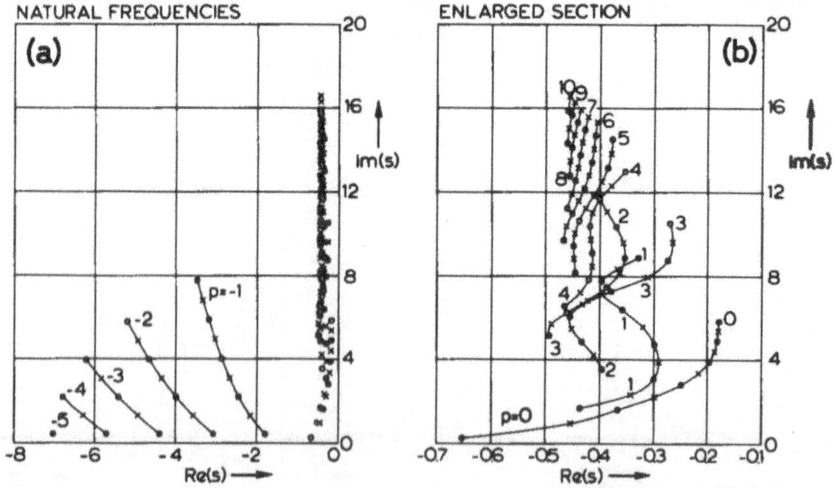

Figure 2.5.8 Natural frequencies s_{mn} with $0 \le m \le 10$, $0 \le n \le 10$ for a dielectric cylinder with a sinusoidal permittivity profile $\varepsilon_r(\rho) = 2.25 + 4\sin^2(\frac{3}{2}\pi\rho)$ and a constant conductivity $\sigma(\rho) = 2$. o: m even; x: m odd. The solid lines connect families of poles with constant $p = n - \mathrm{ent}(m/2)$. In Figure (b), the real s-axis was stretched to bring out the ordering in the poles with $p \ge 0$.

similar manner as their counterparts for the parabolic profile, while the distribution of poles with $p \geq 0$ is more complicated. Nevertheless, the natural frequencies can obviously be ordered according to their index p as well as to their angular order m. In Figure 2.5.9, the first few natural-mode electric-field distributions have been plotted in absolute values for $0 < \rho < 1$ and $m = 7$. As for the parabolic case, the magnitude of the modes with $p < 0$ increases exponentially with increasing ρ, while p minima are observed in the magnitude of those with $p \geq 0$. An interpretation of these observations as well as a more detailed discussion of the form of the distributions displayed in Figure 2.5.9 will be given in Subsection 2.6.3. Time-domain results are shown in Figure 2.5.10. Compared with the signals presented in Figures 2.5.6 and 2.5.7, these signals ex-

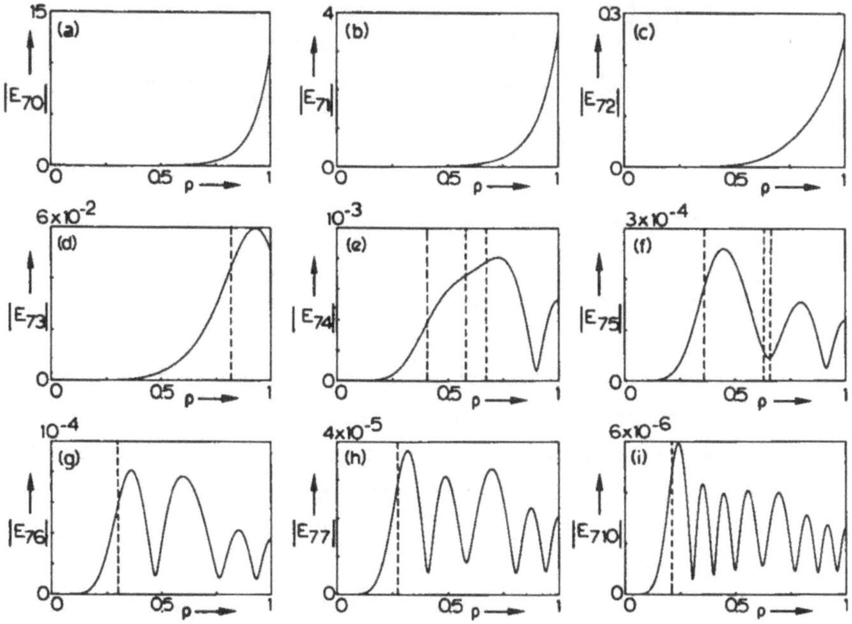

Figure 2.5.9 The magnitude of the natural-mode field distributions $E_{mn}(\rho)$ with $m = 7$, $0 \leq n \leq 7$, and $n = 10$, for the configuration specified in Figure 2.5.8 plotted for $0 < \rho < 1$. The dashed lines mark the real parts of the transition points found from (2.6.59) and tabled in Table 2.6.1.

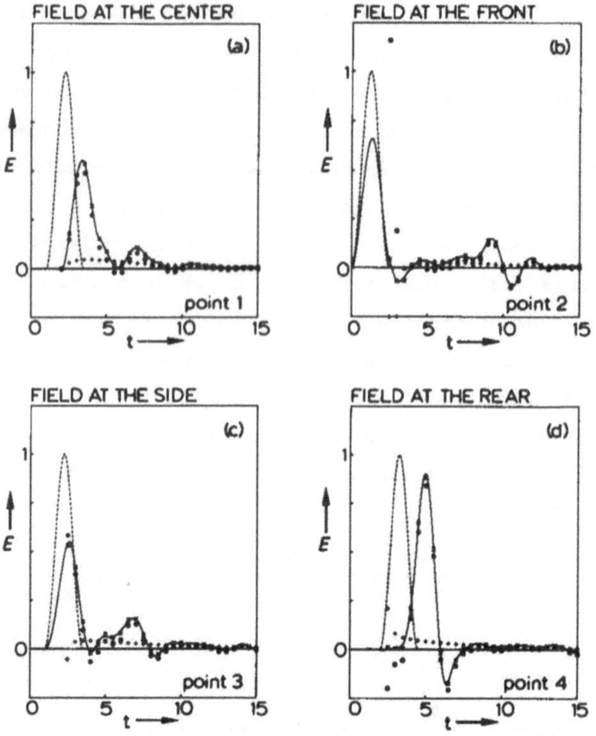

Figure 2.5.10 Time-domain electric fields at the points shown in Figure 2.5.6 for the configuration specified in Figure 2.5.8 excited by an incident pulse of the shape $F(t) = \sin^2(\pi t/T)\,\mathrm{rect}(t - T/2;T)$ with T = 2.5. Symbols as given in Figure 2.5.6.

hibit similar principal wavefronts caused by boundary effects. Furthermore, additional wavefronts show up in between them, which can be explained as being connected with reflection at and propagation along the local maximum in the permittivity at $\rho = \frac{1}{3}$.

Interpretation

A more profound physical interpretation of the results presented above follows from the ordering of the poles in the complex-frequency plane. In accordance with (2.5.36), we consider in this section only the ordering into families $\{s_{mn}\}$ with constant angular order m. The ordering into families with constant index p will be discussed extensively in Section 2.6. Comparing, for a given m, the pole pattern for $|s| \gg |m|$, as obtained in Subsection 2.5.3, with the corresponding one for a dielectric slab (see Subsection 2.3.3), we observe by analogy that the Fourier component

$$\exp(im\phi)E_m(\rho,t) + \exp(-im\phi)E_{-m}(\rho,t) \tag{2.5.55}$$

can be understood in terms of waves repeatedly reflected and diffracted at the cylinder boundary. Each of these waves travels in the radial direction only; the angular field distributions are of the form $\cos(m\phi)$ and, hence, do not change with ρ and t.

From Figures 2.5.6, 2.5.7 and 2.5.10, we see that the branch cut gives rise to a contribution that becomes almost constant in time as $t \gg 1$. Note that this contribution vanishes as $t\to\infty$; this is a direct result of the Riemann-Lebesgue theorem. Obviously, the incident field excites a "stationary" current $\underline{J}(\rho,\phi,t) = \sigma(\rho)E(\rho,\phi,t)\underline{i}_z$ that propagates in the z-direction. For such a current, we have

$$\mathrm{rot}\underline{E}(\rho,\phi,t) = [\underline{i}_\rho\, \rho^{-1}\partial_\phi - \underline{i}_\phi\partial_\rho]E(\rho,\phi,t) \approx 0 \tag{2.5.56}$$

for $t \gg 1$. From (2.5.56), it follows that the electric field E corresponding to this current will be almost constant across the cylinder. In the first place, this implies that this field is mainly produced by the term with $m = 0$. In the second place, this explains the similarity in the late-time behavior of the branch-cut contributions shown in Figures 2.5.6 and 2.5.7 for configuration iv) and in Figure 2.5.10 for configuration v).

Appendix 2.5.A Mean-square convergence of the angular Fourier series as a function of ρ and t

In Subsection 2.5.1, it was argued that for a suitable choice of the incident-pulse shape $F(t)$, the electromagnetic field intensities $E(\rho,\phi,t)$ and $H(\rho,\phi,t)$ can, as far as their dependence on ϕ is concerned, be represented in terms of a Fourier series of the type (2.5.4). For a fixed ρ and t, the series obtained converges in mean square to the relevant quantity. In this appendix, we investigate the convergence as a function of ρ and t. To this end, we consider the local form of the electromagnetic power balance:

$$\sigma(\rho)\underline{E}\cdot\underline{E} + \tfrac{1}{2}\partial_t[\varepsilon(\rho)\underline{E}\cdot\underline{E} + \mu_0\underline{H}\cdot\underline{H}] + \mathrm{div}(\underline{E}\times\underline{H}) = 0, \tag{2.5.57}$$

which holds for any real-valued solution $\{\underline{E}(\rho,\phi,t),\underline{H}(\rho,\phi,t)\}$ of Maxwell's equations for the configuration specified in Subsection 2.5.1.

In the first place, we consider the total field $\{E,H\}$ inside the cylinder. Integrating (2.5.57) over the space-time domain $0 < \rho < a$, $-\pi < \phi < \pi$, $z_0 < z < z_0 + 1$, $0 < t' < t$ and applying Gauss' theorem to the space integral of the last term result in:

$$\begin{aligned}&\int_0^t\int_0^a\int_{-\pi}^{\pi}[\sigma(\rho)\underline{E}\cdot\underline{E}(\rho,\phi,t')]\rho d\rho d\phi dt' \\ &\quad + \tfrac{1}{2}\int_0^a\int_{-\pi}^{\pi}[\varepsilon(\rho)\underline{E}\cdot\underline{E}(\rho,\phi,t) + \mu_0\,\underline{H}\cdot\underline{H}(\rho,\phi,t)]\rho d\rho d\phi \\ &\quad = a\int_0^t\int_{-\pi}^{\pi} E(a,\phi,t')H(a,\phi,t')d\phi dt',\end{aligned} \tag{2.5.58}$$

where E and H denote the tangential field components as specified in (2.5.2). Using the properties of $\sigma(\rho)$ and $\varepsilon(\rho)$ given in Subsection 2.5.1, we arrive at the inequality:

$$\begin{aligned}&\int_0^a\int_{-\pi}^{\pi}[\varepsilon_0 E^2(\rho,\phi,t) + \mu_0 H^2(\rho,\phi,t)]\rho d\rho d\phi \\ &\quad \le 2a\int_0^t\int_{-\pi}^{\pi}E(a,\phi,t')H(a,\phi,t')d\phi dt'.\end{aligned} \tag{2.5.59}$$

Now, we consider the difference

$$\Delta E_M(\rho,\phi,t) \triangleq E(\rho,\phi,t) - \sum_{m=-M}^{M}\exp(im\phi)E_m(\rho,t), \tag{2.5.60}$$

and the similarly defined difference $\Delta H_M(\rho,\phi,t)$. $E(\rho,\phi,t)$ and each linear combination

$$\exp(im\phi)E_m(\rho,t) + \exp(-im\phi)E_{-m}(\rho,t), \tag{2.5.61}$$

along with the corresponding magnetic field, constitute a real-valued solution of Maxwell's equations. By virtue of the superposition principle, the linear combination on the right-hand side of (2.5.60) and the corresponding magnetic field then also meet these conditions and can therefore be substituted in (2.5.59). We arrive at:

$$\int_0^a \int_{-\pi}^{\pi} [\varepsilon_0 \Delta E_M^2(\rho,\phi,t) + \mu_0 \Delta H_M^2(\rho,\phi,t)]\rho d\rho d\phi \leq I_M(t), \tag{2.5.62}$$

with

$$I_M(t) = 2a\int_0^t \int_{-\pi}^{\pi} \Delta E_M(a,\phi,t')\Delta H_M(a,\phi,t')d\phi dt'. \tag{2.5.63}$$

Except for a normalization factor, the left-hand side of (2.5.62) represents the sum of the integrated squared errors in approximating the total fields E and H inside the cylinder at time t by a Fourier series truncated at $|m| = M$. Hence, we are left with the task of estimating the integral $I_M(t)$. To this end, we employ the Cauchy-Schwarz inequality:

$$\begin{aligned}&\int_{-\pi}^{\pi} \Delta E_M(a,\phi,t')\ \Delta H_M(a,\phi,t')d\phi \leq \\ &\left\{\int_{-\pi}^{\pi} \Delta E_M^2(a,\phi,t')d\phi\right\}^{\frac{1}{2}}\left\{\int_{-\pi}^{\pi} \Delta H_M^2(a,\phi,t')d\phi\right\}^{\frac{1}{2}}.\end{aligned} \tag{2.5.64}$$

The integrals between braces on the right-hand side of (2.5.64) are the integrated squared errors in approximating $E(a,\phi,t')$ and $H(a,\phi,t')$ by a truncated Fourier series. Since for a fixed ρ and t', the Fourier series representations of these field quantities converge in mean square, the integral on the left-hand side vanishes for each t' as $M\to\infty$. Moreover, for an incident pulse of finite duration, the integrand of (2.5.63) will become negligible after some finite instant $t' = T_{max}$. From these two facts and (2.5.62), it follows directly that the Fourier-series representations of the total electromagnetic field intensities $E(\rho,\phi,t)$ and $H(\rho,\phi,t)$ converge in mean square on the spatial domain $0 < \rho < a$, $-\pi < \phi < \pi$, uniformly in t for $0 < t < \infty$.

Second, we consider the electromagnetic field outside the cylinder. Clearly, there is no global convergence for the incident pulsed plane wave. Therefore, we restrict ourselves to the scattered field $\{\underline{E}^s,\underline{H}^s\}$, which, in $\mathcal{D}_1$, is by itself a valid solution to Maxwell's equations in vacuo. Substituting this field in (2.5.57), integrating over the space-time domain $a < \rho < R$, $-\pi < \phi < \pi$, $z_0 < z < z_0 + 1$, $0 < t' < t$, and applying Gauss' theorem result in:

$$\begin{aligned} &\tfrac{1}{2}\int_a^R\int_{-\pi}^{\pi}[\varepsilon_0\underline{E}^s\cdot\underline{E}^s(\rho,\phi,t) + \mu_0\underline{H}^s\cdot\underline{H}^s(\rho,\phi,t)]\rho d\rho d\phi \\ &- R\int_0^t\int_{-\pi}^{\pi} E^s(R,\phi,t')H^s(R,\phi,t')d\phi dt' = \\ &-a\int_0^t\int_{-\pi}^{\pi}E^s(a,\phi,t')H^s(a,\phi,t')d\phi dt'. \end{aligned} \tag{2.5.65}$$

For the far-scattered field, we have as $\rho\to\infty$:

$$\begin{aligned} E^s(\rho,\phi,t) &= \rho^{-\frac{1}{2}}e^{\infty}(\phi,t)[1 + O(\rho^{-1})], \\ H^s(\rho,\phi,t) &= \rho^{-\frac{1}{2}}h^{\infty}(\phi,t)[1 + O(\rho^{-1})], \end{aligned} \tag{2.5.66}$$

where the far-field scattering amplitudes e^{∞} and h^{∞} satisfy the relation $h^{\infty} = -Y_0e^{\infty}$. By substituting (2.5.66) in (2.5.65) and taking the limit for $R\to\infty$, we obtain:

$$\begin{aligned} &\int_a^{\infty}\int_{-\pi}^{\pi}\left\{\varepsilon_0[E^s(\rho,\phi,t)]^2 + \mu_0[H^s(\rho,\phi,t)]^2\right\}\rho d\rho d\phi \\ &\leq -2a\int_0^t\int_{-\pi}^{\pi}E^s(a,\phi,t')H^s(a,\phi,t')d\phi dt'. \end{aligned} \tag{2.5.67}$$

By arguments similar to those given above for the total field inside the cylinder, it follows that the Fourier series representations of the scattered field intensities $E^s(\rho,\phi,t)$ and $H^s(\rho,\phi,t)$ converge in mean square on the spatial domain $a < \rho < \infty$, $-\pi < \phi < \pi$, uniformly in t for $0 < t < \infty$. Furthermore, from the form of (2.5.66) and from the occurrence of the factor of ρ on the left-hand side of (2.5.67), it is observed that the global relative accuracy in the far-field region is of the same order of magnitude as the corresponding accuracy in the near-field region.

2.6 Scattering by a lossy, radially inhomogeneous dielectric circular cylinder: complementary interpretation of the singularity expansion method

2.6.1 Formulation of the problem

From the results presented in Subsection 2.5.5, it was observed that the natural modes of a radially inhomogeneous, lossy dielectric circular cylinder can be ordered into families in two complementary ways. In addition, a physical interpretation was given for the total residual contribution from a family of poles with constant angular order m. In this section, we analyze families with constant index p. As we shall see further on, our interpretation can be verified by considering limiting cases where the cylinder either becomes impenetrable, or is surrounded by an impenetrable medium. The latter case is included in the analysis when we replace the configuration described in Subsection 2.5.1 by the more general case, in which the outer medium is a homogeneous, lossless dielectric (see Figure 2.6.1). In $\mathcal{D}_1$, we then have the permittivity $\varepsilon_1 = \varepsilon_{1r}\varepsilon_0 \geq \varepsilon_0$, the conductivity $\sigma_1 = 0$, and the permeability $\mu_1 = \mu_0$, where ε_0 and μ_0

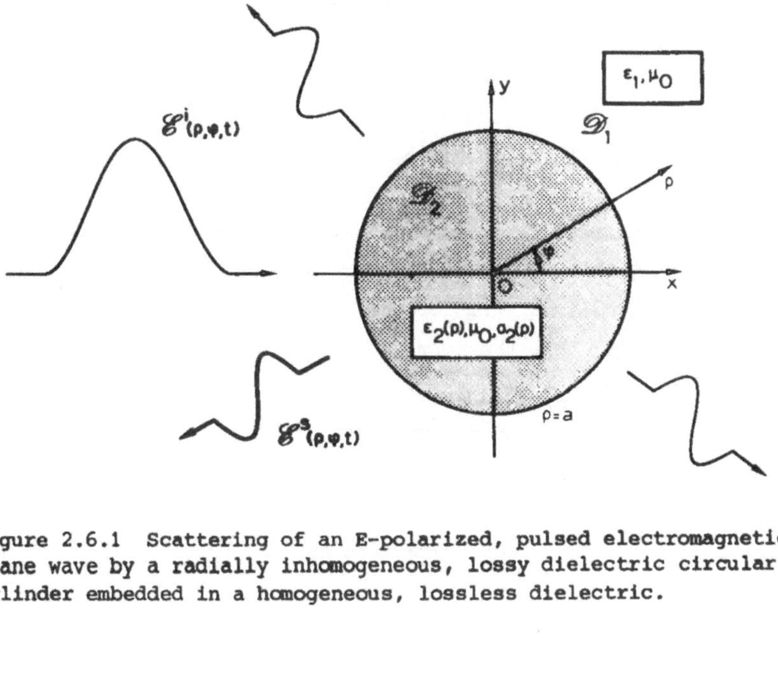

Figure 2.6.1 Scattering of an E-polarized, pulsed electromagnetic plane wave by a radially inhomogeneous, lossy dielectric circular cylinder embedded in a homogeneous, lossless dielectric.

denote the corresponding constitutive coefficients in vacuo. In $\mathcal{D}_2$, we still have $\varepsilon_2 = \varepsilon_{2r}(\rho)\varepsilon_0 \geq \varepsilon_0$, $\sigma_2 = \sigma_2(\rho) \geq 0$, and $\mu_2 = \mu_0$, respectively.

For the incident field, we take

$$\begin{aligned} \underline{E}^i &= F(t - (\rho\cos\phi + a)/c_1)\underline{i}_z, \\ \underline{H}^i &= -Y_1 F(t - (\rho\cos\phi + a)/c_1)(\sin\phi\ \underline{i}_\rho + \cos\phi\ \underline{i}_\phi), \end{aligned} \tag{2.6.1}$$

where $Y_1 = (\varepsilon_1/\mu_0)^{\frac{1}{2}}$ and $c_1 = (\varepsilon_1\mu_0)^{-\frac{1}{2}}$. The coefficient $E_m^i(\rho,s)$ corresponding to the one specified in (2.5.6) follows from applying a Fourier transformation, the shift rule, and a one-sided Laplace transformation to (2.6.1) as

$$E_m^i(\rho,s) = (-1)^m \exp(-sa/c_1)I_m(s\rho/c_1)F(s). \tag{2.6.2}$$

The normalized, source-free electromagnetic field equations given in (2.5.7) and (2.5.8) remain unchanged. It is convenient to rewrite (2.5.8) as

$$[\partial_\rho^2 + \rho^{-1}\partial_\rho - (m^2/\rho^2 + \varepsilon_{1r}s^2)]E_m(\rho,s) = s^2C_1(\rho,s)E_m(\rho,s), \tag{2.6.3}$$

where

$$C_1(\rho,s) = \varepsilon_r(\rho) - \varepsilon_{1r} + \sigma(\rho)/s \tag{2.6.4}$$

denotes the *contrast function* with respect to medium 1, and differs from zero only when $0 \leq \rho < 1$. In $\mathcal{D}_1$, the formal expressions for $E_m(\rho,s)$ and $H_m(\rho,s)$ in terms of the normalized solutions $\{e_m,h_m\}$ defined in (2.5.18) take the form

$$\begin{aligned} E_m(\rho,s) &= (-1)^m \exp(-sN_1)F(s)[I_m(sN_1\rho) + b_m(s)K_m(sN_1\rho)], \\ H_m(\rho,s) &= (-1)^m \exp(-sN_1)F(s)[N_1I_m'(sN_1\rho) + b_m(s)N_1K_m'(sN_1\rho)], \end{aligned} \tag{2.6.5}$$

where $N_1 = \varepsilon_{1r}^{\frac{1}{2}}$. In $\mathcal{D}_2$, we write them as

$$E_m(\rho,s) = (-1)^m \exp(-sN_1)F(s)c_m(s)e_m(\rho,s),$$
$$H_m(\rho,s) = (-1)^m \exp(-sN_1)F(s)c_m(s)h_m(\rho,s). \qquad (2.6.6)$$

Application of the boundary conditions at $\rho = 1$ (continuity of E_m and H_m across the interface) yields

$$b_m(s) = [N_1 I'_m(sN_1)e_m(1,s) - I_m(sN_1)h_m(1,s)]/G_m(s),$$
$$c_m(s) = 1/sG_m(s), \qquad (2.6.7)$$

with

$$G_m(s) = K_m(sN_1)h_m(1,s) - N_1 K'_m(sN_1)e_m(1,s) \qquad (2.6.8)$$

being the characteristic cylinder denominator.

One way to arrive at the SEM representation for this generalized configuration is to start from (2.6.5) - (2.6.8), and go through the same steps that were carried out in Section 2.5, the only difference being that we have to account for the fact that the present outer medium is a homogeneous, lossless dielectric instead of a vacuum. This amounts to judiciously inserting the constant N_1 in some of the expressions obtained in 2.5. In the forthcoming subsections, we will present an alternative way to derive the SEM representation. In doing so, we will arrive in a natural way at the ordering of the natural frequencies according to their index p and its physical interpretation.

2.6.2 Harmonic response

We can also carry out the angular Fourier transformation and the temporal Laplace transformation as defined in Subsection 2.5.1 in reverse order. This leads to

$$\mathcal{E}(\rho,\phi,t) = (2\pi i)^{-1}\int_{L_\beta} \exp(st)E(\rho,\phi,s)ds,$$

with (2.6.9)

$$E(\rho,\phi,s) = \sum_{m=-\infty}^{\infty} \exp(im\phi)E_m(\rho,s).$$

The customary frequency-domain response is obtained from $E(\rho,\phi,s)$ by taking out the spectrum of the incident pulse:

$$E(\rho,\phi,s) = \exp(-sN_1)F(s)\hat{E}(\rho,\phi,s), \tag{2.6.10}$$

and replacing s by $i\omega$. Apart from the complex time factor $\exp(i\omega t)$, $\hat{E}(\rho,\phi,i\omega)$ represents the field excited by an E-polarized, monochromatic plane wave with an angular frequency ω, incident from the direction $\phi = \pi$.

In this subsection, we investigate this harmonic response $\hat{E}(\rho,\phi,s)$. Combining (2.6.5), (2.6.6), (2.6.9) and (2.6.10) results in

$$\hat{E}(\rho,\phi,s) = \sum_{m=-\infty}^{\infty} (-1)^m \exp(im\phi)\hat{E}_m(\rho,s), \tag{2.6.11}$$

with

$$\hat{E}_m(\rho,s) \triangleq \begin{cases} I_m(sN_1\rho) + b_m(s)K_m(sN_1\rho) & \text{in } \mathcal{D}_1, \\ c_m(s)e_m(\rho,s) & \text{in } \mathcal{D}_2. \end{cases} \tag{2.6.12}$$

As is well known from the literature (see e.g. Franz (1957) and Miyazaki (1981)), the convergence of (2.6.11) slows down as s increases in magnitude. In fact, this was the reason for carrying out the two transformations in their original order in Section 2.5. To resolve this problem, we convert the summation over m into an integral in the complex ν-plane by the Watson transformation:

$$\hat{E}(\rho,\phi,s) = \frac{1}{2i}\int_C \frac{d\nu}{\sin\nu\pi} \exp(i\nu\phi)\hat{E}_\nu(\rho,s), \tag{2.6.13}$$

where $\hat{E}_\nu(\rho,s)$ is obtained from $\hat{E}_m(\rho,s)$ by replacing $I_m(z)$, $K_m(z)$ and $e_m(\rho,s)$ with $I_{\sqrt{\nu^2}}(z)$, $K_\nu(z)$ and $e_\nu(\rho,s)$, respectively. The normalized solution $e_\nu(\rho,s)$ is defined by:

$$\begin{gathered} [\partial_\rho^2 + \rho^{-1}\partial_\rho - s^2\varepsilon_r(\rho) - s\sigma(\rho) - \nu^2/\rho^2]e_\nu(\rho,s) = 0, \\ \lim_{\rho\downarrow 0} \rho^{-\sqrt{\nu^2}} e_\nu(\rho,s) = 1. \end{gathered} \tag{2.6.14}$$

In $I_{\sqrt{\nu^2}}$ and in (2.6.14), the square root is chosen according to its principal value, i.e. $\sqrt{\nu^2} = \nu$ when $Re(\nu) > 0$, and $\sqrt{\nu^2} = -\nu$ when $Re(\nu) < 0$. The integration contour consists of two contours C^+ and C^-, which encircle the positive and the negative half of the real ν-axis, respectively, as shown in Figure 2.6.2. Both integrals are defined according to their Cauchy principal value at $\nu = 0$; i.e. half

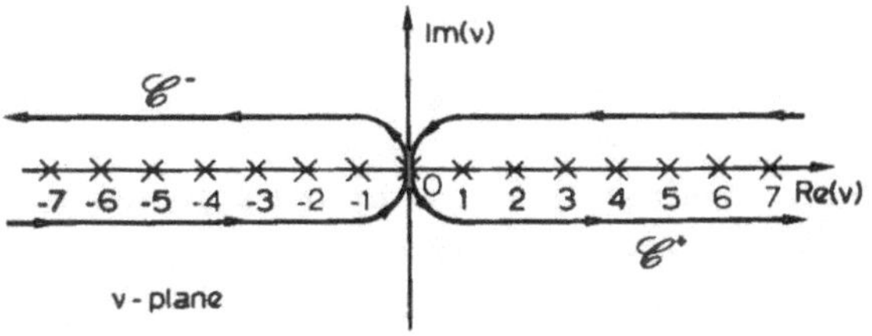

Figure 2.6.2 Integration contour for the Watson transformation (2.6.13) ($C = C^+ + C^-$).

the residue at $\nu = 0$ is attributed to either integral. It should be noted that in (2.6.13) and (2.6.14), we have represented the total field in terms of constituents that remain bounded near the origin.

From definition (2.6.14), it is observed immediately that $e_\nu(\rho,s)$ satisfies two symmetry relations that will be of use further on in the discussion. In the first place, $e_\nu(\rho,s)$ depends on ν^2 only. Hence, we have $e_{-\nu}(\rho,s) = e_\nu(\rho,s)$. In the second place, a complex conjugation of (2.6.14) results in $e_{\nu*}(\rho,s^*) = e^*_\nu(\rho,s)$. From (2.6.7), (2.6.12) and (2.6.13), it is observed that $\hat{E}_\nu(\rho,s)$ only depends on the parameters s and ν via $e_\nu(\rho,s)$ and the modified Bessel functions $K_\nu(sN_1\rho)$ and $I_{\sqrt{\nu^2}}(sN_1\rho)$, which satisfy the same symmetry relations as $e_\nu(\rho,s)$. Hence, we also have:

$$\hat{E}_{-\nu}(\rho,s) = \hat{E}_\nu(\rho,s), \tag{2.6.15}$$

and

$$\hat{E}_{\nu*}(\rho,s^*) = [\hat{E}_\nu(\rho,s)]^*. \tag{2.6.16}$$

At this stage in the analysis, it may be helpful to indicate the relation of the field representation introduced in (2.6.9) - (2.6.14) to the angular transmission representation formulated by Felsen and Marcuvitz (1973, Section 6.2.b). In their formulation, the contour C is the contour enclosing all singularities in the complex ν-plane that correspond to radially propagating waves but no others. Equation (2.6.11) is then the discrete representation of $\hat{E}(\rho,\phi,s)$ in terms of such waves. This is in agreement with the physical interpretation proposed in Subsection 2.5.5 for a Fourier coefficient of the form (2.5.61). Note that (2.6.9) - (2.6.14) could also have been obtained by following the approach of Felsen and Marcuvitz for the case where the field is excited by a properly normalized line source located at $x = x_0$, $y = 0$, and taking the limit of $x_0 \to -\infty$ (see Friedlander (1958)). The main attractions of the present approach are that we avoid this limiting procedure, and that, by starting with an angular Fourier transformation, we have automatically ensured the completeness of the representation (2.6.11) and of any equivalent form described from it. As argued by Felsen and Marcuvitz, an equivalent form in terms of angularly propagating waves is obtained by evaluating the contour integral in (2.6.13) by contour deformation in the complex ν-plane away from the real ν-axis. We will now proceed to derive this representation.

To this end, we supplement C^+ and C^- by circular arcs at infinity in the complex ν-plane as shown in Figure 2.6.3, and apply Cauchy's theorem. From the asymptotic expression for $e_\nu(\rho,s)$ given in (2.6.52) and the Debye approximations for the modified Bessel functions $I_{\sqrt{\nu^2}}$ and K_ν (see Felsen and Marcuvitz (1973, pp. 710 - 719)), it follows that, unless $\arg(\nu) = \pm \pi/2$, $\hat{E}_\nu(\rho,s)$ decays faster than exponentially as $|\nu| \to \infty$. This means that the integrals along the closing contours vanish and that only the singularities of $\hat{E}_\nu(\rho,s)$ in the complex ν-plane contribute to the result. These singularities comprise a denumerable set of poles and a pair of branch cuts from the origin of the complex ν-plane along the positive and negative imaginary axes, as shown in Figure 2.6.3 ($\sqrt{\nu^2}$ may be thought of as $[(i\nu)(-i\nu)]^{\frac{1}{2}}$ with branch points at $\nu = \pm i0$; see also Lewin, Chang and Kuester (1977)).

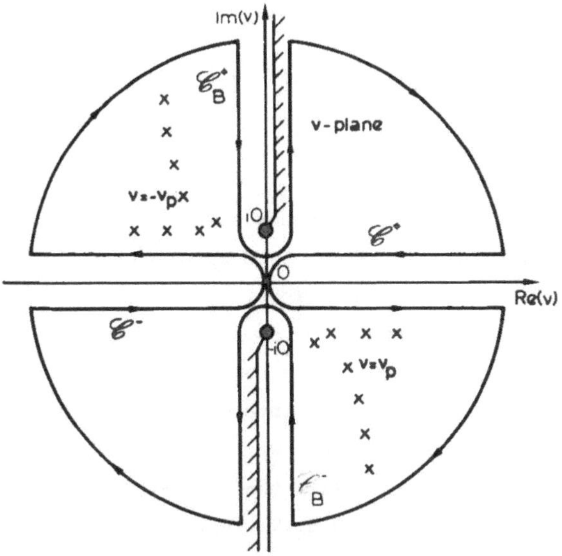

Figure 2.6.3 Contour deformation and singularities in the complex ν-plane (the pole locations apply to s = iω with ω positive real).

We will now first consider the poles in more detail. From (2.6.15), it follows that they occur in pairs $\nu = \pm\,\nu_p(s)$, with $\mathrm{Re}(\nu_p) \geq 0$. The values $\{\nu_p\}$ are found by searching the right half of the ν-plane for roots of the characteristic equation:

$$G_\nu(s) \triangleq K_\nu(sN_1)h_\nu(1,s) - N_1K'_\nu(sN_1)e_\nu(1,s) = 0, \qquad (2.6.17)$$

with $G_\nu(s)$ being the analytic continuation of the characteristic cylinder denominator into the complex ν-plane excepting the imaginary axis. From (2.6.15), it follows that

$$\mathrm{res}\ \hat{E}_\nu(\rho,s)\big|_{\nu=-\nu_p} = -\ \mathrm{res}\ \hat{E}_\nu(\rho,s)\big|_{\nu=\nu_p} \triangleq -r_p(\rho,s). \qquad (2.6.18)$$

With this relation, the residual contribution from a pole pair $\nu = \pm\,\nu_p$ to $\hat{E}(\rho,\phi,s)$ takes the form:

$$(-4\pi i)r_p(\rho,s)\cos(\nu_p\phi)/[2i\,\sin(\nu_p\pi)] = \tag{2.6.19a}$$

$$(-2\pi i)r_p(\rho,s)\sum_{n=0}^{\infty}\Big\{\exp[i\nu_p(\phi - \pi - 2\pi n)] + \exp[i\nu_p(-\phi - \pi - 2\pi n)]\Big\}, \tag{2.6.19b}$$

where, in (2.6.19b), it has been assumed that $\mathrm{Im}(\nu_p) < 0$. As shown in Appendix 2.6.B, $r_p(\rho,s)$ can be determined according to the same procedure that was used in Subsection 2.5.2 to determine the residues of $E_m(\rho,s)$ at poles in the complex s-plane. In this manner, the numerical differentiation of the numerically obtained denominator $G_\nu(s)$ is avoided. The physical interpretation of the residual contribution (2.6.19) follows when we take $s = i\omega$ with ω real and positive and combine (2.6.19b) with the complex time factor $\exp(i\omega t)$. Then, it is observed that (2.6.19) represents this contribution in terms of traveling waves that originate from $\phi = \pi + 2\pi n$ and $\phi = -\pi - 2\pi n$, with $n = 0, 1, 2, \ldots, \infty$, and propagate in the negative and the positive ϕ-direction, respectively. Consequently, the terms with index n represent traveling waves that circumnavigate the central axis n times. This angular behavior is illustrated graphically in Figure 2.6.4. In the radial direction, each wave has a fixed distribution $r_p(\rho,s)$. As shown in Appendix 2.6.B, we have $r_p(\rho,s) = \lambda_p(s)e_{\nu_p}(\rho,s)$ with $\lambda_p(s)$ being a complex coefficient. This interpretation also explains why in Figure 2.6.3 and (2.6.19b) the $\{\nu_p\}$ have been restricted to the fourth quadrant of the complex ν-plane; a pole in the first quadrant would give rise to an expansion of the type (2.6.19b) involving noncausal terms. For ω real and negative,

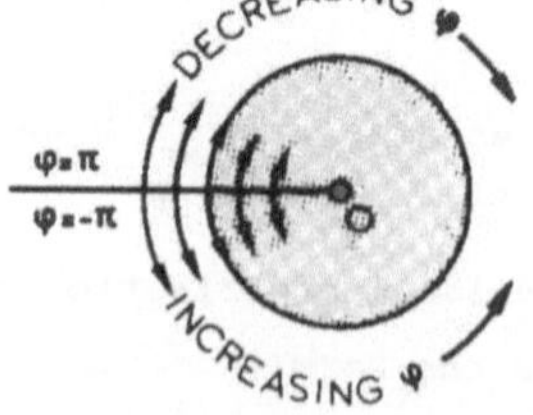

Figure 2.6.4 Angular behavior of the residual contribution from a pole pair $\nu = \pm\,\nu_p(s)$ as given in (2.6.19).

causality requires similarly that the $\{\nu_p\}$ are all in the first quadrant. This is in agreement with the symmetry relation (2.6.16).

For a further interpretation of the poles in the complex ν-plane and the corresponding residual contributions, information is needed on the behavior of $\hat{E}_\nu(\rho,s)$, or equivalently $e_\nu(\rho,s)$, as a function of ρ and ν. Since this distribution can, in general, only be determined numerically, we utilize the asymptotic theory of Appendix 2.6.A. For s large, this theory yields an asymptotic expression for $e_\nu(\rho,s)$ that is valid for each complex ν. As in the appendix, we restrict the analysis to $\mathrm{Re}(\nu) > 0$ and $\mathrm{Im}(s) \geq 0$. The behavior of $E_\nu(\rho,s)$ in the complementary half-planes follows directly from the symmetry relations (2.6.15) and (2.6.16).

Let us first consider combinations of s and ρ that give rise to a transition point as specified in Appendix 2.6.A in the asymptotic field distribution $e^a_\nu(\rho,s)$. Such combinations are encountered for $\arg(\nu) = \arg(s) - \pi/2 + \mathcal{O}(\alpha^{-1}) = \arg(\omega) + \mathcal{O}(\alpha^{-1})$, with $\alpha \triangleq \max\{|\nu|,|s|\}$. Limiting ourselves to choices of $\varepsilon_{2r}(\rho)$ for which we have at most a single transition point, we can then approximate $e_\nu(\rho,s)$ and $h_\nu(\rho,s)$ by the expressions given in (2.6.65) and (2.6.67), with $\varepsilon_r(\rho)$ and $\sigma(\rho)$ replaced by $\varepsilon_{2r}(\rho)$ and $\sigma_2(\rho)$. Substitution of these expressions in the characteristic equation (2.6.17) results in

$$K_\nu(sN_1)[\varepsilon_{2r}(1) + \nu^2/s^2]^{\frac{1}{2}} \sinh[\psi(1,s,\nu)] - N_1K'_\nu(sN_1)\cosh[\psi(1,s,\nu)] = 0, \qquad (2.6.20)$$

with $\psi(\rho,s,\nu)$ as defined in (2.6.64). The asymptotic equation (2.6.20) can be simplified by writing

$$\xi(s,\nu) \triangleq \operatorname{arctanh}\{N_1K'_\nu(sN_1)/K_\nu(sN_1)[\varepsilon_{2r}(1) + \nu^2/s^2]^{\frac{1}{2}}\}, \qquad (2.6.21)$$

where $-\pi/2 \leq \mathrm{Im}[\operatorname{arctanh}(z)] \leq \pi/2$. This reduces (2.6.20) to

$$\sinh[\psi(1,s,\nu) - \xi(s,\nu)] = 0. \qquad (2.6.22)$$

The special form of (2.6.22) induces an ordering by a decomposition of (2.6.22) into the equivalent set of equations

$$\psi(1,s,\nu) - \xi(s,\nu) - p\pi i = 0, \quad p = 0, 1, 2, \ldots, \infty. \qquad (2.6.23)$$

In (2.6.23), we need only consider nonnegative p because it follows from (2.6.21) and (2.6.64) that $\mathrm{Im}[\psi(1,s,\nu) - \xi(s,\nu)] \geq -\frac{1}{4}\pi$. The interpretation of the poles corresponding to the zeros of (2.6.23) follows from the form of the asymptotic field distribution $e_\nu^a(\rho,s)$ inside the cylinder. From $\rho = 0$ up to the transition point $\rho = \rho_1$, e_ν^a increases exponentially in magnitude with increasing ρ. Between the transition point and the boundary, e_ν^a has the form of a standing wave with p minima in its magnitude. In combination with the angular behavior given in (2.6.19) this leads to the visualization in Figure 2.6.5a, which suggests that a pole of this type corresponds to a *whispering-gallery mode*. The term $\xi(s,\nu)$ is then identified as being half the phase shift due to a reflection at the cylinder boundary, and the term $-i\pi/4$ in (2.6.64) as half the phase shift due to a total reflection at the surface $\rho = \rho_1$ (see also Wait (1967)). Consequently, the exponentially increasing part of the modal field distribution up to $\rho = \rho_1$ can be understood as being the broken wave

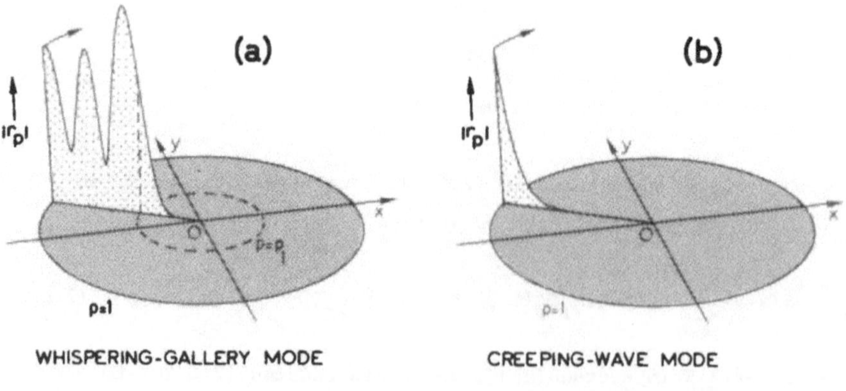

Figure 2.6.5 Behavior of a natural mode inside the cylinder. (a): whispering-gallery mode; (b) surface-wave part of a creeping-wave mode. In both cases, a similar mode moves in the opposite ϕ-direction.

caused by this total reflection. The interpretation given above is confirmed by the dependence of $\psi(1,s,\nu)$ on the constitutive coefficients: $\varepsilon_{2r}(\rho)$ and $\sigma_2(\rho)$ only occur in integrals from $\rho = \rho_0 \approx \rho_1$ to $\rho = 1$. A final verification is obtained by letting the outer medium become impenetrable, i.e. by taking the limit for $N_1 \to \infty$. In that limit, $\xi(s,\nu)$ approaches the value $\pi i/2$. With this result and (2.6.64), (2.6.23) reduces to

$$s\int_{\rho_0}^{1}\left\{[\varepsilon_{2r}(\rho) + \nu^2/\rho^2 s^2]^{\frac{1}{2}} + \frac{\sigma_2(\rho)}{2s[\varepsilon_{2r}(\rho) + \nu^2/\rho^2 s^2]^{\frac{1}{2}}}\right\}d\rho = (p + \tfrac{1}{4})\pi i, \tag{2.6.24}$$

$p = 0, 1, 2, \ldots, \infty$. For the case of a homogeneous, lossless inner medium, Wasylkiwskyj (1975) obtained a phase equation of the same kind, using a first-order uniform asymptotic approximation of the Bessel function in terms of the Airy function Ai. In the terminology of the present section, his result reads

$$s\int_{i\nu/s\varepsilon_{2r}^{\frac{1}{2}}}^{1}(\varepsilon_{2r} + \nu^2/s^2\rho^2)^{\frac{1}{2}}d\rho = \tfrac{2}{3}\sigma_{p+1}^{3/2}i, \tag{2.6.25}$$

$p = 0, 1, 2, \ldots, \infty$, with σ_n being minus the zeros of the Airy function, viz. $Ai(-\sigma_n) = 0$, $n = 1, 2, \ldots, \infty$. Note that the lower boundary of the integration interval is exactly the transition point as found from (2.6.59). Hence, the only difference between (2.6.25) and (2.6.24) is that in the former, σ_n has been replaced by its asymptotic approximation $\sigma_n \approx [3\pi(4n - 1)/8]^{2/3}$ (see Abramowitz and Stegun (1965), formula 10.4.94). Thus, we observe that the general asymptotic equation (2.6.23) indeed comprises the special whispering-gallery-mode equations discussed in Wasylkiwskyj (1975) and in Ishihara, Felsen and Green (1978).

In the absence of transition points, i.e. for all combinations of s and ν that are not covered by the preceding paragraph, we can use the simpler asymptotic expressions given in (2.6.52) and (2.6.54). Substitution in (2.6.17) leads to

$$K_\nu(sN_1)[s^2\varepsilon_{2r}(1) + \nu^2]^{\frac{1}{2}} - sN_1K'_\nu(sN_1) = 0. \qquad (2.6.26)$$

Unlike (2.6.20), this equation cannot be decomposed directly. Since we do require such a decomposition to arrive at an ordering, we consider the related equation

$$K_\nu\{sN_1 - sN_1/[s^2\varepsilon_{2r}(1) + \nu^2]^{\frac{1}{2}}\} = 0, \qquad (2.6.27)$$

which reduces to (2.6.26) when $K_\nu(z)$ is replaced by the first two terms of its Taylor expansion around $z = sN_1$. As $|\nu|\to\infty$, (2.6.27) approximates (2.6.26) up to $\mathcal{O}(s^2/\nu^2)$. For $|\nu + iz| < \mathcal{O}(z^{1/3})$, it follows from the asymptotic expansions for Bessel functions in the transition regions (see Abramowitz and Stegun (1965), formulas 9.3.23 and 9.3.24) that $(\partial_z)^n K_\nu(z) = \mathcal{O}(z^{-n/3})\mathcal{O}(K_\nu) = \mathcal{O}(\nu^{-n/3})\mathcal{O}(K_\nu)$. Therefore, (2.6.27) approximates, in that range of z and ν, (2.6.26) up to $\mathcal{O}(\nu^{-2/3}) = \mathcal{O}(z^{-2/3})$. Since the zeros of $K_\nu(z)$ in the complex ν-plane are located near a contour that stretches from $\nu = -iz$ to infinity (with $\arg(\nu)\to-\pi/2$ as $|\nu|\to\infty$ (Felsen and Marcuvitz (1973), pp. 710 - 719)), this suggests that the roots of (2.6.26) are located near the roots of

$$\nu = \gamma_{-p}\{sN_1 - sN_1/[s^2\varepsilon_{2r}(1) + \nu^2]^{\frac{1}{2}}\}, \quad p = -1, -2, \ldots, -\infty, \qquad (2.6.28)$$

with $\gamma_p(z)$ being the p-th zero of $K_\nu(z)$ in the right half of the complex ν-plane, as discussed in that reference. (2.6.28) introduces an ordering that links up with the one introduced in (2.6.23), where $p = 0, 1, 2, \ldots, \infty$. As in the case of the whispering-gallery modes, the physical interpretation of the present pole contributions follows from the asymptotic field distributions inside the cylinder. From (2.6.51) and (2.6.52), it is observed that e_ν^a increases exponentially in magnitude in the entire interval $0 < \rho < 1$. In combination with (2.6.19), this leads to a residual contribution that, in $\mathcal{D}_2$, behaves as a surface wave propagating along the cylinder boundary, as visualized in Figure 2.6.5b. Like the part of the whispering-gallery mode up to $\rho = \rho_1$, such a wave can be understood as being a broken wave caused by the total reflection of a part of the incident field at the surface. The dominant part of the propagation takes place in $\mathcal{D}_1$, where this reflection gives rise to a creeping wave, i.e. a wave that creeps along the outer boundary of the cylinder, gradually radiating away its energy. This explanation is confirmed by the fact that $\varepsilon_{2r}(1)$ is the only obstacle parameter that occurs in the characteristic equation. It therefore seems appropriate to designate a mode of the present type as a *creeping-wave mode*. An additional verification of this explanation can be obtained by letting the inner medium become impenetrabl , i.e. by taking the limit for $\varepsilon_{2r}(1)\to\infty$. In this limit, either of the equations (2.6.26) and (2.6.27) reduces directly to

$$K_\nu(sN_1) = 0, \tag{2.6.29}$$

which is the exact characteristic equation for an impenetrable cylinder in a lossless dielectric, as discussed by Heyman and Felsen (1983).

As argued above, we have in addition to the residual contributions from the poles, a contribution from a pair of branch cuts along the imaginary ν-axis. From (2.6.13), (2.6.15), and the location of the branch cuts, it is observed that this contribution consists of a continuous spectrum of terms of the form (2.6.19), with ν being negative imaginary. In the angular directions, these constituents do not propagate at all; they simply decay exponentially.

Solutions of the equations (2.6.17), (2.6.20), (2.6.26) and related forms were obtained numerically. The roots were obtained by the same two-step zero-finding procedure that was used in Subsections 2.3.4 and 2.5.4 to search for zeros in the complex s-plane. The field distributions were computed by numerically integrating the pair of first-order equations (2.5.7), with m replaced by $\sqrt{\nu^2}$, with a fifth- and sixth-order Runge-Kutta-Verner method (IMSL (1982)). For details, the reader is referred to Subsection 2.5.4. The only new element was the evaluation of the Bessel functions of complex order and argument. For that purpose, a special set of subroutines was developed. As an illustration, Figure 2.6.6 shows results for a

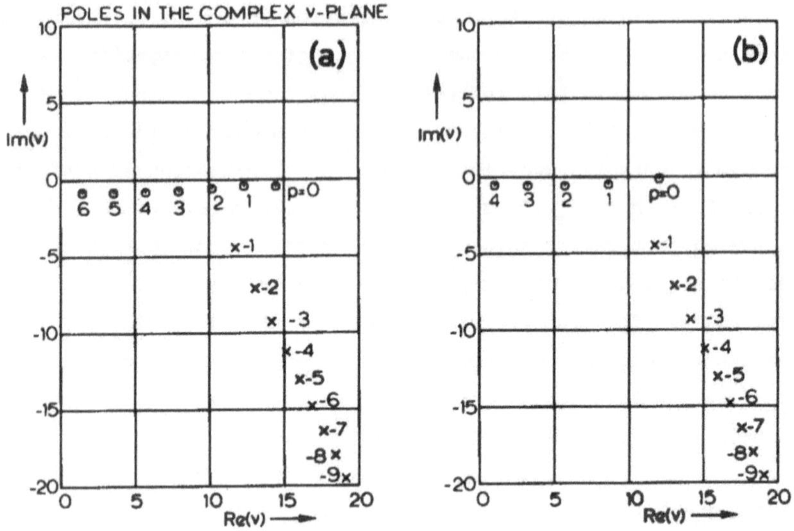

Figure 2.6.6 Poles in the complex ν-plane for s = 10i for two configurations with $\varepsilon_{1r} = 1$ and $\varepsilon_{2r}(1) = 2.25$. (a): $\varepsilon_{2r}(\rho) = 6.25 - 4\rho^2$, $\sigma_2(\rho) = 2$; (b): $\varepsilon_{2r}(\rho) = 2.25$, $\sigma_2(\rho) = 0$; o: whispering-gallery modes; x: creeping waves.

real-valued frequency $\omega = 10$ ($s = 10i$) for two configurations with the same permittivity at the cylinder boundary and a vacuum outside the cylinder. In the first configuration (referred to as Configuration A), we have a parabolic permittivity profile $\varepsilon_{2r}(\rho) = 6.25 - 4\rho^2$ and a constant conductivity $\sigma_2(\rho) = 2$. This is the same configuration that was defined as case iv) in Subsection 2.5.5, and whose results were shown in Figures 2.5.3 - 2.5.7. The second configuration (Configuration B) is case i) of Subsection 2.5.5, i.e. a homogeneous, lossless cylinder with $\varepsilon_{2r}(\rho) = 2.25$ and $\sigma_2(\rho) = 0$. Only the exact poles have been plotted. On the scale of Figure 2.6.6, the asymptotic poles found from (2.6.20) and (2.6.26) cannot be distinguished from the exact ones. For Configuration A, the maximum distance between corresponding values was 0.18 for the poles related to whispering-gallery modes and 0.04 for those related to creeping waves. For Configuration B, these distances were 0.09 and 0.08, respectively. The results presented in Figure 2.6.6 were also compared with the corresponding approximate values obtained from

(2.6.27). It turns out that these approximations are less accurate: for both configurations, the distances between corresponding values ranged from 0.6 for $p = -9$ to 2.3 for $p = -1$. Nevertheless, the asymptotic poles were sufficiently close to the actual ones to allow an ordering of the latter according to (2.6.28). From Figure 2.6.6, it is observed that the creeping-wave poles for both configurations are nearly identical, while the poles corresponding to whispering-gallery modes differ considerably. This is in agreement with the observations made above about the dependence of the location of the poles on the configuration: the creeping-wave poles depend on $\varepsilon_{2r}(1)$ and ε_{1r} only, while the whispering-gallery-mode poles also depend on the interior properties of the cylinder.

2.6.3 Transient response

From (2.6.9) and (2.6.10), it follows directly that the transient response caused by the incident pulse specified in (2.6.1) is found from $\hat{E}(\rho,\phi,s)$ by evaluating the integral

$$E(\rho,\phi,t) = (2\pi i)^{-1}\int_{L_\beta} \exp(st - sN_1)F(s)\hat{E}(\rho,\phi,s)ds. \qquad (2.6.30)$$

The SEM representation is obtained when this evaluation is effectuated by deforming the contour L_β into the negative half of the complex s-plane. Obviously, the modal representation only holds for space-time points where the contribution from the closing contour vanishes, i.e. when the closing conditions derived in Subsection 2.5.3 are met.

The alternative derivation of this representation follows when $\hat{E}(\rho,\phi,s)$ in (2.6.30) is replaced by the result of the evaluation of the integral in (2.6.13) by contour deformation in the complex ν-plane. As discussed in Subsection 2.6.2, this procedure leads to a representation of $\hat{E}_\nu(\rho,\phi,s)$ in terms of a series of residues of the form (2.6.19a) and a contour integral around the branch cuts. Since the factor $\exp(st - sN_1)F(s)$ is analytic in the entire complex s-plane, the poles in that plane must, in this derivation, correspond to the zeros of the factor $\sin(\nu_p(s)\pi)$ in the denominator of (2.6.19a). This leads to

$$\nu_p(s) = m, \quad m\in\mathbb{Z}\,, \tag{2.6.31}$$

which is exactly the equation obtained by Heyman and Felsen (1983) for the case of a perfectly conducting cylinder. In the terminology of the SEM approach, the solutions $s = s_{mp}$ of (2.6.31) are then the complex natural frequencies. It should be observed that the pair of equations (2.6.17) and (2.6.31) provides a description of these natural frequencies which is dual to the description given in Section 2.5. In that section, the angular order was immediately restricted to $\nu = m$, with $m\in\mathbb{Z}$, which led to the definition of the natural frequencies as being the zeros of the characteristic cylinder denominators $G_m(s)$ in the complex s-plane (see (2.5.21)).

The connection found above between the poles in the ν-plane and those in the s-plane directly provides a physical interpretation of the total residual contribution from a family of the latter with constant p. By (2.6.30) and (2.6.13), that contribution equals an integral of the type of (2.6.30) with $\beta = 0$ and with $\hat{E}(\rho,\phi,s)$ replaced by (2.6.19b) for $\mathrm{Im}(s) > 0$ and by a similar form for $\mathrm{Im}(s) < 0$. In view of the interpretation of (2.6.19b) given in Subsection 2.6.2, it is clear that this integral, and hence also the equivalent residual contribution, represents a pair of angularly propagating waves. The nature of these waves depends on the index p: they are of the whispering-gallery and of the creeping-wave type for $p \geq 0$ and $p < 0$, respectively.

The special form of (2.6.31) allows us to derive a related approximate equation which is more accurate than the one given in (2.6.27). From (2.6.28), it is observed that (2.6.26) can only have a root on the real ν-axis when $|sN_1 - i\nu| < O(s^{1/3})$. For s large, we can therefore expand the factor $[s^2\varepsilon_{2r}(1) + \nu^2]^{\frac{1}{2}}$ in (2.6.26) in powers of $(s^2\varepsilon_{1r} + \nu^2)/s^2$, which is of order $s^{-2/3}$. For $\varepsilon_{2r}(1) > \varepsilon_{1r}$, $\nu = m$ and $\mathrm{Re}(s) < 0$, this results in

$$K_m(sN_1)\left\{1 + \frac{s^2\varepsilon_{1r} + m^2}{2s^2[\varepsilon_{2r}(1) - \varepsilon_{1r}]} + O(s^{-4/3})\right\}$$
$$+ \frac{N_1}{[\varepsilon_{2r}(1) - \varepsilon_{1r}]^{\frac{1}{2}}} K'_m(sN_1) = 0. \tag{2.6.32}$$

Using the identity

$$\frac{s^2\varepsilon_{1r} + m^2}{s^2} K_m(sN_1) = \frac{N_1}{s} K'_m(sN_1) + \varepsilon_{1r}K''_m(sN_1), \tag{2.6.33}$$

which immediately follows from the second-order differential equation for the modified Bessel functions, we arrive at

$$K_m(sN_1) + \frac{N_1}{[\varepsilon_{2r}(1) - \varepsilon_{1r}]^{\frac{1}{2}}} K'_m(sN_1)$$
$$+ \frac{\varepsilon_{1r}}{2[\varepsilon_{2r}(1) - \varepsilon_{1r}]} K''_m(sN_1) = 0, \tag{2.6.34}$$

which holds up to $O(s^{-1})$. Recognizing in the left-hand side of (2.6.34) the first three terms of a Taylor expansion, and using the estimate $(\partial_z)^n K_m(z) = O(z^{-n/3})O(K_m)$, we arrive at

$$K_m\{sN_1 + N_1/[\varepsilon_{2r}(1) - \varepsilon_{1r}]^{\frac{1}{2}}\} = 0, \tag{2.6.35}$$

which holds again up to $O(s^{-1})$. The roots of (2.6.35) are equal to the corresponding roots for the case of a perfectly conducting cylinder, as discussed by Heyman and Felsen, shifted by $-[\varepsilon_{2r}(1) - \varepsilon_{1r}]^{-\frac{1}{2}}$ along the real s-axis. This correspondence once more confirms the physical interpretation given in Subsection 2.6.2. In the residual contributions, the shift has the effect of an additional damping, which is identical for all of them. This damping is explained from the fact that energy is lost in preserving the surface-wave part of the modes on the inside of the cylinder. For $\varepsilon_{2r}(1) < \varepsilon_{1r}$, $\nu = m$ and $\mathrm{Im}(s) > 0$, we obtain by a similar argument:

$$K_m\{sN_1 - iN_1/[\varepsilon_{1r} - \varepsilon_{2r}(1)]^{\frac{1}{2}}\} = 0. \tag{2.6.36}$$

Now the roots are obtained by shifting the corresponding ones for the perfect conductor over $[\varepsilon_{1r} - \varepsilon_{2r}(1)]^{-\frac{1}{2}}$ along the imaginary s-axis. In the residual contributions, this shift has the effect of an overall increase in velocity, which is explained physically from the fact that the surface wave on the inside of the cylinder travels faster in the ϕ-direction than its counterpart on the outside.

In order to verify (2.6.31), we have repeated the computation of Figure 2.6.6 for the complex natural frequencies $s = s_{mp}$ with $m = 8$ and $p = \pm 2$, which were available from the numerical computations described in Subsection 2.5.5. The results are shown in Figures 2.6.7 and 2.6.8. Clearly, they are in agreement with (2.6.31). In each case, the asymptotic equations (2.6.20) and (2.6.26) provided good approximations. For the poles occurring in Figures 2.6.7 and 2.6.8, the maximum distance between corresponding asymptotic and actual poles was 0.16. In Figure 2.6.9, we reconsider the natural frequencies of Configuration A, which were already plotted in Figure

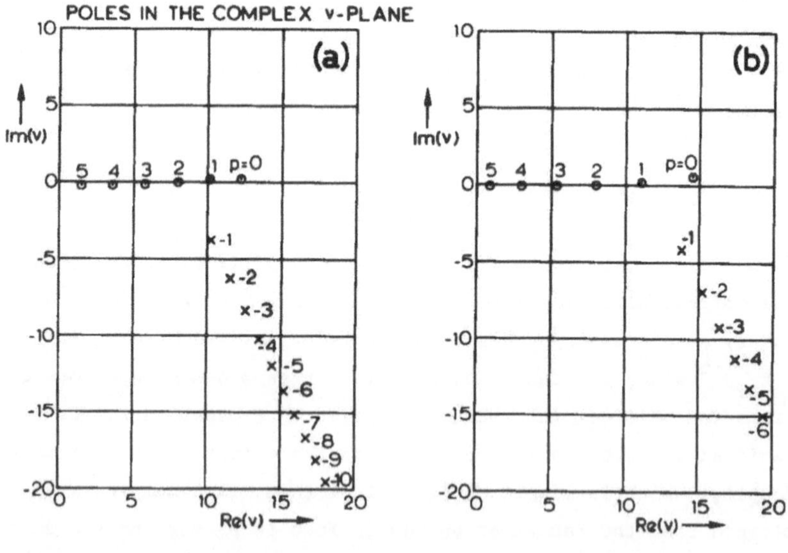

Figure 2.6.7 Result of the same computation as in Figure 2.6.6 for the complex natural frequency $s = s_{mp}$ with $m = 8$ and $p = 2$.

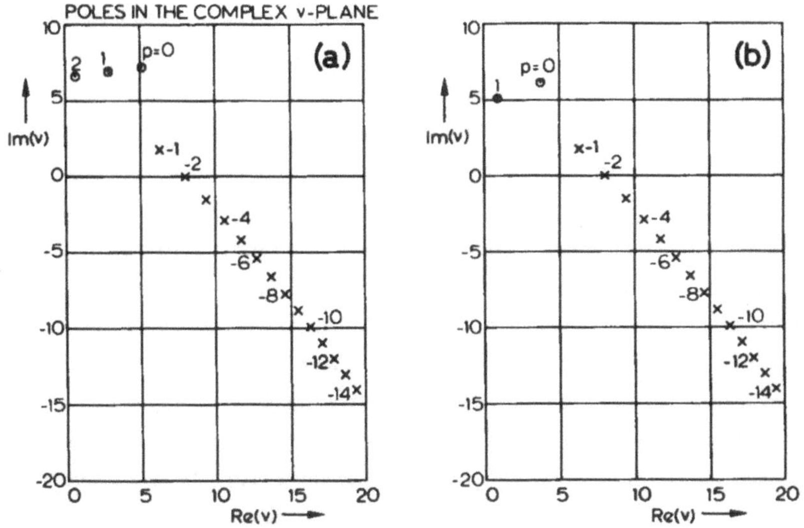

Figure 2.6.8 Result of the same computation as in Figure 2.6.7 with $p = -2$.

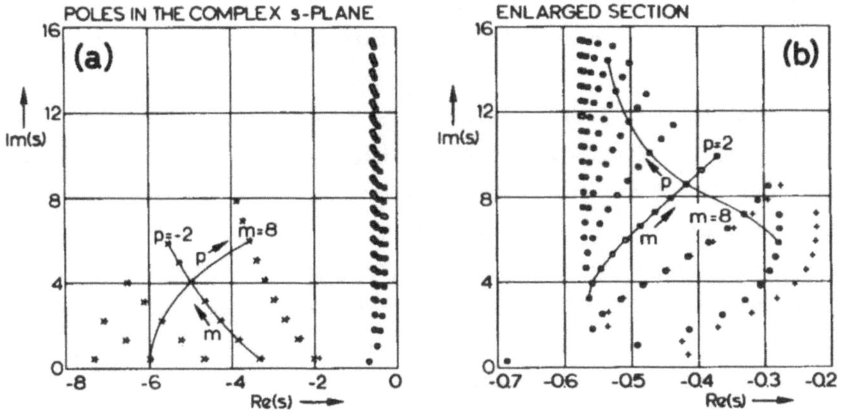

Figure 2.6.9 Natural frequencies as specified in Figure 2.5.3 ordered according to type and compared with their asymptotic approximations. o: whispering-gallery modes; x: creeping waves; +: asymptotic poles in so far as distinguishable from the actual ones.

2.5.3. The solid lines connect families of natural frequencies with $m = 8$ and $p = \pm 2$. The intersections of these lines mark the complex frequencies for which Figures 2.6.7a and 2.6.8a were computed. The corresponding asymptotic natural frequencies were obtained by substituting $\nu = m$ in (2.6.20) and (2.6.26), and numerically searching their solutions in the complex s-plane. For $p \leq 1$, the frequencies thus obtained are also indicated. For $p \geq 2$, the results are indistinguishable on the scale of Figure 2.6.9. The exact natural frequencies have also been compared with the approximations obtained from (2.6.35). As could be expected, these "shifted poles" are about as accurate as the results obtained from (2.6.26).

Finally, we investigate whether the interpretation given in this section also applies to the results obtained for the last configuration treated in Subsection 2.5.5, i.e. configuration v) as specified in (2.5.54). Since, for this configuration, both $\varepsilon_{2r}(\rho)$ and $\sigma_2(\rho)$ are continuous functions of ρ, the asymptotic expressions (2.6.52) and (2.6.54), and hence also the asymptotic characteristic equations (2.6.26) and (2.6.35), apply. To verify this, we repeated the numerical experiment described above for the present case. The agreement between the exact frequencies with $p < 0$ shown in Figure 2.5.8a and the corresponding asymptotic ones obtained from (2.6.26) and (2.6.35) turned out to be even better than for the related results shown in Figure 2.6.9a, possibly due to the larger value of $\varepsilon_{2r}(1)$. Therefore, the families of natural frequencies with $p < 0$ observed in Figure 2.5.8 and the corresponding natural modes can be identified as being of the creeping-wave type.

The interpretation of the natural frequencies with $p \geq 0$ is less straightforward. In particular, for the present permittivity profile, which has minima at $\rho = 0$, $2/3$ and maxima at $\rho = 1/3$, 1, there exist combinations of values of s and ν for which equation (2.6.59) has more than one solution near the real ρ-axis on the interval $0 < \rho < 1$, i.e. more than one transition point. For such combinations, the asymptotic expressions (2.6.65) and (2.6.67) are no longer valid. In order to investigate whether this complication arises for any of the natural modes, we searched, for all the natural frequencies $s = s_{mn}$ indicated in Figure 2.5.8, for complex solutions $\rho_0(s_{mn},m)$ in the vicinity of the interval $0 < \rho < 1$. Some representative results can be found in Table 2.6.1, where, for each of the natural modes considered in Figure 2.5.9, the two or three solutions with the smallest imaginary parts have been listed. For the creeping-wave-type natural frequencies, we need only consider the two solutions that are closest to the real ρ-axis. These two form an almost complex conjugate pair with real parts of approximately $\mathrm{Re}(\rho) = 2/3$. Their imaginary parts are so large in magnitude that we need not associate them with transition points. This is in agreement with the success of the asymptotic characteristic equations (2.6.26) and (2.6.35) described above. As far as the interpretation of the modes is concerned, we can therefore designate these solutions as being spurious. For all other natural frequencies, it suffices to look at the three solutions with the smallest imaginary part. It turns out that one of these three solutions is always so close to the real ρ-axis that it can be interpreted as being associated with a transition point, while the additional pair is located on opposite

Table 2.6.1 Complex solutions of (2.6.59) nearest to the real ρ-axis on the interval $0 < \rho < 1$ for $\varepsilon_r(\rho) = 2.25 + 4\sin^2(\frac{3}{2}\pi\rho)$, $\nu = m$ and $s = s_{mn}$, with $m = 7$ and $0 \le n \le 10$.

indices n	p	frequency s_{mn}	"transition points" $\rho_0(s_{mn},m)$	"spurious" $\rho_0(s_{mn},m)$
0	-3	-4.937 + 1.308i	not present	0.672 + 0.219i 0.718 - 0.199i
1	-2	-4.343 + 3.078i	not present	0.636 + 0.216i 0.740 + 0.168i
2	-1	-3.039 + 4.945i	not present	0.594 + 0.191i 0.746 - 0.117i
3	0	-0.187 + 4.392i	0.812 + 0.012i	0.482 + 0.214i 0.462 - 0.214i
4	1	-0.395 + 7.127i	0.399 - 0.036i 0.576 + 0.060i 0.669 - 0.035i	not present
5	2	-0.312 + 7.958i	0.352 - 0.015i 0.625 + 0.081i 0.648 - 0.073i	not present
6	3	-0.357 + 9.598i	0.294 - 0.009i	0.645 + 0.110i 0.653 - 0.107i
7	4	-0.414 + 11.008i	0.263 - 0.008i	0.652 + 0.121i 0.657 - 0.119i
8	5	-0.397 + 12.492i	0.239 - 0.006i	0.656 + 0.127i 0.659 - 0.126i
9	6	-0.412 + 14.054i	0.218 - 0.005i	0.659 + 0.132i 0.661 - 0.131i
10	7	-0.422 + 15.588i	0.202 - 0.004i	0.660 + 0.135i 0.662 - 0.132i

sides of the real ρ-axis. For some small p, the imaginary parts of all three solutions are comparable in magnitude, and, consequently, the latter pair should also have transition points attributed to them (e.g. n = 4 in Table 2.6.1). In that case, there is a considerable difference between the real parts of the two additional solutions. When, for a fixed m, p or, equivalently, n increases, the additional pair of solutions rapidly moves away from the real ρ-axis and the two real parts approach 2/3 (e.g. $n \geq 5$ in Table 2.6.1). In that limit, the pair of solutions resembles the one found for the creeping-wave modes, and no longer need be associated with a pair of transition points. In Figure 2.5.9, the dashed lines mark the real parts of those solutions $\rho_0(s_{mn},m)$ from Table 2.6.1 that give rise to transition points. Clearly, these suffice to mark all sudden transitions in the relevant natural-mode field distributions, which comfirms our conclusion. In both situations that arise for $p \geq 0$, we have an odd number of transition points and, hence, a standing-wave-type modal field distribution near $\rho = 1$. Because of both this observation and the gradual change-over between the situations with one and three transition points, it still seems justified to designate the modes with $p \geq 0$ as being of the whispering-gallery type.

Finally, we consider the interpretation of the variations that are observed in the period of the standing-wave part of the modes shown in Figures 2.5.9g, h, i. The results listed in Table 2.6.1 indicate that these variations are caused by a variation in the local value of $\varepsilon_{2r}(\rho)$ rather than by the presence of transition points. This explanation also corresponds with the functional form of the asymptotic field distribution specified in (2.6.64) and (2.6.65).

In conclusion, the only difference between the results for the present configuration and the corresponding ones for Configurations A and B seems to be the presence of the additional pair of solutions in the complex ρ-plane. Because of the limiting value of $\mathrm{Re}(\rho) = 2/3$, it seems fair to attribute this presence to the occurrence of a minimum in $\varepsilon_{2r}(\rho)$ at $\rho = 2/3$. Even though the additional solutions create complications, they do not prevent the interpretation of the natural modes as being either of the creeping-wave or of the whispering-gallery type.

Appendix 2.6.A Asymptotic expressions for the fields inside an inhomogeneous dielectric cylinder

In this appendix, we derive the asymptotic expressions for the electric and magnetic field strengths in the interior of a radially inhomogeneous cylinder which are used to obtain the asymptotic characteristic equations. To this end, we consider, for $0 \leq \rho < 1$, the solution of the second-order differential equation

$$[\partial_\rho^2 + \rho^{-1}\partial_\rho - s^2\varepsilon_r(\rho) - s\sigma(\rho) - \nu^2/\rho^2]e_\nu(\rho,s) = 0 \qquad (2.6.37)$$

that satisfies the normalization condition

$$\lim_{\rho \downarrow 0} \rho^{-\sqrt{\nu^2}} e_\nu(\rho,s) = 1. \qquad (2.6.38)$$

In (2.6.37), $\varepsilon_r(\rho)$ and $\sigma(\rho)$ are real-valued, continuous functions with $\varepsilon_r(\rho) \geq 1$ and $\sigma(\rho) \geq 0$, whose derivatives are discontinuous at no more than a finite number of points in the interval of interest. The complex frequency s and the complex angular order ν occur as parameters; at least one of them is assumed to be large. In (2.6.38), the square root is chosen according to its principal value, i.e. $\mathrm{Re}\sqrt{\nu^2} \geq 0$. From (2.6.37) and (2.6.38), which uniquely define the solution, it is observed that $e_\nu(\rho,s)$ satisfies the symmetry relations:

$$e_{-\nu}(\rho,s) = e_\nu(\rho,s), \qquad e_{\nu^*}(\rho,s^*) = [e_\nu(\rho,s)]^*. \qquad (2.6.39)$$

Therefore, we may, in what follows, restrict ourselves to $\mathrm{Re}(\nu) \geq 0$ and $\mathrm{Im}(s) \geq 0$. For other values of s and ν, the solution can be obtained directly with the aid of the relations (2.6.39).

Since we want to analyze the differential equation (2.6.37) by a WKB-type technique, it is convenient to eliminate the first derivative with respect to ρ in (2.6.37). This is achieved by the substitution of $e_\nu(\rho,s) \stackrel{\Delta}{=} \rho^{-\frac{1}{2}}u(\rho,s,\nu)$ (see e.g. Erdélyi (1956)). We arrive at

$$[\partial_\rho^2 - s^2\varepsilon_r(\rho) - s\sigma(\rho) - (\nu^2 - \tfrac{1}{4})/\rho^2]u(\rho,s,\nu) = 0. \qquad (2.6.40)$$

We will now construct an asymptotic solution to this equation. Following Murray (1984), we write formally

$$u(\rho,s,\nu) = \exp\{\textstyle\sum_{n=0}^{\infty} U_n(\rho,s,\nu)\}. \qquad (2.6.41)$$

Substitution of (2.6.41) in (2.6.40) yields

$$\begin{aligned}\textstyle\sum_{n=0}^{\infty} \partial_\rho^2 U_n(\rho,s,\nu) + \sum_{n=0}^{\infty} \sum_{m=0}^{\infty} [\partial_\rho U_n(\rho,s,\nu)][\partial_\rho U_m(\rho,s,\nu)] = \\ s^2\varepsilon_r(\rho) + s\sigma(\rho) + [\nu^2 - \tfrac{1}{4}]/\rho^2.\end{aligned} \qquad (2.6.42)$$

For a general combination of s and ν, we have for each fixed ρ with $0 < \rho < 1$:

$$s^2\varepsilon_r(\rho) + \nu^2/\rho^2 = \mathcal{O}(\alpha^2) \quad \text{as } \alpha\to\infty, \qquad (2.6.43)$$

with $\alpha \triangleq \max\{|s|,|\nu|\}$. This term dominates the right-hand side of (2.6.42) since the remaining terms are at most of $\mathcal{O}(\alpha)$. Consequently, (2.6.42) can be solved by requiring

$$U_n(\rho,s,\nu) = \mathcal{O}(\alpha^{1-n}), \qquad (2.6.44)$$

and enforcing the equality sign in (2.6.42) for terms of equal order in α. For U_0, we obtain:

$$[\partial_\rho U_0(\rho,s,\nu)]^2 = s^2\varepsilon_r(\rho) + \nu^2/\rho^2, \qquad (2.6.45)$$

which is solved by

$$U_0^{\pm}(\rho,s,\nu) = \pm\int_0^\rho\{[s^2\varepsilon_r(\rho') + \nu^2/\rho'^2]^{\frac{1}{2}} - \nu/\rho'\}d\rho' \pm \nu\ln\rho. \qquad (2.6.46)$$

In (2.6.46), the term $\nu\ln\rho$ has been extracted in order to remove the singularity of the integrand at $\rho' = 0$. The second term of the sum in (2.6.41) is found from the transport equation

$$\partial_\rho^2 U_0(\rho,s,\nu) + 2[\partial_\rho U_1(\rho,s,\nu)][\partial_\rho U_0(\rho,s,\nu)] = s\sigma(\rho), \tag{2.6.47}$$

which leads to

$$U_1^{\pm}(\rho,s,\nu) = -\tfrac{1}{4}\ln[s^2\varepsilon_r(\rho) + \nu^2/\rho^2] \pm \int_0^\rho \frac{s\rho'\sigma(\rho')}{2[s^2\rho'^2\varepsilon_r(\rho') + \nu^2]^{\frac{1}{2}}}\, d\rho'. \tag{2.6.48}$$

In principle, terms of higher order in α^{-1} can be obtained by continuing as outlined above. Since we are only interested in a first-order asymptotic approximation, we terminate the procedure here. Combination of (2.6.41), (2.6.44), (2.6.46) and (2.6.48) yields

$$u(\rho,s,\nu) = [A^+(s,\nu)u_a^+(\rho,s,\nu) + A^-(s,\nu)u_a^-(\rho,s,\nu)][1 + O(\alpha^{-1})], \tag{2.6.49}$$

with

$$u_a^{\pm}(\rho,s,\nu) = \frac{\rho^{\pm\nu}}{[s^2\varepsilon_r(\rho) + \nu^2/\rho^2]^{\frac{1}{4}}} \exp[\pm\phi(\rho,s,\nu)], \tag{2.6.50}$$

where

$$\phi(\rho,s,\nu) = \int_0^\rho \{[s^2\varepsilon_r(\rho') + \nu^2/\rho'^2]^{\frac{1}{2}} - \nu/\rho' + \frac{s\rho'\sigma(\rho')}{2[s^2\rho'^2\varepsilon_r(\rho') + \nu^2]^{\frac{1}{2}}}\}\, d\rho'. \tag{2.6.51}$$

Finally, we select the proper linear combination by imposing the normalization condition (2.6.38). This results in $A^+(s,\nu) = \nu^{\frac{1}{2}}$ and $A^-(s,\nu) = 0$. We end up with $e_\nu(\rho,s) = e_\nu^a(\rho,s)[1 + O(\alpha^{-1})]$, where

$$e_\nu^a(\rho,s) = \frac{\rho^\nu}{[1 + s^2\rho^2\varepsilon_r(\rho)/\nu^2]^{\frac{1}{4}}} \exp[\phi(\rho,s,\nu)]. \tag{2.6.52}$$

The magnetic-field distribution $h_\nu(\rho,s)$ that corresponds to a given electric-field distribution $e_\nu(\rho,s)$ is found from Maxwell's equation

$$h_\nu(\rho,s) = s^{-1}\partial_\rho e_\nu(\rho,s). \tag{2.6.53}$$

Now, we utilize the fact that $e^a_\nu(\rho,s)$ is locally just a first-order WKB-approximation to $e_\nu(\rho,s)$. Differentiation of such a WKB approximation of a given accuracy directly yields the derivative of the solution to the same order of accuracy (see Erdélyi (1956)). With the aid of this property and (2.6.53), we find $h_\nu(\rho,s) = h^a_\nu(\rho,s) \times [1 + O(\alpha^{-1})]$, with

$$h^a_\nu(\rho,s) = s^{-1}[s^2\varepsilon_r(\rho) + \nu^2/\rho^2]^{\frac{1}{2}} \, e^a_\nu(\rho,s). \tag{2.6.54}$$

A key point in the analysis presented up to now is the estimate (2.6.43), which holds for almost all combinations of s and ν. It is violated only when both these parameters are of the same order of magnitude and $\arg(s) \approx \arg(\nu) + \pi/2$. Then, there may be some ρ in the interval $0 < \rho < 1$ where the left-hand side of (2.6.43) is smaller than $O(\alpha^2)$. For those values of ρ, we then have

$$|\varepsilon_r^{\frac{1}{2}}(\rho)s - i\nu/\rho| = O(\alpha). \tag{2.6.55}$$

Clearly, this condition can only hold in a nonvanishing ρ-interval if $\varepsilon_r(\rho)$ is a multiple of ρ^{-2} in that interval. Since this is a rather special type of permittivity profile, we leave it out of consideration. Then, such an estimate only holds for a finite number of isolated points ρ_i, $i = 1, 2, \ldots, N$, with $0 < \rho_i < \rho_{i+1} < 1$. Since we expect the characteristic equation to have a number of solutions especially for combinations of s and ν for which (2.6.55) holds at some ρ_i, we have to include this case in our analysis. We will distinguish between two cases, namely

$$|\varepsilon_r^{\frac{1}{2}}(\rho_i)s - i\nu/\rho_i| \geq O(\alpha^\delta) \tag{2.6.56}$$

for some δ with $0 < \delta < 1$, and

$$|\varepsilon_r^{\frac{1}{2}}(\rho_i)s - i\nu/\rho_i| \leq O(1). \tag{2.6.57}$$

As both $\varepsilon_r^{\frac{1}{2}}(\rho)$ and ρ^{-1} are real-valued functions, it follows that the same condition must hold for all ρ_i.

Asymptotic expressions for $|\varepsilon_r^{\frac{1}{2}}(\rho)s - i\nu/\rho| \geq O(\alpha^\delta)$

First, we consider the case in which (2.6.56) is satisfied for all ρ_i. Then, we also have

$$s^2\varepsilon_r(\rho_i) + \nu^2/\rho_i^2 \geq O(\alpha^{1+\delta}), \tag{2.6.58}$$

with $0 < \delta < 1$. In view of this estimate, the terms separated off in (2.6.45) still dominate (2.6.42). Hence, the representation (2.6.41) also applies for $\rho = \rho_i$, although the derivatives of the functions U_n at these points are not of the same order in α as the corresponding values outside them. Nevertheless, the asymptotic expressions (2.6.52) and (2.6.54) remain valid because the value of the integrals in (2.6.46), (2.6.48) and related terms of higher order are not affected by the value of the integrands at a finite number of points.

Asymptotic expressions in the presence of a single transition point

Next, we consider the case in which (2.6.57) is satisfied at the points $\rho = \rho_i$. Then, the analysis leading to (2.6.52) and (2.6.54) breaks down since for $\rho = \rho_i$, the terms $\partial_\rho^2 U_0(\rho,s,\nu)$ and $s\sigma(\rho)$ in (2.6.42) are at least of the same order of magnitude as the terms $[\partial_\rho U_0(\rho,s,\nu)]^2$ and $s^2\varepsilon_r(\rho) + \nu^2/\rho^2$. At such so-called transition points, it is therefore no longer justified to separate off the latter pair as carried out in (2.6.45). This implies that the expressions (2.6.52) and (2.6.54), which were derived starting from (2.6.45), are only valid up to $\rho = \rho_1$. In the present discussion, we restrict ourselves to the case where there is only one transition point inside the cylinder. The extension to the case of more transition points is straightforward.

As mentioned above, the expression (2.6.52) remains a valid approximation in the subdomain $0 < \rho < \rho_1$. In the complementary subdomain $\rho_1 < \rho < 1$, $u(\rho,s,\nu) = \rho^{\frac{1}{2}} e_\nu(\rho,s)$ is also represented asymptotically by a linear combination of the type (2.6.49). The proper coefficients in this linear combination follow from the local behavior of u near $\rho = \rho_1$. In view of the restrictions imposed on $\varepsilon_r(\rho)$ and $\sigma(\rho)$ at the beginning of this appendix, we can at least continue these functions analytically into the complex ρ-plane in some non-vanishing domain enclosing $\rho = \rho_1$. From a first-order Taylor expansion around $\rho = \rho_1$, we then observe that there is always a complex zero $\rho_0(s,\nu)$ with $|\rho_0(s,\nu) - \rho_1(s,\nu)| = O(\alpha^{-1})$, such that

$$s^2\varepsilon_r(\rho_0(s,\nu)) + \nu^2/\rho_0^2(s,\nu) = 0. \tag{2.6.59}$$

Following standard transition-point theory (Murray (1984)), we approximate $s^2\varepsilon_r(\rho) + \nu^2/\rho^2$ near ρ_0 by a linear approximation, and $\sigma(\rho)$ by a constant. This results in an asymptotic expression for $u(\rho,s,\nu)$ in terms of the Airy functions Ai and Bi of the form

$$u(\rho,s,\nu) \approx A(s,\nu)\mathrm{Ai}(\xi) + B(s,\nu)\mathrm{Bi}(\xi). \tag{2.6.60}$$

In (2.6.60), we have

$$\begin{aligned} &\xi = -(-is\beta)^{2/3}[\rho - \rho_0 + \sigma(\rho_0)/s\beta^2], \quad \text{with} \\ &\beta = \{\partial_\rho[\varepsilon_r(\rho) + \nu^2/\rho^2 s^2]|_{\rho=\rho_0}\}^{\frac{1}{2}}, \quad \mathrm{Re}(\beta) \geq 0. \end{aligned} \tag{2.6.61}$$

The coefficients $A(s,\nu)$ and $B(s,\nu)$ are obtained by comparing, for a fixed ρ on the real ρ-axis immediately to the left of $\rho = \mathrm{Re}(\rho_0)$, the behavior of the expansions (2.6.49) and (2.6.60) as $\alpha\to\infty$. We find:

$$A(s,\nu) = \frac{2\pi^{\frac{1}{2}}\nu^{\frac{1}{2}}}{(-i\beta s)^{1/3}}\rho_0^\nu \exp[\phi(\rho_0,s,\nu)], \qquad B(s,\nu) = 0. \tag{2.6.62}$$

In (2.6.62), the function ϕ should be determined by evaluating the integral in (2.6.51) along a contour on or near the real ρ-axis, where $\varepsilon_r(\rho)$ and $\sigma(\rho)$ or their analytic continuations are well defined. By a similar comparison for ρ to the right of $\rho = \mathrm{Re}(\rho_0)$, we now obtain in the subdomain $\rho_1 < \rho < 1$ the asymptotic approximation $u(\rho,s,\nu) = u_a(\rho,s,\nu)[1 + \mathcal{O}(\alpha^{-1})]$, with

$$u_a(\rho,s,\nu) = \frac{A(s,\nu)\beta^{1/3}}{\pi^{1/2}(-is)^{1/6}}[\varepsilon_r(\rho) + \nu^2/\rho^2 s^2]^{-1/4} \cosh[\psi(\rho,s,\nu)]. \quad (2.6.63)$$

In (2.6.63), the complex phase integral $\psi(\rho,s,\nu)$ is defined by

$$\psi(\rho,s,\nu) \triangleq s\int_{\rho_0}^{\rho} \{[\varepsilon_r(\rho') + \nu^2/\rho'^2 s^2]^{1/2} + \frac{\sigma(\rho')}{2s[\varepsilon_r(\rho') + \nu^2/\rho'^2 s^2]^{1/2}}\}\, d\rho' - \frac{i\pi}{4}. \quad (2.6.64)$$

In these two formulas, the singularity in the integrand at $\rho' = 0$ has no longer been extracted explicitly since this point is located outside the integration interval. The combination of (2.6.62), (2.6.63) and the relation between e and u then yields $e_\nu(\rho,s) = e_\nu^a(\rho,s) \times [1 + \mathcal{O}(\alpha^{-1})]$, with

$$e_\nu^a(\rho,s) = C(s,\nu)\rho^{-1/2}[\varepsilon_r(\rho) + \nu^2/\rho^2 s^2]^{-1/4} \cosh[\psi(\rho,s,\nu)]. \quad (2.6.65)$$

In (2.6.65), the coefficient $C(s,\nu)$ is given by:

$$C(s,\nu) = 2(i\nu/s)^{1/2}\rho_0^{\nu} \exp[\phi(\rho_0,s,\nu)]. \quad (2.6.66)$$

As in the first part of this appendix, we determine the asymptotic magnetic-field distribution that corresponds to $e_\nu^a(\rho,s)$ by a direct differentiation according to (2.6.53). This results in $h_\nu(\rho,s) = h_\nu^a(\rho,s)[1 + \mathcal{O}(\alpha^{-1})]$, with

$$h_\nu^a(\rho,s) = C(s,\rho)\rho^{-1/2}[\varepsilon_r(\rho) + \nu^2/\rho^2 s^2]^{1/4} \sinh[\psi(\rho,s,\nu)]. \quad (2.6.67)$$

With (2.6.52), (2.6.54), (2.6.65) and (2.6.67), we now have at our disposal the set of asymptotic field distributions that allows us to derive the asymptotic characteristic equations discussed in the text for s and ν lying in the half-planes as specified and with at least one of them being large in magnitude.

Appendix 2.6.B The residue of $\hat{E}_\nu(\rho,s)$ at a pole $\nu = \nu_p$

The residues of $\hat{E}_\nu(\rho,s)$ at poles in the complex ν-plane can be determined by the same method that was used in Subsection 2.5.2 to determine the residues of $E_m(\rho,s)$ at poles in the complex s-plane. The method is based on a procedure that has been outlined by Marin (1973). The starting point of the analysis is the integral relation in operator form:

$$L_\nu(s)\hat{E}_\nu(\rho,s) = I_{\sqrt{\nu^2}}(sN_1\rho), \tag{2.6.68}$$

in which the right-hand side originates from the incident field, and where

$$L_\nu(s)f(\rho) = \int_0^\infty [\delta(\rho - \rho') + s^2 C_1(\rho',s)G_\nu(\rho,\rho';s)]f(\rho')d\rho'. \tag{2.6.69}$$

In (2.6.69), $C_1(\rho',s)$ denotes the *contrast function* with respect to medium 1, as defined in (2.6.4), and $G_\nu(\rho,\rho';s)$ the Green's function

$$G_\nu(\rho,\rho';s) = \rho' K_\nu(sN_1\rho_>)I_{\sqrt{\nu^2}}(sN_1\rho_<), \tag{2.6.70}$$

where $\rho_> = \max(\rho,\rho')$ and $\rho_< = \min(\rho,\rho')$. The integral relation (2.6.68) is equivalent to (2.6.3) with m^2 replaced by ν^2. Let $\hat{E}_\nu(\rho,s)$ have a simple pole at $\nu = \nu_p$. Then, we can write

$$\hat{E}_\nu(\rho,s) = \frac{r_p(\rho,s)}{\nu - \nu_p(s)} + \hat{E}_p^e(\rho,s,\nu), \tag{2.6.71}$$

where $\hat{E}_p^e(\rho,s,\nu)$ is a function which is analytic in a finite domain around $\nu = \nu_p$. Substituting (2.6.71) in (2.6.68), and using the analytic properties of $L_\nu(s)$, $I_{\sqrt{\nu^2}}(sN_1\rho)$ and $\hat{E}_p^e(\rho,s,\nu)$, we then find

$$L_{\nu_p}(s)r_p(\rho,s) = 0. \tag{2.6.72}$$

It now follows from the characteristic equation (2.6.17) that for $\nu = \nu_p$ the normalized solution $e_\nu(\rho,s)$ defined by (2.6.14) can, in $\mathcal{D}_1$, be written as:

$$e_{\nu_p}(\rho,s) = K_{\nu_p}(sN_1\rho)e_{\nu_p}(1,s)/K_{\nu_p}(sN_1). \tag{2.6.73}$$

From (2.6.73), it is observed that $e_{\nu_p}(\rho,s)$ satisfies the homogeneous equation (2.6.72). Hence, we can identify

$$r_p(s) = \lambda_p(s)e_{\nu_p}(\rho,s), \tag{2.6.74}$$

where $\lambda_p(s)$ is an as yet unknown complex constant. Next, we define the *object product* of two functions f and g as

$$\langle f(\rho,s,\nu_1)|g(\rho,s,\nu_2)\rangle = \int_0^1 f(\rho,s,\nu_1)C_1(\rho,s)g(\rho,s,\nu_2)\rho d\rho. \tag{2.6.75}$$

From (2.6.69), we find that the operator $L_\nu(s)$ has the following property:

$$\langle f(\rho,s,\nu_1)|L_{\nu_2}(s)g(\rho,s,\nu_2)\rangle = \langle L_{\nu_2}(s)f(\rho,s,\nu_1)|g(\rho,s,\nu_2)\rangle. \tag{2.6.76}$$

Using (2.6.68), (2.6.76), and the fact that $e_{\nu_p}(\rho,s)$ satisfies (2.6.72), we find

$$\langle [L_\nu(s) - L_{\nu_p}(s)]|e_{\nu_p}(\rho,s)\rangle = \langle I_{\sqrt{\nu^2}}(sN_1\rho)|e_{\nu_p}(\rho,s)\rangle. \tag{2.6.77}$$

The unknown constant $\lambda_p(s)$ is determined by substituting in (2.6.77) the formulas (2.6.71) and (2.6.74) and the expansion

$$L_\nu(s) = L_{\nu_p}(s) + L'_{\nu_p}(s)(\nu - \nu_p) + O[(\nu - \nu_p)^2], \tag{2.6.78}$$

where $L'_\nu(s)$ is the derivative operator of L(s):

$$L'_\nu(s)f(\rho) = s^2\int_0^1 C_1(\rho',s)\partial_\nu G_\nu(\rho,\rho';s)f(\rho')d\rho'. \tag{2.6.79}$$

Taking the limit $\nu \to \nu_p$, we then determine $\lambda_p(s)$ as

$$\lambda_p(s) = \frac{\langle I_{\sqrt{\nu_p^2}}(sN_1\rho)|e_{\nu_p}(\rho,s)\rangle}{\langle L'_{\nu_p}(s)e_{\nu_p}(\rho,s)|e_{\nu_p}(\rho,s)\rangle}. \tag{2.6.80}$$

2.7 Conclusions

In this chapter, two different frequency-domain techniques have been applied to resolve the transient-scattering problem for an inhomogeneous, lossy dielectric slab and a lossy, radially inhomogeneous dielectric circular cylinder. Both these scatterers were either embedded in vacuo or surrounded by homogeneous dielectric media, and were illuminated by an E-polarized, plane electromagnetic pulse of finite duration. For each solution technique, the analytical as well as the numerical aspects have been investigated. The common starting point for both techniques is the representation of the transient field in terms of a Laplace inversion integral; the main difference lies in the procedure, by which this integral is evaluated.

In the *singularity expansion method*, this evaluation is carried out by contour deformation in the complex-frequency plane. Conventionally, only the contributions from the singularities in that plane are taken into account. This results in a representation of the excited field solely in terms of natural-mode contributions. The completeness of this representation depends on the possible contribution from a closing contour at infinity. For both configurations under consideration, this well-known completeness problem has been resolved. It was shown that there exists a nonzero region in the space-time domain where the scattered field cannot be expressed in terms of natural-mode contributions only and where the conventional SEM representation does not hold. On the other hand, for each space point a finite initial instant can be found, after which the natural-mode representation holds. These results are in agreement with the principle of causality: an incident pulse cannot excite natural modes before it enters the scattering obstacle. Furthermore, the validity of the SEM representation requires restrictions on the shape of the incident pulse. For instance, often-employed incident pulses like the delta-function or the Gaussian pulse violate these restrictions.

For the cylinder problem, the conclusions stated above were obtained by decomposing it into a series of one-dimensional scattering problems, each of them similar to the one for the slab. It was shown that, for the total field inside the cylinder and for the scattered field outside, the representation obtained converges globally in space and time. From a computational point of view, this has the advantage that the truncated series approximates the relevant field quantities to the same global accuracy in the entire space-time domain. The main difference between this and previous applications of the SEM is the presence of a continuous part in the natural-frequency spectrum.

In order to verify our analytical results, we have also implemented the SEM theory numerically. It turns out that the SEM approach allows an accurate computation of the transient scattering by both configurations. For the slab, an acceleration of the convergence allowed us to handle even discontinuous or fluctuating incident pulses. Moreover, a second acceleration of the convergence was applied in the computation of the field at space-time points where the original SEM representation is not valid. Nevertheless, the computation remained time consuming. In the case of the cylinder, the occurrence of a factor of $s^{-\frac{1}{2}}$ in the asymptotic expressions prevents such accelerations.

The second technique that has been applied to evaluate the Laplace inversion integral is its *direct computation* with the aid of an *FFT algorithm*. Globally, the potentialities of this method appear to be about the same as those of the SEM. As in the case of the latter method, discontinuous and fluctuating pulses can be handled by using the high-frequency approximations to accelerate the convergence by one order. Moreover, the total computation time consumed in determining the desired time-domain solutions is of the same order of magnitude for both methods. This was observed for either of the two configurations that have been analyzed in this chapter. Most of this computation time is spent on solving the frequency-domain scattering problem for a sizable number of frequencies. For the larger of them, in particular, the numerical effort required increases linearly with their magnitudes. In the SEM, such computations form part of the search process for the natural frequencies; in the FFT evaluation, the solutions occur as factors in the integrand. The main advantages of the direct computation are that it yields the excited field at all instants, and that a Gaussian incident pulse can be handled. On the other hand, the SEM provides more insight into the physical aspects of the scattering process. Since the computational effort involved in applying both techniques is comparable, it should be these advantages that decide the choice between them.

In the present study, the *physical interpretation* of the excited fields has been obtained primarily from asymptotic expressions. It turns out that the propagation inside the slab as well as the radial behavior of the modes in the cylinder problem can be analyzed with the aid of the high-frequency WKB expressions derived in investigating the role of the closing contours. In both configurations, the excited fields travel up and down the scattering obstacle, repeatedly reflected and diffracted at the interface with the surrounding medium. The analysis of the angular behavior of the cylinder modes turns out to be more complicated. In Section 2.6, an alternative derivation of the SEM representation for the relevant fields that explains this behavior was presented. This derivation interprets the additional ordering that is observed in the natural frequencies: each family with constant index p corresponds to a pair of waves with a weakly time-independent radial distribution that propagate in opposite angular directions. With the aid of special high-frequency high-order asymptotic expressions for the radial field distributions, the occurrence of two different types of natural modes can also be understood. These types can be identified as creeping-wave and whispering-gallery modes. For each type, an asymptotic characteristic equation was derived which holds as long as at least one of the relevant parameters (the complex frequency s and the complex angular order ν) is large in magnitude. These equations are simple enough to allow a straightforward physical interpretation of their solutions and the corresponding modal field distributions and yet lead to surprisingly accurate approximate results. It was verified that the formulation includes results obtained previously by other authors for limiting cases where either the inner or the outer medium becomes impenetrable. As a by-product of the derivation, a representation was obtained in Subsection 2.6.2 for the harmonic response in terms of contributions from the singularities in the complex ν-plane. Both from a physical and a mathematical point of view, this representation turns out to be more consistent than related representations that have appeared in the literature.

References

Abramowitz, M., and Stegun, I.A. (1965), *Handbook of mathematical functions*, Dover Publications, New York.
Baum, C.E. (1971), On the singularity expansion method for the solution of electromagnetic interaction problems, *Interaction Note 88*, Air Force Weapons Laboratory, Albuquerque, New Mexico.
Baum, C.E. (1976a). The singularity expansion method. In: *Transient electromagnetic fields*, Felsen, L.B. (Ed.), Springer Verlag, Berlin, Chap. 3.
Baum, C.E. (1976b), Emerging technology for transient and broad-band analysis and synthesis of antennas and scatterers, *Proc. IEEE 64*, 1598-1616.
Baum, C.E. (1978), Toward an engineering theory of scattering: the singularity and eigenmode expansion methods. In: *Electromagnetic scattering*, Uslenghi, P.L.E. (Ed.), Academic Press, New York, Chap. 15.
Berg, P.W., and McGregor, J.L. (1966), *Elementary partial differential equations*, Holden-Day, San Francisco, pp. 156-163.
Blok, H. (1970), *Diffraction theory of open resonators*, Ph.D. thesis, Delft University of Technology, Delft, The Netherlands.
Bollig, G., and Langenberg, K.J. (1983), The singularity expansion method as applied to the elastodynamic scattering problem, *Wave Motion 5*, 331-354.
Brekovskikh, L.M. (1980), *Waves in layered media*, Academic Press, New York, pp. 164-180.
Chadan, K., and Sabatier, P.C. (1977), *Inverse problems in quantum scattering theory*, Springer Verlag, New York, pp. 295-304.
Copson, E.T. (1976), *An introduction to the theory of functions of a complex variable*, Oxford University Press, Oxford, 13th impression, p. 119.
De Hoop, A.T. (1965), A note on the propagation of waves in a continuous layered medium, *Appl. Sci. Res. B 12*, 74-80.
Dil, J.G., and Blok, H. (1973), Propagation of electromagnetic surface waves in a radially inhomogeneous optical waveguide, *Optoelectronics 5*, 415-428.
Dudley, D.G. (1985), Comments on SEM and the parametric inverse problem, *IEEE Trans. Antennas Propagat. 33*, 119-120.
Erdélyi, A. (1956), *Asymptotic expansions*, Dover Publications, New York, Chap. 4.
Felsen, L.B. (1984), Progressing and oscillatory waves for hybrid synthesis of source excited propagation and diffraction, *IEEE Trans. Antennas Propagat. 32*, 775-796.
Felsen, L.B. (1985), Comments on early time SEM, *IEEE Trans. Antennas Propagat. 33*, 118-119.
Felsen, L.B., and Marcuvitz, N. (1973), *Radiation and scattering of waves*, Prentice-Hall, Englewood Cliffs, N.J.
Fiorito, R., Madigosky, W., and Überall, H. (1981), Acoustic resonances and the determination of the material parameters of a viscous fluid layer, *J. Acoust. Soc. Am. 69*, 897-903.
Flax, L., Caunaurd, G.C., and Überall, H. (1981), Theory of resonance scattering, In: *Physical Acoustics 15*, Mason, W.P., and Thurston, R.N. (Eds.), Academic Press, New York, Chap. 3.

Franz, W. (1957), *Theorie der Beugung elektromagnetischer Wellen*, Springer Verlag, Berlin, Chap. 11.

Friedlander, F.G. (1958), *Sound pulses*, Cambridge University Press, Cambridge, pp. 149-166.

Gaunaurd, G.C., and Werby, M.F. (1985), Resonance response of submerged, acoustically excited thick and thin shells, *J. Acoust. Soc. Am. 77*, 2081-2093.

Heyman, E., and Felsen, L.B. (1983), Creeping waves and resonances in transient scattering by smooth convex objects, *IEEE Trans. Antennas Propagat. 31*, 426-437.

Heyman, E., and Felsen, L.B. (1985a), Wavefront interpretation of SEM resonances, turn-on times and entire functions. In: *Proceedings of the Nato Advanced Research Workshop on Hybrid Formulation of Wave Propagation and Scattering, IAFE, Castel Gandolfo, Italy, August 30 - September 3, 1983*, Felsen, L.B. (Ed.), Martinus Nijhoff Publishers, Dordrecht.

Heyman, E., and Felsen, L.B. (1985b), A wavefront interpretation of the singularity expansion method, *IEEE Trans. Antennas Propagat. 33*, 706-718.

Howell, W.E., and Überall, H. (1984), Complex frequency poles of radar scattering from coated conducting spheres, *IEEE Trans. Antennas Propagat. 32*, 624-627.

IMSL Library Reference Manual (1982), Int. Math. Stat. Libraries, Houston, Subroutine DVERK.

Ishihara, T., Felsen, L.B., and Green, A. (1978), High-frequency fields excited by a line source located on a perfectly conducting concave cylindrical surface, *IEEE Trans. Antennas Propagat. 26*, 757-767.

Langenberg, K.J. (Ed.) (1983), Special issue on transient fields, *Wave Motion 5*.

Lewin, L., Chang, D.C., and Kuester, E.F. (1977), *Electromagnetic waves and curved structures*, Peter Peregrinus Ltd., Stevenage, pp. 132-139.

Marin, L. (1973), Natural-mode representation of transient fields, *IEEE Trans. Antennas Propagat. 21*, 809-818.

McKelvey, R. (1959), Solution about a singular point of a linear differential equation involving a large parameter, *Trans. Am. Math. Soc. 91*, 410-424.

Michalski, K.A. (1981), Bibliography of the singularity expansion method and related topics, *Electromagnetics 1*, 493-511.

Miyazaki, Y. (1981), Scattering and diffraction of electromagnetic waves by inhomogeneous dielectric cylinder, *Radio Sci. 16*, 1009-1014.

Morgan, M.A. (1984), Singularity expansion representations of fields and currents in transient scattering, *IEEE Trans. Antennas Propagat. 32*, 466-473.

Morgan, M.A. (1985), Response to comments regarding SEM representation, *IEEE Trans. Antennas Propagat. 33*, 120.

Murray, J.D. (1984), *Asymptotic analysis*, Springer Verlag, Berlin, second edition, Chap. 6.

Nussenzveig, H.M. (1972), *Causality and dispersion relations*, Academic Press, New York, p. 223.

Pearson, L.W. (1983), Present thinking on the use of the singularity expansion in electromagnetic scattering computation, *Wave Motion 5*, 355-368.

Pearson, L.W. (1984), A note on the representation of scattered fields as a singularity expansion, *IEEE Trans. Antennas Propagat. 32*, 520-524.

Pearson, L.W., and Marin, L. (Eds.) (1981), Special issue on the singularity expansion method, *Electromagnetics 1*.

Singaraju, B.K., Giri, D.V., and Baum, C.E. (1976), Further developments in the application of contour integration to the evaluation of the zeros of analytic functions and relevant computer programs, *Mathematics Note 42*, Air Force Weapons Laboratory, Albuquerque, New Mexico.

Stratton, J.A. (1941), *Electromagnetic theory*, McGraw-Hill, New York, Chap. 9.

Streifer, W., and Kodis, R.D. (1964a), On the solution of a transcendental equation arising in the theory of scattering by a dielectric cylinder, *Quart. Appl. Math. 21*, 285-298.

Streifer, W., and Kodis, R.D. (1964b), On the scattering of electromagnetic waves by a dielectric cylinder, *Quart. Appl. Math. 22*, 193-206.

Streifer, W., and Kodis, R.D. (1965), On the solution of a transcendental equation in scattering theory-Part II, *Quart. Appl. Math. 23*, 27-38.

Streifer, W., and Kurz, C.N. (1967), Scalar analysis of radially inhomogeneous guiding media, *J. Opt. Soc. Am. 57*, 779-786.

Tijhuis, A.G. (1986), Angularly propagating waves in a radially inhomogeneous, lossy dielectric cylinder and their connection with the natural modes, *IEEE Trans. Antennas Propagat. 34*, 813-824.

Tijhuis, A.G., and Blok, H. (1980), SEM approach to the transient scattering by an inhomogeneous, lossy dielectric slab, *Proceedings of the International URSI Symposium on Electromagnetic Waves, Munich, August 26-29*, pp. 221B/1-4.

Tijhuis, A.G., and Blok, H. (1984a), SEM approach to the transient scattering by an inhomogeneous slab; Part I: the homogeneous case, *Wave Motion 6*, 61-78.

Tijhuis, A.G., and Blok, H. (1984b), SEM approach to the transient scattering by an inhomogeneous slab; Part II: the inhomogeneous case, *Wave Motion 6*, 167-182.

Tijhuis, A.G., and Van der Weiden, R.M. (1986), SEM approach to transient scattering by a lossy, radially inhomogeneous dielectric circular cylinder, *Wave Motion 8*, 43-63.

Titchmarsh, E.C. (1950), *The theory of functions*, Oxford University Press, London, second edition, Chap. 13.

Traub, J.F. (1964), *Iterative methods for the solution of equations*, Prentice-Hall, Englewood Cliffs, N.J., Chap. 10.

Überall, H., and Gaunaurd, G.C. (1981), The physical content of the singularity expansion method, *Appl. Phys. Lett. 39*, 362-364.

Überall, H., and Gaunaurd, G.C. (1984), Relation between the ringing of resonances and surface waves in radar scattering, *IEEE Trans. Antennas Propagat. 32*, 1071-1079.

Wait, J.R. (1967), Electromagnetic whispering gallery modes in a dielectric rod, *Radio Sci.* *2*, 1005-1017.
Wasylkiwskyj, W. (1975), Diffraction by a concave perfectly conducting circular cylinder, *IEEE Trans. Antennas Propagat.* *23*, 480-492.
Weinstein, L.A. (1969), *Open resonators and open waveguides*, The Golem Press, Boulder, Chap. 10-11.
Whittaker, E.T., and Watson, G.N. (1952), *A course of modern analysis*, Cambridge University Press, Cambridge, fourth edition, pp. 286-288.

3. TIME-DOMAIN TECHNIQUES

3.1 Introduction

In this chapter, we take leave of the idea of solving time-domain electromagnetic direct-scattering problems with the aid of frequency-domain techniques. Instead, we consider solving such problems directly in the time domain by utilizing the common property in the relevant integral equations that the scattered field at each space-time point is expressed in terms of one or more integrals of field values at previous instants. The spatial domain of these integrals is either the boundary of the scattering obstacle (for homogeneous or impenetrable scatterers) or its interior (for inhomogeneous scatterers). In the *marching-on-in-time method*, the integral equations mentioned above are discretized in space and time. This results in a system of linear equations for the approximate field values at the discrete space-time points that can be solved by a step-by-step updating procedure involving only a linear combination of known field values. This procedure is schematically illustrated in Figure 3.1.1, which shows a flow diagram of a calculation performed according to this method for the same problem as considered in Figures 2.1.1 and 2.1.2. The most striking difference between those figures and Figure 3.1.1 is that, in the present case, no intermediate results which can be used in the calculation of the response of more than one incident field are available. In this chapter, we consider some general aspects of the marching-on-in-time method as well as its application to a set of specific scattering problems, similar to the ones solved in Chapter 2.

Before we specify the contents of this chapter in more detail, some historical background should be provided. The first applications of the marching-on-in-time method pertained to transient scattering problems that can be formulated in terms of boundary integral equations. As far as acoustic scattering problems are concerned, the method was applied to two- and three-dimensional impenetrable obstacles with a variety of boundary conditions (Friedman and Shaw (1962), Shaw (1967, 1968, 1975), Mitzner (1967), Neilson et al. (1978), Bennett and Mieras (1981a, 1983)). In electromagnetics, two

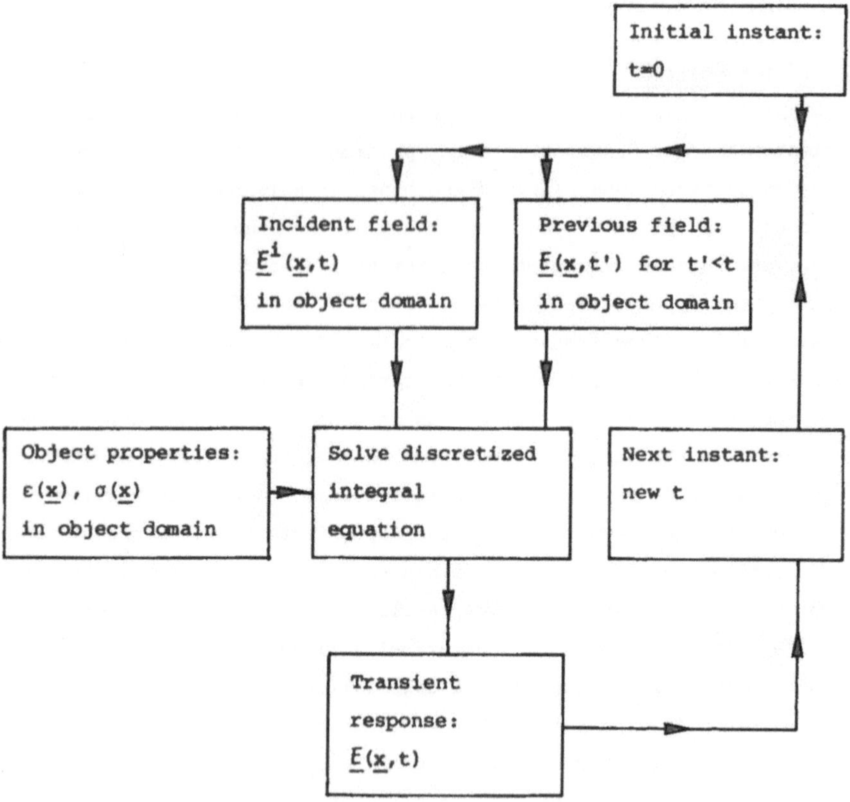

Figure 3.1.1 Flow diagram of the calculation of a transient field via the marching-on-in-time method. In the diagram, $\underline{x}$ denotes a Cartesian position vector, t the time cooridinate, $\varepsilon(\underline{x})$ a permittivity, $\sigma(\underline{x})$ a conductivity, and $\underline{E}$ and $\underline{E}^i$ electric-field strengths.

types of scattering problems that give rise to a boundary-integral equation can be distinguished. The analog of the acoustic scattering problems cited above is the scattering by a perfectly conducting two- or three-dimensional target. Hence, it is not surprising that such scattering problems were analyzed during the same period as their acoustic counterparts (Bennett (1968), Bennett and Weeks (1970), Bennett and Mieras (1981b), Damaskos et al. (1985)). For this class of problems, the development of the marching-on-in-time method has also been summarized in review papers by Poggio and Miller (1973), Mittra (1976), Bennett and Ross (1978) and Bennett (1978). A specifically electromagnetic application that falls within this class is the transient scattering from wires and wire-grid models (Miller and Landt (1980)). The second category of electromagnetic scattering problems that was solved by applying the marching-on-in-time method is the transient scattering by homogeneous, lossless dielectric targets. Such problems were formulated and solved in Mieras and Bennett (1982) and Marx (1984).

When the scatterer is inhomogeneous, boundary integral equations no longer suffice, and domain integral equations have to be resolved. For a number of three-dimensional acoustic and elastodynamic scattering problems, this was carried out by Herman (1981a,b, 1982). In addition, the two-dimensional acoustic case was dealt with by Lesselier (1983). As far as electromagnetic scattering problems are concerned, to best of the author's knowledge, only the inhomogeneous lossy dielectric slab (Bolomey et al. (1978), Lesselier (1982), Tijhuis (1981)), and the inhomogeneous, lossy dielectric cylinder (Tijhuis (1984a,b)) have been investigated.

The most important difficulty in the application of the marching-on-in-time method is the occurrence of exponentially increasing instabilities in the solutions obtained. These instabilities can be explained from the accumulation of discretization errors in the time-recursive solution procedure. The majority of the references cited above does not address this difficulty. Generally, the unwanted instabilities are eliminated by some ad hoc method that is applicable to the specific problem under consideration. The more fundamental approaches to the instability problem can be subdivided into two categories. The first category avoids the instabilities by modifying the method of solution. To this end, the numerical solution is redefined as the set of approximate field values that minimizes a squared error in the equality sign of the discretized integral equation. This solution is then obtained by a gradient (Herman (1981b), Herman and Van den Berg (1982)) or conjugate-gradient method (Van den Berg (1984), Sarkar (1984), Sarkar and Rao (1984), Rao et al.

(1984)). A possible disadvantage of these methods is that, because of the vast number of field values to be determined, it may take a large number of iteration steps to arrive at an acceptable solution. Since each iteration step requires about as much computation time as two complete marching-on-in-time computations, this may take up an unacceptably large amount of computation time. The second category attempts to control the unstable behavior in the solutions obtained by improving the discretization. A first example is the work of Marx (1985a,b), who calculated more accurate approximations to the dominant parts of the discretized integrals in the integral equation for scattering by a three-dimensional perfect conductor. A second example was described by Tijhuis (1981, 1984a,b), who demonstrated that, for a certain type of scattering problem, both the accuracy and the stability of a marching-on-in-time result can be controlled by reducing the mesh size in a systematically generated discretization. Work along these lines was also carried out by Rynne (1985), who generalized the analysis to a different type of discretization. This type of method, too, has the disadvantage that it becomes computationally expensive for multi-dimensional scattering problems. This is due to the fact that the number of operations required is proportional to some large inverse power of the mesh size.

In the present chapter, a proposal is to combine the best ideas of both approaches. To this end, the marching-on-in-time method is interpreted as the recursive solution of a lower-triangular matrix equation of a large dimension. The instabilities can then be envisaged as the inaccurate components of the solution vector along a few problematic eigenvectors of the system matrix associated with a small eigenvalue. Since these components are located in a small subspace of the total solution space they may be designated as "local" errors. By drawing up a simplified model of a discretized time-domain integral equation, it will be shown that such "local" errors may appear regardless of the accuracy of the discretization. On the other hand, the marching-on-in-time method will determine the components of the solution vector along the majority of the eigenvectors of the system matrix up to some small "global" accuracy. An iterative method will be described that uses this property as well as the lower-triangular structure of the matrix equation in determining a regularized solution that does not have unwanted components along the problematic eigenvectors. In eliminating these components, the regularized inversion has the same effect as minimizing a squared error by a gradient-type method. The essential difference is that our method *partially* improves the values of the components along *all* the well-behaved eigenvectors in each iteration step, while a conjugate-gradient method *totally* improves the component in a *single* direction. Hence, the former seems better suited to determine a solution vector of a large dimension. The iterative procedure leaves us with a gradually varying solution whose "global" error may still increase gradually with increasing time. To that error, we then apply the analysis of Tijhuis (1984a,b), which argues that the increase in error per time step will be of the order of magnitude of the discretization error, i.e. the error in approximating the integrals in the time-domain integral equation by the corresponding discretized forms for the *exact* field. The "global" error can then be controlled by taking care that this discretization error is at most proportional to the square of the space-time mesh size and by choosing that mesh size sufficiently small.

The organization of this chapter is as follows. In Section 3.2, we first formulate the marching-on-in-time method in its general form. Next, we carry out the stability analysis summarized above and discuss various stabilization methods for the marching-on-in-time method. In Sections 3.3 and 3.4, we reconsider two problems that were already solved in Tijhuis (1981, 1984a,b), namely the transient scattering of a plane, pulsed electromagnetic wave by an inhomogeneous, lossy dielectric slab and by a perfectly conducting cylinder. For both problems, the stability problem is reinterpreted in the light of the stability considerations of Section 3.2. In Section 3.5, we describe a successful attempt along the lines of Section 3.2 to resolve a problem for which up to now only partial success has been achieved, namely that of plane-wave scattering by an inhomogeneous, lossy dielectric cylinder. Finally, the conclusions are stated in Section 3.6.

3.2 General aspects of the marching-on-in-time method

3.2.1 Formulation of the method

In this section, we consider the marching-on-in-time method from a formal point of view. This allows us to analyze some general properties of this method that show up in its application to various specific time-domain scattering problems without considering these problems in detail. With the applications in Sections 3.3 - 3.5 in mind, we restrict ourselves to a scalar wave field $\phi(\underline{x},t)$. However, the scalar nature of $\phi(\underline{x},t)$ is not essential for the present discussion. For a general, one-, two- or three-dimensional linear time-domain scattering problem, the field $\phi(\underline{x},t)$ satisfies an integral equation of the form

$$\phi(\underline{x},t) = \phi^{i}(\underline{x},t) + \int_{\mathcal{D}} d\underline{x}' \int_0^{t-R/c} dt' \; K(\underline{x},\underline{x}';t - t')\phi(\underline{x}',t') \qquad (3.2.1)$$

(see also Herman and Van den Berg (1982), Van den Berg (1984)). In (3.2.1), $\underline{x}$ is a Cartesian position vector and $R = |\underline{x} - \underline{x}'|$. Furthermore, $\mathcal{D}$ denotes a finite domain, $K(\underline{x},\underline{x}';t - t')$ a linear time-invariant operator acting on $\phi(\underline{x}',t')$, $\phi^{i}(\underline{x},t)$ a known incident field that for $\underline{x}\in\mathcal{D}$ vanishes when $t \le 0$, and c a wave speed parameter. Further, we have $K = 0$ when $\underline{x}'\in\mathcal{D}'$, where $\mathcal{D}'$ is the complement of the closure of $\mathcal{D}$ in the space under consideration. When $\underline{x}\in\mathcal{D}$, $0 \le t < \infty$, (3.2.1) is an integral equation that allows the numerical determination of ϕ with the *marching-on-in-time method*. In this method, we discretize in space and time, approximate the second term on the right-hand side of (3.2.1) accordingly, and invoke the equality sign in (3.2.1) at the relevant space-time points. To this end, we construct a uniform spatial grid $\{\underline{x}_\alpha\}$ with mesh size h and take $t = n\Delta t$, where $n = 0,1,2,\ldots,\infty$. The time step Δt is chosen such that $\Delta t = \min(R_{\alpha\alpha'})/c$, where $R_{\alpha\alpha'} = |\underline{x}_\alpha - \underline{x}_{\alpha'}|$. Then we end up with algebraic equations of the type

$$\tilde{\phi}(\alpha,n) = \tilde{\phi}^{i}(\alpha,n) + \sum_{\alpha'}\sum_{n'=0}^{n} \tilde{K}(\alpha,\alpha';n - n')\tilde{\phi}(\alpha',n'), \qquad n = 0,1,2,\ldots,\infty, \qquad (3.2.2)$$

where $\tilde{\phi}^i(\alpha,n) = \phi^i(\underline{x}_\alpha, n\Delta t)$. Formally, the system of equations (3.2.2) can be resolved as follows. First we rewrite (3.2.2) as

$$\sum_{\alpha'} W(\alpha,\alpha')\tilde{\phi}(\alpha',n) = \tilde{\phi}^i(\alpha,n) + \sum_{\alpha'}\sum_{n'=0}^{n-1} \tilde{K}(\alpha,\alpha';n-n')\tilde{\phi}(\alpha',n'). \tag{3.2.3}$$

In (3.2.3), we have introduced the "weighting matrix" $W(\alpha,\alpha')$ given by

$$W(\alpha,\alpha') = \delta_{\alpha\alpha'} - \tilde{K}(\alpha,\alpha';0), \tag{3.2.4}$$

with $\delta_{\alpha\alpha'}$ being the Kronecker symbol. Next, we define the "inverse weighting matrix" $W^{-1}(\alpha,\alpha')$ in the usual manner, i.e.

$$\sum_{\alpha'} W^{-1}(\alpha,\alpha')W(\alpha',\alpha'') = \delta_{\alpha\alpha''} \tag{3.2.5}$$

for all relevant α and α''. Combining (3.2.3) and (3.2.5) then yields

$$\tilde{\phi}(\alpha,n) = \sum_{\alpha'} W^{-1}(\alpha,\alpha')\left\{\tilde{\phi}^i(\alpha',n) + \sum_{\alpha''}\sum_{n'=0}^{n-1} \tilde{K}(\alpha',\alpha'';n-n')\tilde{\phi}(\alpha'',n')\right\}. \tag{3.2.6}$$

Equation (3.2.6) is the basis of the *marching-on-in-time method.* This equation expresses, for each fixed instant n, the field values $\{\tilde{\phi}(\alpha,n)\}$ in terms of the given incident field at the instant $t = n\Delta t$ and the field values $\{\tilde{\phi}(\alpha,n') \mid n' < n\}$, i.e. the numerical solution at previous instants. Hence, (3.2.6) can be solved by a step-by-step updating procedure involving only the linear combination, at each space point, of known field values.

In terms of matrix calculus, the system of equations (3.2.2) can be envisaged as a matrix equation of infinite dimension for the unknown vector $\{\tilde{\phi}(\alpha,n)\}$. In that perspective, Equation (3.2.6) can be understood as being the result of reducing (3.2.2) to lower-triangular form. The marching-on-in-time method is then just the recursive solution of that lower-triangular matrix equation. It should be observed that the dimension of the weighting matrix $W(\alpha,\alpha')$, which has to be inverted in the triangularization of (3.2.2), depends on the space discretization only. Furthermore, the upper limit of the time integration in (3.2.1) is always less that t, unless R = 0, which occurs only if $\alpha = \alpha'$. Hence, owing to the choice of Δt, the interpolations in the discretization can generally be organized such that $\tilde{K}(\alpha,\alpha';0) = 0$ if $\alpha' \neq \alpha$. In that case, the weighting matrices reduce to

$$W(\alpha,\alpha') = [1 - \tilde{K}(\alpha,\alpha;0)]\delta_{\alpha\alpha'}, \tag{3.2.7a}$$

$$W^{-1}(\alpha,\alpha') = [1 - \tilde{K}(\alpha,\alpha;0)]^{-1}\delta_{\alpha\alpha'}, \tag{3.2.7b}$$

which eliminates the necessity for a numerical inversion of $W(\alpha,\alpha')$. Whether (3.2.7a) can be met in the discretization, depends on the structure of the specific integral equation under consideration. We will investigate this aspect in more detail when the relevant equations come up for discussion. Even when the discretization cannot be organized as specified above, the weighting matrices will still closely resemble the ones given in (3.2.7). A configuration for which this happens will be discussed in Section 3.5. For passive obstacles, either of the properties

$$\tilde{K}(\alpha,\alpha;0) \leq 0 \quad \text{or} \tag{3.2.8a}$$

$$\tilde{K}(\alpha,\alpha;0) = O(1) \tag{3.2.8b}$$

can be shown to hold as $h \to 0$. Hence, for a sufficiently small mesh size h, we need not expect computational difficulties in the determination of $W^{-1}(\alpha,\alpha')$.

Difficulties can be expected due to the discretization of the multiple integral in (3.2.1). Because of the error made in this discretization, $\tilde{\phi}(\alpha,n)$ will only be approximately equal to the actual field value $\phi(\underline{x}_\alpha, n\Delta t)$. Since, in the numerical solution of (3.2.6), each field value $\tilde{\phi}(\alpha,n)$ is computed from field values $\tilde{\phi}(\alpha',n')$ at previous instants, the computational errors caused in this manner will accumulate. As a consequence, the solution obtained may be unstable. In the upcoming two subsections we will analyze this error accumulation.

3.2.2 Error accumulation

For scattering by passive obstacles, the field $\phi(\underline{x},t)$ with $\underline{x}\epsilon\mathcal{D}$ generally becomes negligible after some finite instant T_{max}. Hence, it suffices to organize the numerical computation such that the deviation between the actual field values $\phi(\underline{x}_\alpha,n\Delta t)$ with $0 < n\Delta t < T_{max}$ and the corresponding marching-on-in-time results $\tilde{\phi}(\alpha,n)$ can be controlled. In this subsection, we investigate the conditions under which this is possible. To this end, we consider the deviation mentioned above:

$$\Delta\phi(\alpha,n) \overset{\Delta}{=} \phi(\underline{x}_\alpha,n\Delta t) - \tilde{\phi}(\alpha,n). \tag{3.2.9}$$

Combining (3.2.1) and (3.2.2), we find that this deviation is governed by the equation

$$\Delta\phi(\alpha,n) = D(\alpha,n) + \sum_{\alpha'}\sum_{n'=0}^{n} \tilde{K}(\alpha,\alpha';n - n')\Delta\phi(\alpha',n'), \tag{3.2.10}$$

with

$$\begin{aligned} D(\alpha,n) \overset{\Delta}{=} & \int_{\mathcal{D}}d\underline{x}'\int_0^{n\Delta t-R/c}dt'\ K(\underline{x}_\alpha,\underline{x}';n\Delta t - t')\phi(\underline{x}',t') \\ & - \sum_{\alpha'}\sum_{n'=0}^{n} \tilde{K}(\alpha,\alpha';n - n')\phi(\underline{x}_{\alpha'},n'\Delta t). \end{aligned} \tag{3.2.11}$$

In (3.2.10), the term $D(\alpha,n)$ denotes the discretization error resulting from approximating the integral in (3.2.1) by the discrete sum in (3.2.2) for the *exact* field $\phi(\underline{x},t)$; the second term on the right-hand side of (3.2.10) prescribes the effect of the errors made in previous updating steps.

With (3.2.10) and (3.2.11), we have isolated the behavior of the marching-on-in-time error from that of the actual solution. Note that (3.2.10) is of the same form as the discretized equation (3.2.2), where the discretization error $D(\alpha,n)$ plays the role of the incident field. The special form of (3.2.10) already provides us with a first impression of the type of contributions that will show up in $\Delta\phi(\alpha,n)$. As remarked in Subsection 3.2.1, an equation of the form (3.2.10) can be envisaged as a matrix equation of infinite dimension. Moreover, because of the upper bound on the summation index n', a finite subsystem of linear equations can be selected for the field in the time interval $0 < t < T_{max}$. In Appendix 3.2.A, it will be argued that the numerical condition of such a matrix equation depends on the magnitudes of the eigenvalues of its system matrix. When these eigenvalues are all of $O(1)$ or larger, the components of the known vector (i.e. $D(\alpha,n)$ in (3.2.10)) along the corresponding eigenvectors will not be amplified in the solution vector (i.e. $\Delta\phi(\alpha,n)$). On the other hand, when the system matrix of (3.2.10) has some eigenvectors with small eigenvalues, the components of $D(\alpha,n)$ along these eigenvectors will be amplified and the solution obtained will be unstable. Now we should keep in mind that (3.2.2) is a, presumably accurate, discretization of the well-conditioned integral equation (3.2.1). The sampled exact solution of (3.2.1) will therefore meet in good approximation the equality sign in (3.2.2). This indicates that the system matrix of (3.2.2) (and (3.2.10)) will have at most a few "problematic" eigenvectors with small eigenvalues. Interpreting the eigenvectors of that system matrix as a set of basis vectors for the space of possible $\{D(\alpha,n)\}$ and $\{\Delta\phi(\alpha,n)\}$, we can then distinguish between two types of error. The components of $D(\alpha,n)$ within the subspace spanned by the "problematic" eigenvectors will result in magnified components of $\Delta\phi(\alpha,n)$ within that same subspace. In view of the localization of this effect within the subspace, an error of this type may be called a *local* error. Note that this qualification need not imply a local behavior of the error in the space-time domain. Rather, it refers to an error behavior of a special, unwanted character. The components of $D(\alpha,n)$ along the remaining majority of the eigenvectors will not be amplified in $\Delta\phi(\alpha,n)$. Consequently, the resulting component in the marching-on-in-time error $\Delta\phi(\alpha,n)$ will be of the same order of magnitude as the discretization error $D(\alpha,n)$. In line with the designation proposed above, such an error may be named a *global* error.

Now that we know what type of errors to expect, let us consider

(3.2.10) and (3.2.11) in more detail. If, in the solution of (3.2.2), only global errors were excited, it would be sufficient to organize the discretization such that $D(\alpha,n) = O(1)$ as $h \to 0$. This is essentially the result obtained by Rynne (1985). However, in view of the time-recursive character of the marching-on-in-time method, it seems overly optimistic to ignore the possibility of a deviation $\Delta\phi(\alpha,n)$ that increases gradually with increasing n. The origin of such a deviation can be understood as follows. Generally, the discretization error $D(\alpha,n)$ consists entirely of interpolation errors. Each of these errors is proportional to some power of the mesh size h and to some higher-order space or time derivative at a space-time point that lies within a specified space or time interval of length $O(h)$. For a smoothly varying $\phi(\underline{x},t)$ and a sufficiently small h, these errors will contain a systematic part that need not be canceled out and, hence, may result in a deviation $\Delta\phi(\alpha,n)$ that grows with n.

In view of this observation, it seems more realistic to consider the "error growth" per update step, i.e. $\Delta\phi(\alpha,n) - \Delta\phi(\alpha,n-1)$. From (3.2.10), we have

$$\begin{aligned}\Delta\phi(\alpha,n) - \Delta\phi(\alpha,n-1) &= D(\alpha,n) - D(\alpha,n-1) \\ &+ \sum_{\alpha'}\sum_{n'=0}^{n} \tilde{K}(\alpha,\alpha';n-n')[\Delta\phi(\alpha',n') - \Delta\phi(\alpha',n'-1)],\end{aligned} \tag{3.2.12}$$

which is of the same form as (3.2.2) and (3.2.10). In (3.2.12) the role of the incident field is played by the error difference $D(\alpha,n) - D(\alpha,n-1)$. This eliminates the major part of the systematic behavior signaled in the previous paragraph. Since $D(\alpha,n)$ may also have a randomly varying part, we have to estimate

$$D(\alpha,n) - D(\alpha,n-1) = O(D(\alpha,n)). \tag{3.2.13}$$

Equations (3.2.12) and (3.2.13) suggest that the error accumulation in the marching-on-in-time method can be controlled when the following two criteria are met. In the first place, the discretization error should meet *criterion I*:

$$D(\alpha,n) = O(h^2) \quad \text{as } h\to 0, \tag{3.2.14}$$

for all relevant α and n. In the second place, the error difference $D(\alpha,n) - D(\alpha,n-1)$ should be either so smoothly varying or so randomly distributed over the relevant space-time points that the solution of (3.2.12) meets *criterion II*:

$$\Delta\phi(\alpha,n) - \Delta\phi(\alpha,n-1) = O(D(\alpha,n)). \tag{3.2.15}$$

In the terminology introduced at the beginning of this subsection, the latter criterion may be reformulated as: the error difference $D(\alpha,n) - D(\alpha,n-1)$ should only excite a "global" error growth.

With $\Delta\phi(\alpha,0) = 0$, which holds in view of the initial conditions formulated in Subsection 3.2.1, and with (3.2.14) - (3.2.15), it follows by induction that

$$\Delta\phi(\alpha,n) = O(nh^2). \tag{3.2.16}$$

Because the time step Δt is proportional to h, the error at $t = T_{max}$ will then at most be proportional to

$$(T_{max}/h)h^2 = T_{max}h,$$

and, hence, it can be controlled by choosing h sufficiently small.

As we will see in the upcoming sections, in the discretization of an actual scattering problem, it may be hard to meet both the criteria formulated above simultaneously. With sufficient ingenuity, criterion I can generally be met. The main hindrance to achieving this seems to be the possible singular behavior of the operator $K(\underline{x},\underline{x}';t-t')$ near the point of observation, i.e. for $\underline{x}' \approx \underline{x}$ and $t' \approx t$. In the verification of this criterion, we have the advantage that it refers to the *exact* field. In the error estimation, we may therefore utilize the continuity and/or differentiability properties of that field. Criterion II, on the other hand, may be more difficult to verify. This requires some knowledge of the behavior of the unknown exact solution as well as some insight into the nature of the "problematic" eigenvectors, i.e. the instabilities that may show up in the marching-on-in-time results. In particular, criterion II may be violated when the $\tilde{\phi}(\alpha,n)$ contain a component that alternates with n. Such an (unwanted) component may increase exponentially as n increases and, eventually, dominate the numerical solution. Moreover, it cannot be removed by refining the discretization since the argument given above no longer serves. In the next subsection, we will elaborate on the origin of such instabilities or, equivalently, local errors. In addition, we will indicate two ways these can be avoided by modifying the marching-on-in-time procedure.

3.2.3 Instabilities

Origin

Greater comprehension of the instabilities that show up in marching-on-in-time results can only be obtained when the operator $K(\underline{x},\underline{x}';t - t')$ is specified in more detail. In general, this operator involves one or two time differentiations acting directly on $\phi(\underline{x}',t')$. In the discretization, the resulting derivatives have to be approximated by suitable difference formulas. For example, in compliance with criterion I, we can approximate $\partial_{t'}\phi(\underline{x}',t')$ by a three-term backward difference formula based on quadratic time interpolation:

$$\partial_{t'}\phi(\underline{x}',t') = [\tfrac{3}{2}\phi(\underline{x}',t') - 2\phi(\underline{x}',t' - \Delta t) + \tfrac{1}{2}\phi(\underline{x}',t' - 2\Delta t)]/\Delta t, \tag{3.2.17}$$

which holds up to $O(\Delta t^2)$ (see Hildebrand (1956)). Moreover, as mentioned above, $K(\underline{x},\underline{x}';t - t')$ is singular at the point of observation for most multi-dimensional scattering problems. Restricting ourselves to operators K involving only a single time derivative, we can then come up with the following simplified model of the discretized integral equation (3.2.2). Let $\mathfrak{f}[n]$ and $g[n]$, with $n\in\mathbb{Z}$, be real-valued time sequences that vanish for $n \leq 0$ and satisfy the difference equation

$$\begin{aligned} f[n] &+ w_0 C\left\{\tfrac{3}{2}f[n] - 2f[n-1] + \tfrac{1}{2}f[n-2]\right\} \\ &+ w_1 C\left\{\tfrac{3}{2}f[n-1] - 2f[n-2] + \tfrac{1}{2}f[n-3]\right\} = g[n]. \end{aligned} \tag{3.2.18}$$

In (3.2.18), w_0 and w_1 are real-valued, positive weighting factors with $w_1 > w_0$, and C is a positive configuration parameter. This parameter represents either the local contrast (for penetrable obstacles) or the local curvature (for impenetrable obstacles). The first term on the left-hand side of (3.2.18) may be regarded as the field at a point of observation that shows up on the left-hand side of (3.2.1) and (3.2.2). The terms with C may be viewed in two complementary ways. The first interpretation holds when the operator $K(\underline{x},\underline{x}';t - t')$ is singular at the point of observation. In that case, the term with w_0 in (3.2.18) may be envisaged as the "self term" from the discretized integral in (3.2.2) and the one with w_1 as the contribution from the "nearest neighbors" in space and time. Since there is always more than one neighbor, it seems fair to assume that the contribution from these neighbors is larger than the "self term", i.e. $w_1 > w_0$. The second interpretation holds in the absence of a singularity. In that case, the second and third terms in (3.2.18) can be interpreted as a systematic contribution from space-time points with n - n' even and odd, respectively. Then, both contributions will have approximately the same magnitude. Hence, we may still have the situation $w_1 > w_0$. The sequence $g[n]$ in (3.2.18) stands for the incident field and the remainder of the discretized integral. Here, it is assumed to be gradually varying with n.

It should be noted that the model proposed above may be an oversimplification in two respects. In the first place, it presupposes a fixed relation between the field values represented by the left-hand side of (3.2.18). In the second place, assuming $g[n]$ to be smoothly varying restricts the instability analysis to the interaction between field values at a few selected space-time points. However, the only consequence of these simplifications is that certain types of instabilities may be excluded from the analysis. This implies that at least the instabilities predicted by our model can be regarded as being realistic.

To obtain an impression of these instabilities, we introduce the bilateral z-transform of a time sequence $f[n]$ as

$$F(z) \triangleq \sum_{n=-\infty}^{\infty} f[n]z^{-n}. \tag{3.2.19}$$

Taking the z-transform of (3.2.18) and defining the system function H(z) as $H(z) \triangleq F(z)/G(z)$, with G(z) being the z-transform of the input sequence $g[n]$, we arrive at

$$H(z) = z^3/P(z), \quad \text{with} \tag{3.2.20a}$$

$$P(z) = z^3 + w_0 C(\tfrac{3}{2}z^3 - 2z^2 + \tfrac{1}{2}z) + w_1 C(\tfrac{3}{2}z^2 - 2z + \tfrac{1}{2}). \tag{3.2.20b}$$

As in the time-continuous case discussed in Chapter 2, the homogeneous solutions of the time-discrete equation (3.2.18) can be found by searching for the poles of the system function H(z), i.e. the zeros of the characteristic polynomial P(z). To each zero $z = z_m$ of that polynomial, there corresponds a solution of the homogeneous equation of the form

$$f_m[n] = (z_m)^n, \quad n \in \mathbb{Z}. \tag{3.2.21}$$

In the numerical solution of (3.2.18), difficulties can be expected when one or more of the $\{z_m\}$ lie outside the unit circle $|z| = 1$. In that case, the corresponding homogeneous solutions $\{f_m[n]\}$ increase exponentially as $n \to \infty$. Even though the input sequence $g[n]$, which vanishes for $n \le 0$, does not excite these solutions, numerical round-off errors will cause them to creep in. As a result, the numerically obtained solution $f[n]$ will be unstable.

For the cubic polynomial P(z) defined in (3.2.20b), the conditions for the presence of a zero $z = z_m$ with $|z_m| > 1$ can be obtained in closed form. Analyzing P(z) and its z-derivative shows that P(z) always has a single zero on the negative real z-axis. This zero is located on the subinterval $-\infty < z < -1$ when $P(-1) > 0$, i.e. when

$w_1 > w_0$ and

$$C > (4w_1 - 4w_0)^{-1}. \tag{3.2.22}$$

Moreover, from the sign distribution of P(z) on the positive half of the real z-axis, it is observed that, for the two remaining zeros $z = z_2$ and $z = z_3$, we have either of the following situations. For a sufficiently large C, both z_2 and z_3 are located on the real z-interval $\frac{1}{3} < z < 1$, where the term proportional to C in (3.2.20b) is negative. Otherwise, z_2 and z_3 form a complex-conjugate pair off the real z-axis. In the former case, z_2 and z_3 are obviously located within the unit disk $|z| < 1$. In the latter case, their magnitudes can be estimated with the aid of the identity

$$|z_2|^2 = |z_3|^2 = z_2 z_3 = -\tfrac{1}{2}w_1 C/z_1(1 + \tfrac{3}{2}w_0 C) \tag{3.2.23}$$

(see Abramowitz and Stegun (1965), formula 3.8.2). By direct substitution, it follows that

$$P(-\tfrac{1}{2}w_1 C/(1 + \tfrac{3}{2}w_0 C)) > 0. \tag{3.2.24}$$

Since P(z) becomes negative as $z \to -\infty$, we then have

$$z_1 < -\tfrac{1}{2}w_1 C/(1 + \tfrac{3}{2}w_0 C), \tag{3.2.25}$$

which, in combination with (3.2.23), implies that $|z_2| = |z_3| < 1$. Combining the results derived above, we arrive at the following conclusions. Under the assumption that the operator K contains only a single time derivative, numerically solving (3.2.2) will resemble solving the simplified equation (3.2.18). Hence, a stable result will only be produced for a finite range of the configuration parameter C. When C lies inside that range, the discretization error $D(\alpha,n)$ defined in (3.2.11) and the numerical round-off errors made during the computation only excite exponentially decaying terms. Hence, criterion II as formulated in Subsection 3.2.2 is met and the error accumulation in the marching-on-in-time method can be controlled. When C lies outside the stability range, the numerically obtained solution will contain an exponentially exploding, unstable component that alternates with n. Finally, it should be pointed out that the instability result derived above also applies when $K(\underline{x},\underline{x}';t - t')$ involves second-order time derivatives. Such derivatives can be modeled by including in (3.2.18) a four-point backward difference formula based on cubic time interpolation. Repeating the analysis for the thus extended difference equation leads to a quartic characteristic polynomial similar to (3.2.20b) with a similar zero on the negative real z-axis.

Modifying the discretization

The conclusions formulated in the preceding paragraph also provide us with the key to the solution of the instability problem: within the limitations imposed by criterion I, the discretized equation should be modified such that, in effect, the weighting factors w_0 and w_1 are reduced. We will now describe two ways to accomplish this. The first method that comes to mind is to modify the discretization of the integral in (3.2.1). For example, when w_0 and w_1 are both of $O(1)$ as $h \to 0$, a reduction in the mesh size will eliminate the instability. An example where this technique takes effect will be discussed in Section 3.4. Similarly, w_0 and w_1 can be reduced by replacing the approximations to the first- and second-order time derivatives with difference formulas having the double time step (e.g. by taking $2\Delta t$ instead of Δt in (3.2.17)). At the cost of a minor increase in the discretization error, this extends the stability range of the relevant configuration parameters by a factor of two and four, respectively. Moreover, doubling the time step shifts the stability problem to the combination of the "self term" and the contributions from the "next-nearest neighbors", which may have a smaller weighting factor. The effectiveness of doubling the time step was first reported by Rynne (1985), who dealt with the scattering by a three-dimensional perfect conductor. A second example of a successful application will be given in Section 3.5. A disadvantage of modifying the discretization of the integral might be that it still leaves us with a finite range of applicability. For the doubling of the time step in the difference formulas, this limitation is obvious. Reducing the mesh size results in a sharp increase in computation time.

Modifying the method of solution

To some extent, both these limitations are overcome in the second method of solving the instability problem: leave the discretization as it is and modify the method of solution. One way to achieve this is to redefine $\tilde{\phi}(\alpha,n)$ as the solution of minimizing a squared error of the type

$$\sum_{\alpha}\sum_{n=0}^{\infty}\Big\{\tilde{\phi}(\alpha,n) - \tilde{\phi}^{i}(\alpha,n) - \sum_{\alpha'}\sum_{n'=0}^{n} \tilde{K}(\alpha,\alpha';n-n')\tilde{\phi}(\alpha',n')\Big\}^2. \tag{3.2.26}$$

This solution can be obtained numerically by applying a gradient-type method (Herman and Van den Berg (1982), Van den Berg (1984), Sarkar (1984), Sarkar and Rao (1984), Rao et al. (1984)). If desired, the individual squared errors can further be weighted such that the sum over α and n corresponds to a space-time integral over $\mathcal{D}$ and $0 < t < \infty$. By assuming that the sum (3.2.26) exists, one implicitly imposes the restriction that $\tilde{\phi}(\alpha,n)$ vanishes as $n\to\infty$. When the initial estimate for $\tilde{\phi}(\alpha,n)$ exhibits this behavior, all subsequent approximations will do so as well. In matrix terminology, this means that we have restricted the space of possible solutions $\{\tilde{\phi}(\alpha,n)\}$ to a subspace spanned by well-behaved eigenvectors. Hence, we need only expect "global" errors in the results obtained. This interpretation also shows the principal disadvantage of gradient-type methods. In each iteration step of such a method, at most the component of $\tilde{\phi}(\alpha,n)$ along a single direction in the space of possible solutions receives its proper value. In view of the vast dimension of that solution space, it may therefore take a large number of iteration steps to obtain a sufficiently accurate solution. Since each iteration step requires twice the computational effort involved in carrying out a direct marching-on-in-time computation, applying a gradient-type method may take up an excessive amount of computation time. This conclusion was verified by applying the conjugate-gradient method to the example problems that will be discussed in Sections 3.3 and 3.5. In both these problems, hundreds of iteration steps are required to arrive at an acceptable solution.

The inefficiency observed above can be explained by the fact that, in applying a gradient-type method, we also throw overboard the favorable properties of the marching-on-in-time method. In particular, this applies to the numerical efficiency owing to the lower-triangular form of the system matrix, and to the property that the "global" components of $\tilde{\phi}(\alpha,n)$ do receive their proper value. We will now describe an iterative procedure that exploits these properties. Our modification of the marching-on-in-time method can again be understood from the difference equation (3.2.18). Suppose that we have available some sufficiently smooth estimate $\bar{\delta}[n]$ of the time sequence $\delta[n]$. Then we can replace (3.2.18) by

$$\begin{aligned}(p + 1)\delta[n] &+ w_0C\left\{\tfrac{3}{2}\delta[n] - 2\delta[n-1] + \tfrac{1}{2}\delta[n-2]\right\} \\ &+ w_1C\left\{\tfrac{3}{2}\delta[n-1] - 2\delta[n-2] + \tfrac{1}{2}\delta[n-3]\right\} = g[n] + p\bar{\delta}[n],\end{aligned} \tag{3.2.27}$$

with p being a real-valued, positive parameter. Repeating the stability analysis carried out above for the modified equation (3.2.27), we find that that equation allows a stable numerical solution for

$$0 \le C \le (p + 1)/(4w_1 - 4w_0). \tag{3.2.28}$$

For a given configuration parameter C, we can therefore always choose p in (3.2.27) such that instabilities do not show up in its numerical solution. Obviously, the time sequence obtained from (3.2.27) is only an approximate solution of the actual equation (3.2.18). Its quality depends on the quality of the initial estimate $\bar{\delta}[n]$ and on the magnitude of the parameter p. An impression of this quality can be obtained from the magnitude of $p\{\delta[n] - \bar{\delta}[n]\}$, which constitutes the error in the equality sign of (3.2.18) for the solution obtained from (3.2.27). The improvement in the stability found above forms the basis of the following procedure for the iterative solution of the system of equations (3.2.2). Starting from a sufficiently smooth initial estimate $\bar{\phi}^{(0)}(\alpha,n)$, we carry out a number of iteration steps. In step number j, the approximate solution $\tilde{\phi}^{(j)}(\alpha,n)$ is obtained as the result of minimizing, for each fixed n, the squared error

$$\sum_\alpha\left\{\tilde{\phi}^{(j)}(\alpha,n) - \tilde{\phi}^{i}(\alpha,n) - \sum_{\alpha'}\sum_{n'=0}^{n} \tilde{K}(\alpha,\alpha';n-n')\tilde{\phi}^{(j)}(\alpha',n')\right\}^2 + p^2\sum_\alpha\left\{\tilde{\phi}^{(j)}(\alpha,n) - \bar{\phi}^{(j-1)}(\alpha,n)\right\}^2 \tag{3.2.29}$$

for known values of $\tilde{\phi}^{i}(\alpha,n)$, $\bar{\phi}^{(j-1)}(\alpha,n)$, $\tilde{\phi}^{(j)}(\alpha,n')$ with $n' < n$, and p. For $j > 1$, we choose $\bar{\phi}^{(j-1)}(\alpha,n)$ to be a smoothed version of the approximate solution $\tilde{\phi}^{(j-1)}(\alpha,n)$ obtained in the previous step. The smoothing operation is applied after each iteration step to remove small, unwanted components from the approximation obtained in that step which may cause an instability in subsequent steps. It amounts to replacing $\tilde{\phi}^{(j)}(\alpha,n)$ with

$$\bar{\phi}^{(j)}(\alpha,n) = [\tilde{\phi}^{(j)}(\alpha,n-1) + 2\tilde{\phi}^{(j)}(\alpha,n) + \tilde{\phi}^{(j)}(\alpha,n+1)]/4, \tag{3.2.30}$$

for $0 < n < \infty$, and is based on the assumption that the actual solution $\phi(\underline{x},t)$ varies approximately linearly within the time interval $(n-1)\Delta t < t < (n+1)\Delta t$. Note that this operation is consistent with criterion I as well as with approximating the time derivatives of $\phi(\underline{x}',t')$ by three- and four-term difference formulas. If necessary, the smoothing operation can be repeated several times. This method, too, will be applied in Section 3.5.

In terms of matrix calculus, the iterative procedure formulated in Equations (3.2.29) and (3.2.30) is just the relaxation method explained in Appendix 3.2.A. This is a method for solving a matrix equation, which, in the solution procedure, avoids components along the unwanted eigenvectors associated with small eigenvalues. In view of the infinite dimension of the system matrix, we cannot apply this method to the full matrix equation. Instead, we utilize the lower-triangular structure of the system matrix to apply the relaxation method to subsystems of equations pertaining to field values

$\{\tilde{\phi}(\alpha,n)\}$ at a fixed instant n. This interpretation also allows us to use the convergence results derived in Appendix 3.2.A. These results ensure that the iteration scheme is strictly convergent even when the component along an unwanted eigenvector of the system matrix is systematically removed after each iteration step. Moreover, the convergence is obtained regardless of the initial estimate. It should be noted that each iteration step is in fact the solution of a regularized version of the linear algebraic equation (3.2.2) (see Tikhonov and Arsenin (1977) and Sarkar et al. (1981)). The new element in our solution technique is that we have incorporated this regularized inversion in an iterative scheme.

The iterative procedure should not be confused with solving, for a fixed n, (3.2.29) repeatedly until convergence for known (converged) values of $\tilde{\phi}(\alpha',n')$ with $n' < n$. The latter procedure is equivalent to solving (3.2.6) and, hence, will produce the unstable result that we aimed at avoiding. The improvement in our iterative procedure is just that, by computing in each iteration step $\tilde{\phi}^{(j)}(\alpha,n)$ for *all* n, we are able to apply the smoothing procedure that removes the component along the problematic eigenvector of the system of equations (3.2.2).

The only remaining question is the choice of a suitable value for the parameter p. This value must be established by trial and error in the first iteration step. When the instabilities do not show up in that step, we may expect them not to turn up in subsequent steps either. Only when the initial guess is totally different from the eventual solution, will it perhaps become necessary to adjust p during the iterative procedure. An additional lead to the proper value of p is given by Equation (3.2.28), which suggests that for large values of the configuration parameter C, p should be linearly proportional to that parameter.

The essential difference between our method and a gradient-type method is that our method produces, in each iteration step, a *partial* improvement to the components along *all* well-behaved eigenvectors. A gradient-type method, on the other hand, produces a *total* improvement in a *single* direction. The common element of both methods is that they will produce similar global errors. To control these errors, the theory of Subsection 3.2.2 must be put into operation.

Appendix 3.2.A A relaxation method for the solution of an ill-conditioned system of linear equations

We consider a system of N linear equations for N unknown variables $\{x_n\}$, components of a vector $\underline{x}$. This system can be written in matrix form as

$$\sum_{n=1}^{N} A_{mn}\, x_n = b_m, \qquad m = 1,2,\ldots,N, \tag{3.2.31}$$

where A denotes the system matrix and where b_m is a component of a known vector $\underline{b}$. We assume the matrix A to be nonsingular. In that case, A has N nonvanishing eigenvalues λ_ℓ, $\ell = 1,2,\ldots,N$, with N corresponding independent eigenvectors $\underline{v}^{(\ell)}$. Expanding both the unknown vector $\underline{x}$ and the known vector $\underline{b}$ in terms of these eigenvectors, we formally have

$$\underline{b} = \sum_{\ell=1}^{N} \beta_\ell\, \underline{v}^{(\ell)}, \tag{3.2.32a}$$

$$\underline{x} = \sum_{\ell=1}^{N} \xi_\ell\, \underline{v}^{(\ell)} = \sum_{\ell=1}^{N} \lambda_\ell^{-1}\, \beta_\ell\, \underline{v}^{(\ell)} \tag{3.2.32b}$$

In (3.2.32) and in the remainder of this appendix, we have used Greek symbols to denote the components along the eigenvectors $\{\underline{v}^{(\ell)}\}$. This notation is used to avoid confusion with the components in the original coordinate frame. In the numerical solution of (3.2.31), problems may be expected when one or more of the eigenvalues $\{\lambda_\ell\}$ are small in magnitude. As (3.2.32b) indicates, a small deviation in the known vector $\underline{b}$ along the corresponding eigenvectors may then cause a considerable error in the solution vector $\underline{x}$ obtained.

When the components of $\underline{x}$ along the problematic eigenvectors of A are known from a priori information (for example because they should vanish), the following iterative procedure may be employed to solve (3.2.31). Starting from an initial estimate $\bar{\underline{x}}^{(0)}$, the j'th approximation $\underline{x}^{(j)}$ is defined as the solution of the system of equations

$$px_m^{(j)} + \sum_{n=1}^{N} A_{mn} x_n^{(j)} = b_m + p\bar{x}_m^{(j-1)}, \tag{3.2.33}$$

with m = 1,2,...,N. In (3.2.33), p is a real-valued, positive parameter and $\bar{\underline{x}}^{(j-1)}$ is the result of resetting the components of the previous approximation $\underline{x}^{(j-1)}$ along the problematic eigenvectors to their a priori known values. The convergence of the iteration scheme can be analyzed by also expanding the iteration results $\underline{x}^{(j)}$ in terms of the eigenvectors $\underline{v}^{(\ell)}$. In analogy with (3.2.32), we have

$$\underline{x}^{(j)} = \sum_{\ell=1}^{N} \xi_\ell^{(j)} \underline{v}^{(\ell)}, \quad \bar{\underline{x}}^{(j)} = \sum_{\ell=1}^{N} \bar{\xi}_\ell^{(j)} \underline{v}^{(\ell)}. \tag{3.2.34}$$

Substituting (3.2.34) in (3.2.33) directly yields the recurrence relation

$$\xi_\ell^{(j)} = (\beta_\ell + p\, \bar{\xi}_\ell^{(j-1)})/(p + \lambda_\ell). \tag{3.2.35}$$

In order to establish whether $\xi_\ell^{(j)}$ is a better approximation to the actual solution $\xi_\ell = \lambda_\ell^{-1}\beta_\ell$ than $\bar{\xi}_\ell^{(j-1)}$, we consider how it differs from that solution. From (3.2.35), we have

$$\xi_\ell^{(j)} - \lambda_\ell^{-1}\beta_\ell = [\bar{\xi}_\ell^{(j-1)} - \lambda_\ell^{-1}\beta_\ell]p/(p + \lambda_\ell). \tag{3.2.36}$$

When $\xi_\ell^{(j)}$ is not tampered with in the reset step, i.e. when $\bar{\xi}_\ell^{(j)} = \xi_\ell^{(j)}$, it follows from (3.2.36) that we have strict convergence if and only if $\lambda_\ell > 0$. Moreover, this convergence property holds regardless of the initial estimate and for each individual component $\xi_\ell^{(j)}$. The latter observation is especially important, since it shows that resetting a particular component $\xi_\ell^{(j)}$ to its a priori value does not affect the convergence of those remaining. Finally, (3.2.36) indicates that the components corresponding to the smallest eigenvalues digress the least from their initial estimates. This suggests that, when only a few iteration steps are required to arrive at an acceptable result, we may leave out the reset step. In subsequent steps, however, we will then gradually obtain the exact solution (3.2.32b), whose inaccuracy initially motivated our taking up the iteration scheme.

The result (3.2.36) rules out the possibility of applying our scheme when one or more of the eigenvalues λ_ℓ are negative. In the application that we have in mind, this restriction is not acceptable, since we are not even able to determine all the eigenvalues. However, negative eigenvalues can be accounted for by redefining $\underline{x}^{(j)}$ as the vector that minimizes the squared error

$$\sum_{m=1}^{N}\left\{\sum_{n=1}^{N} A_{mn}\, x_n^{(j)} - b_m\right\}^2 + p^2\sum_{m=1}^{N}\left\{x_m^{(j)} - \bar{x}_m^{(j-1)}\right\}^2. \qquad (3.2.37)$$

Differentiating this error with respect to the individual components $x_k^{(j)}$, $k = 1,2,\ldots,N$, reveals that its minimum is obtained when the linear equation

$$p^2\, x_k^{(j)} + \sum_{n=1}^{N}\sum_{m=1}^{N} A_{mk}\, A_{mn}\, x_n^{(j)} = \sum_{m=1}^{N} A_{mk}\, b_m + p^2\, \bar{x}_k^{(j-1)} \qquad (3.2.38)$$

is met for all k. The system of equations (3.2.38) is of the same form as the one given in (3.2.33), the only difference being that the system matrix A is replaced by the product resulting from a left multiplication of A with its transpose. As is well known, such a product matrix has only positive eigenvalues.

Finally, we are left with naming our solution procedure. Since the deviation between the approximation obtained after j iteration steps and the eventual solution decays exponentially with increasing j, it seems appropriate to designate our technique as a *relaxation method*.

3.3 Scattering by an inhomogeneous, lossy dielectric slab

3.3.1 Slab in vacuum: formulation of the problem

In this section, we consider the application of the marching-on-in-time method to the scattering by an inhomogeneous, lossy dielectric slab as specified in Subsection 2.2.1. In the first three subsections, we will treat the problem of the slab embedded in vacuum, as formulated in Subsection 2.2.1. Further on in this section, we will investigate the case where the slab is situated in between two homogeneous, lossless half-spaces, as specified in Subsection 2.4.1. Throughout this section, we will use the same normalized quantities as in the frequency-domain analysis of Sections 2.2 - 2.4 (see p. 37).

The relevant time-domain integral equations can be obtained from their frequency-domain equivalents with the aid of Lerch's theorem (Widder (1946)). This theorem prescribes that a causal function is uniquely determined by its known Laplace transform on the positive real s-axis. Now it follows from Table 2.3.2 that, for each fixed z, there exists a finite initial instant before which the incident as well as the scattered field vanishes. Hence, both fields may be regarded as causal signals to which Lerch's theorem applies.

As remarked above, we first consider the slab embedded in vacuum as analyzed in Section 2.3. Substituting the definitions (2.2.8) and (2.2.10) in the integral relation (2.2.9) and taking into account that the contrast ranges over the interval $0 < z < 1$ results in

$$E(z,s) = E^{i}(z,s) - \tfrac{1}{2}\int_0^1 [s\chi(z') + \sigma(z')] \times \exp(-s|z - z'|)E(z',s)dz'. \tag{3.3.1}$$

In (3.3.1) the factor of s can be associated with a time differentiation; the exponent can be identified with a time-delay factor. With these identifications and Lerch's theorem, (3.3.1) is directly carried over to the time domain as:

$$\mathcal{E}(z,t) = \mathcal{E}^{i}(z,t) - \tfrac{1}{2}\int_0^1 [\chi(z')\partial_t + \sigma(z')]\mathcal{E}(z',t')dz', \tag{3.3.2}$$

with $t' \triangleq t - |z - z'|$. The integral relation (3.3.2) holds for all z and t. For $0 \le z \le 1$ and $0 \le t < \infty$, it is an integral equation of the second kind for $E(z,t)$. As an illustration, Figure 3.3.1 shows the space-time points that occur in (3.3.2) for a specific choice of (z,t). From Figure 3.3.1, it is observed that this integral equation is indeed suited to the application of the marching-on-in-time method.

The second term on the right-hand side of (3.3.2) is identified with the reflected field $E^r(z,t)$ when $z < 0$, while the total field in $z > 1$ is identified with the transmitted field $E^t(z,t)$. From (3.3.2), we have the relations

$$\begin{aligned} E^r(z,t) &= E^r(0,t+z) && \text{for } z < 0, \\ E^t(z,t) &= E^t(1,t-(z-1)) && \text{for } z > 1, \end{aligned} \tag{3.3.3}$$

by which the field outside the slab can be obtained from the fields at the slab's interfaces. The relations (3.3.3) indicate that, when we restrict ourselves to observations outside the slab, all the retrievable information about the configuration must be present in the reflected and transmitted fields at those interfaces. As the final

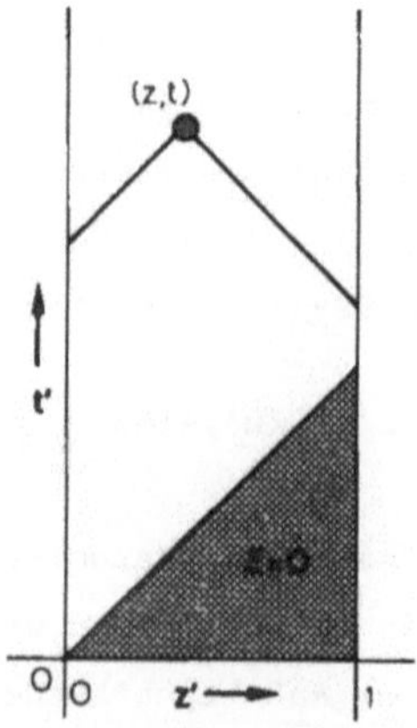

Figure 3.3.1 The space-time points that are relevant in the integral equation (3.3.2).

goal of our study is to retrieve such information, we will therefore, in the presentation of the numerical results, concentrate on those fields.

In the upcoming two subsections, we will now investigate the details of applying the marching-on-in-time method to the integral equation (3.3.2).

3.3.2 Slab in vacuum: discretization

In order to solve the integral equation (3.3.2) numerically, we discretize in space and time. When the dimensionless space step and the dimensionless time step are $h = \Delta t = 1/N$, with N being the number of space steps inside the slab, this discretization results in

$$\begin{aligned}\tilde{E}(m,n) &= \tilde{E}^{i}(m,n)\\ &- \tfrac{1}{2}\sum_{m'=0}^{N} w_{m'} \frac{\tilde{\chi}(m')}{2h}\left[\tfrac{3}{2}\tilde{E}(m',n') - 2\tilde{E}(m',n'-2) + \tfrac{1}{2}\tilde{E}(m',n'-4)\right]\\ &- \tfrac{1}{2}\sum_{m'=0}^{N} w_{m'}\, \tilde{\sigma}(m')\tilde{E}(m',n'),\end{aligned} \tag{3.3.4}$$

where $m = 0,1,2,\ldots,N$ and $n = 0,1,2,\ldots,\infty$. In (3.3.4), we have $n' \triangleq n - |m - m'|$, $\tilde{E}^{i}(m,n) \triangleq E^{i}(mh,n\Delta t)$, $\tilde{\chi}(m) \triangleq \chi(mh)$, $\tilde{\sigma}(m) \triangleq \sigma(mh)$, $w_m \triangleq h$ for $0 < m < N$ and $w_0 = w_N \triangleq h/2$. The spatial integral in (3.3.2) has been approximated by a repeated trapezoidal rule and the time derivative by a three-point backward interpolation formula with the double time step. As an illustration, the discrete space-time points that occur in (3.3.4) are indicated in Figure 3.3.2 for a specific choice of (m,n). This figure may be envisaged as a discretized version of Figure 3.3.1.

Note that, for the discretization (3.3.4), the assumptions made in connection with (3.2.8) turn out to be correct. In the right-hand side of (3.3.4), the field value at the point of observation $\tilde{E}(m,n)$ occurs with a weighting factor of

$$-\tfrac{1}{2}w_m[\tfrac{3}{4}\tilde{\chi}(m)/h + \tilde{\sigma}(m)], \tag{3.3.5}$$

which is negative and at most of $O(1)$ as $h \to 0$. Moreover, as Figure

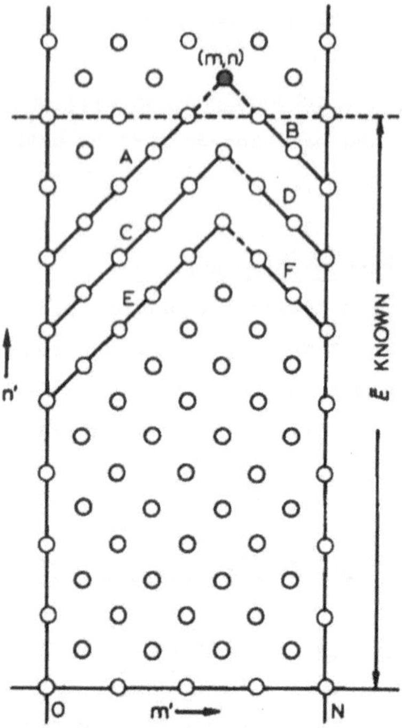

Figure 3.3.2 Space-time diagram of the discretized integral equations (3.3.4) and (3.3.8). The field at the point (m,n) can be computed from known values of $\tilde{E}(m',n')$ with $n' < n$. The solid skew lines connect the points that occur in each of the sums in (3.3.8): A: $S^1_{\sigma,\chi}(m-1,n-1)$; B: $S^2_{\sigma,\chi}(m,n)$; C: $S^1_{\chi}(m,n-2)$; D: $S^2_{\chi}(m,n-2)$; E: $S^1_{\chi}(m,n-4)$; F: $S^2_{\chi}(m,n-4)$.

3.3.2 shows, the right-hand side of (3.3.4) contains no other field values $\tilde{E}(m',n')$ with $n' = n$. Hence the weighting matrices corresponding to the discretization (3.3.4) indeed assume the form (3.2.7). In view of these properties, we may solve the discretized integral equation (3.3.4) with the marching-on-in-time method.

With the aid of the error estimates for the approximation of an integral by a repeated trapezoidal rule and for the approximation of a time derivative by a three-point backward interpolation formula (Hildebrand (1956), Stoer (1972)), it follows directly that the discretization meets criterion I. If, as in Bolomey et al. (1978) or Lesselier (1982), the time derivative were approximated by a two-point formula, the error in that approximation would be of $O(h)$, and, hence, criterion I would be violated. Now that we have checked criterion I, we only have to analyze the possible presence of an instability of the type discussed in Subsection 3.2.3. In the absence of such an instability, a smoothly varying incident field will result in a smoothly varying marching-on-in-time result that satisfies criterion II. Clearly, the integrand in the integral equation (3.3.2) is not singular at the point of observation. This rules out the first interpretation of (3.2.18). Therefore, we need only consider the second interpretation, where a number of errors in the field values $\tilde{E}(m',n')$ occurring on the right-hand side of (3.3.4) add up systematically. In order to show rigorously that this does not happen, we need additional information on the variation of

$$\Delta E(m',n') \overset{\Delta}{=} E(m'h,n'\Delta t) - \tilde{E}(m',n') \tag{3.3.6}$$

with m' and n'. Since such information is not available, we resort to the following intuitive argument. The error growth per updating step in the marching-on-in-time method is governed by Equation (3.2.12), in which ϕ and α can be recognized as E and m of the present problem. As Equation (3.2.15) shows, this error growth is incited by the discretization error $D(\alpha,n)$ defined in (3.2.11). Since the approximations leading to the discretization (3.3.4) are all based on interpolation in z and t, this discretization error consists entirely of such interpolation errors. Moreover, the distribution of these errors is determined by the higher-order space and time derivatives of the *exact* field occurring in the corresponding error estimates. Now $E(z,t)$ represents a pulsed wave which, due to repeated reflection at the slab's interfaces, travels backwards and forwards across the slab. For such a field, the characteristics of the traveling waves run in approximately the same direction as the skew lines in Figure 3.3.2. Hence, it may be imagined that the discretization error $D(\alpha,n)$ adds up along one of these lines. However, the differences $D(\alpha,n) - D(\alpha,n-1)$ will tend to average out along them. In view of the uniform magnitude of the weighting factors for the field values in (3.3.4), it therefore seems plausible that the errors $D(\alpha,n)$ will not cause an unwanted systematic behavior in the resulting $\Delta\phi(\alpha,n)$, i.e. $\Delta E(m,n)$. With this conclusion, the second possible type of instability discussed in Subsection 3.2.3 has also been ruled out, and criterion II is met.

The choice of the time interval of double length in the discretization of the time derivative in (3.3.4) was made with the numerical efficiency in mind, and not to eliminate instabilities as suggested in Section 3.2.3. This choice allows a restriction of the numerical computation to space-time points $(mh,n\Delta t)$ where $m + n$ is even. A further reduction in computation time can be achieved by using recurrence relations. Two types of recurrence relations can be distinguished. In Bolomey et al. (1978) and in Tijhuis (1981) recurrence relations were used between the approximate field values $\tilde{E}(m,n)$ and $\tilde{E}(m - 1,n - 1)$. From Figure 3.3.2, it is observed that the discretized integral equation (3.3.4) pertaining to these points contains a common contribution from space-time points with $m' \le m - 1$. This property is used to its full advantage when the computation is organized such that, systematically, the relevant field values are computed along lines where $m + n$ is fixed. For more details, the reader is referred to the publications cited above. An even more efficient scheme is obtained by decomposing the summation in (3.3.4) according to

$$\sum_{m'=0}^{N} = \sum_{m'=0}^{m} + \sum_{m'=m+1}^{N} \tag{3.3.7}$$

and separating off the terms containing $\tilde{E}(m,n)$. This results in

$$\begin{aligned}\tilde{E}(m,n) = \Big\{1 + \tfrac{1}{2}w_m[\tfrac{1}{4}\tilde{\chi}(m)/h + \tilde{\sigma}(m)]\Big\}^{-1} \Big\{\tilde{E}^i(m,n) &+ \tfrac{3}{2}[S_\chi^1(m - 1,n - 1) + S_\chi^2(m,n)] \\ &- 2[S_\chi^1(m,n - 2) + S_\chi^2(m,n - 2)] \\ &+ \tfrac{1}{2}[S_\chi^1(m,n - 4) + S_\chi^2(m,n - 4)] \\ &+ S_\sigma^1(m - 1,n - 1) + S_\sigma^2(m,n)\Big\},\end{aligned} \tag{3.3.8}$$

where

$$S_{\sigma}^{1}(m,n) = -\tfrac{1}{2}\sum_{m'=0}^{m} w_{m'}\tilde{\sigma}(m')\tilde{E}(m',n-m+m'),$$
$$S_{\sigma}^{2}(m,n) = -\tfrac{1}{2}\sum_{m'=m+1}^{N} w_{m'}\tilde{\sigma}(m')\tilde{E}(m',n+m-m'), \tag{3.3.9}$$

and where similar definitions hold for S_{χ}^{1} and S_{χ}^{2}. If $\tilde{E}(m',n')$ is known for $0 \le m' \le N$ and $0 \le n' < n$, $\tilde{E}(m,n)$ can be determined by computing the right-hand side of (3.3.8). In the computation of the field at the next instant, the sums S_{σ}^{1}, S_{σ}^{2}, S_{χ}^{1} and S_{χ}^{2} need not be evaluated again but can be obtained from the recurrence relations

$$S_{\sigma}^{1}(m,n) = S_{\sigma}^{1}(m-1,n-1) - \tfrac{1}{2}w_{m}\tilde{\sigma}(m)\tilde{E}(m,n),$$
$$S_{\sigma}^{2}(m-1,n+1) = S_{\sigma}^{2}(m,n) - \tfrac{1}{2}w_{m}\tilde{\sigma}(m)\tilde{E}(m,n), \tag{3.3.10}$$

and similar relations for S_{χ}^{1} and S_{χ}^{2}. At the slab's boundaries, we have the initial conditions

$$S_{\sigma}^{1}(-1,n) = S_{\chi}^{1}(-1,n) = 0,$$
$$S_{\sigma}^{2}(N,n) = S_{\chi}^{2}(N,n) = 0, \tag{3.3.11}$$

which complete the recurrence scheme. If recurrence relations of either type are used, the computation time for the numerical solution is proportional to N^2 as $N\to\infty$, while a direct evaluation of (3.3.4) leads to a computation time of order N^3.

3.3.3 Slab in vacuum: numerical results

Numerical results were obtained for several incident-pulse shapes $F(t)$ and various susceptibility and conductivity profiles. Numerical instabilities were not observed, even for rapid variations in the shape of the incident pulse and/or for small values of N, for which the results become inaccurate. As an illustration, we present in Figure 3.3.3 and in Table 3.3.1 results for a sine-squared pulse incident on a homogeneous, lossless slab. For that configuration, the

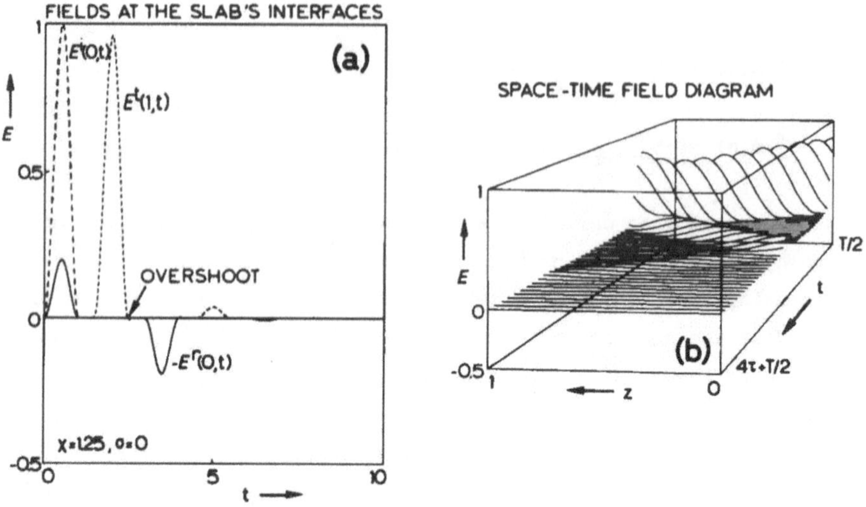

Figure 3.3.3 Results of applying the marching-on-in-time method to the scattering of a sine-squared incident pulse given by $F(t) = \sin^2(\pi t/T)\text{rect}(t - T/2;T)$ with $T = 1$ by a homogeneous, lossless dielectric slab with $\chi = 1.25$. (a): Incident, reflected and transmitted fields, $N = 160$; (b): three-dimensional space-time plot of the field inside the slab for 4 runs of the pulse through the slab, $N = 100$. The travel time for a single traverse through the slab is denoted by τ with, in this case, $\tau = 1.5$. In Figure (a), the arrow marks the location of the overshoot mentioned in the text.

Table 3.3.1 Computational data from the numerical solution of the scattering problem specified in Figure 3.3.3 for increasing N. Computation times 1 and 2 refer to the recurrence schemes described in Tijhuis (1981) and in Subsection 3.3.2. These computation times and the outgoing energy correspond to the time interval $0<t<20$.

	IP^{out}	error (%)	$E(1,\tau+T/2)$	$E(0,2\tau+T/2)$	CPU time 1	CPU time 2
exact	0.375	-	0.96	0.192	-	-
N=10	0.291	28.7	0.665	0.116	0.14s	0.03s
N=20	0.344	9.1	0.842	0.155	0.56s	0.12s
N=40	0.370	1.4	0.946	0.186	2.24s	0.50s
N=80	0.3743	0.18	0.958	0.1916	8.81s	1.96s
N=160	0.37492	0.02	0.9598	0.191993	35.6s	7.89s

solution is known in closed form in terms of repeatedly reflected and transmitted waves. In fact, it is obtained directly by substituting $\sigma = 0$ in the asymptotic expression (2.2.46) and the relevant quantities defined in (2.2.25) and (2.2.47). Figure 3.3.3a shows the computed reflected and transmitted fields at the ends of the slab while Figure 3.3.3b gives the field distribution inside the slab for the first four runs of the pulse across it. In Figure 3.3.3a, a slight overshoot is observed at the instant where the directly transmitted wave has completely emerged from the slab. This can be explained from the fact that at that space-time point, the derivative $\partial_t^3 E(z,t)$, which occurs in the error estimate for the discretization of the time derivative, becomes unbounded. We have not taken special measures to prevent this effect since it turns out to vanish as N increases. In Table 3.3.1, we provide results of an accuracy test. For a lossless medium ($\sigma = 0$), we have the identity $IP^{in} = IP^{out}$, with

$$\begin{aligned} IP^{in} &\triangleq \int_0^\infty P^{in}(t)dt = \int_0^\infty F^2(t)dt, \\ IP^{out} &\triangleq \int_0^\infty P^{out}(t)dt = \int_0^\infty [E^r(0,t)^2 + E^t(1,t)^2]dt, \end{aligned} \tag{3.3.12}$$

where IP^{in} and IP^{out} denote the total energy that, per unit surface, flows in and out of the slab, respectively. In Table 3.3.1, we have, for increasing N, checked this identity, compared the actual maximum values of $E(0,t)$ and $E(1,t)$ with the corresponding numerical values and listed the computation times for the numerical determination of $E(z,t)$ over a fixed time interval with both the recurrence schemes mentioned in the previous subsection. It is observed that the computation times are indeed proportional to N^2. Furthermore, the error in the computed fields decreases considerably faster than was predicted by the worst-case estimate of $O(N^{-1})$. Similar numerical experiments were carried out for the discretization used in Bolomey et al. (1978) and Lesselier (1982), where a two-point formula is employed to approximate the time derivative. It turns out that for incident pulses of short durations ($T \le 1$) computational problems arise that cannot be removed by increasing N.

Results for the scattering of the same incident pulse by an inhomogeneous slab are given in Figures 3.3.4 and 3.3.5. Figure 3.3.4 shows the reflected and transmitted fields at the slab's interfaces for a lossless slab with a linear and a quadratic susceptibility profile, respectively. In Figure 3.3.5, results are presented for a slab that has the same, constant permittivity as the one analyzed in Figure 3.3.3 and Table 3.3.1. In Figures 3.3.5a,b,c we had a linear, a quadratic, and a sine-squared conductivity profile, respectively. The contrasts of the slabs treated in Figures 3.3.3 - 3.3.5 were all chosen within the same range so that the influence of the constitutive parameters could be estimated by comparing the fields excited in the various configurations. Carrying out this comparison reveals that the influence of the configuration is noticed primarily in the reflected field at the front of the slab. In particular, this influence is centered in between the arrivals of the directly reflected pulses originating from the reflection of the primary wave at both the slab's interfaces. For moderate contrasts, it turns out

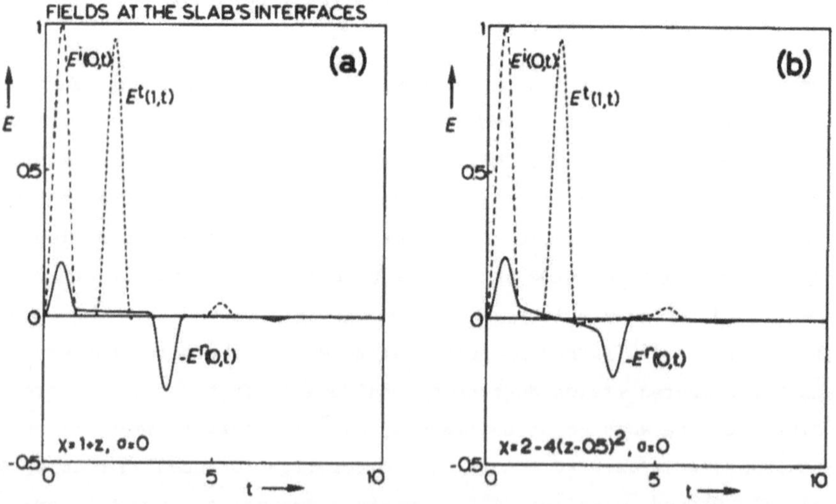

Figure 3.3.4 Results of repeating the computation specified in Figure 3.3.3a for two inhomogeneous lossless dielectric slabs. (a): linearly increasing profile; (b): parabolic profile.

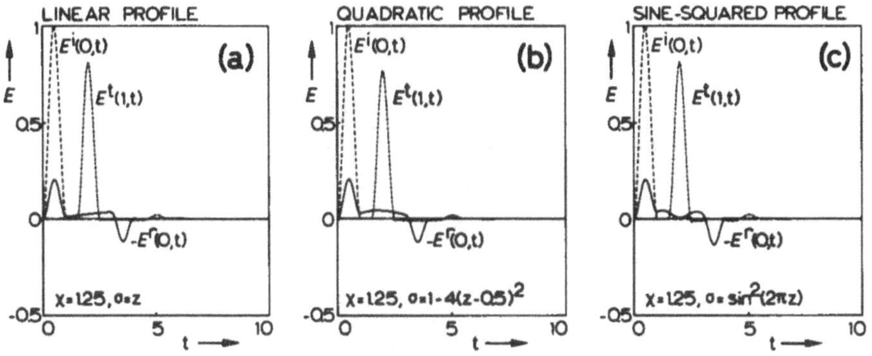

Figure 3.3.5 Results of repeating the computation specified in Figure 3.3.3a for three slabs with the same, constant susceptibility $\chi = 1.25$ and three different conductivity profiles as specified in the figures.

that the negative of the field in that time interval can be envisaged as a convolution of the incident pulse with a linear combination of $\sigma(z)$ and $d_z\chi(z)$. The direct reflections originating from both ends of the slab can be incorporated in this interpretation by understanding the derivative $d_z\chi(z)$ in the sense of generalized functions. For higher contrasts, distortion effects appear due to the fact that a pulse experiences a variation in the local wave speed as well as an attenuation, when traveling across the slab. In that case, the correspondence seems to be nonlinear. The correspondence signaled above forms the basis of the inverse-scattering methods investigated in Chapters 5 and 6. It will be analyzed in more detail in Sections 5.2 and 5.3.

Numerical results were also obtained for most of the configurations specified in Subsection 2.3.5. For these configurations, the marching-on-in-time method behaved as well as for the examples given above. The only problem that required special attention was discontinuity of the permittivity profiles specified in Figure 2.3.11. At these discontinuities, a minor adjustment in the definition of $\tilde{\chi}(m)$ was needed. In fact, for a discontinuity at $z = mh$, the repeated trapezoidal rule yields

$$\tilde{\chi}(m) = \lim_{\delta \downarrow 0}[\chi(mh + \delta) + \chi(mh - \delta)]/2, \tag{3.3.13}$$

which is in agreement with the proper principal-value definition. Furthermore, replacing the location of a discontinuity to the nearest multiple of h causes an error of $O(h)$ in the excited field. Since the marching-on-in-time error is of the same order of magnitude, combining such a replacement with the definition (3.3.13) provides an acceptable solution to the discontinuity problem. Results for the slabs specified in Figures 2.3.4 and 2.3.11b will be encountered in Subsection 4.3.3, where they are used as input data for a Prony-type calculation.

3.3.4 Slab in between half-spaces: formulation of the problem

In the second part of this section, we consider the dielectric slab situated in between two homogeneous, lossless dielectric half-spaces as specified in Subsection 2.4.1. It seems tempting to solve this problem by forming an integral equation of the type (3.3.2), in which a homogeneous, lossless dielectric with $\varepsilon = \varepsilon_1$ is treated as a reference medium. Such an integral equation is obtained directly from the frequency-domain integral relation (2.4.33), which holds for $z \le 1$. Substituting $C(z,s)$ as defined in (2.4.18) and $G_1(z,s)$ as defined in (2.4.31), and recognizing similar terms as in (3.3.1), we end up with

$$E(z,t) = E^i(z,t) - \frac{1}{2N_1}\int_0^1 \{[\varepsilon_r(z') - \varepsilon_{1r}]\partial_t + \sigma(z')\}E(z',t')dz' \\ + [(N_1 - N_3)/2N_1]E(1,t - N_1(1 - z)), \tag{3.3.14}$$

where $t' \triangleq t - N_1|z - z'|$, and where

$$E^i(z,t) = F(t - N_1 z). \tag{3.3.15}$$

As usual, $F(t)$ represents the time distribution of the incident field at $z = 0$. For $0 \le z \le 1$ and $0 \le t < \infty$, the relation (3.3.14) constitutes an integral equation of the second kind for $E(z,t)$. If it can be solved, the field outside the slab can be obtained directly from the relations

$$\begin{aligned} E^r(z,t) &= E^r(0,t + N_1 z) && \text{for } z < 0, \\ E^t(z,t) &= E^t(1,t - N_3(z - 1)) && \text{for } z > 1, \end{aligned} \qquad (3.3.16)$$

which follow directly by converting (2.4.4) to the time domain. The relations (3.3.16) should be regarded as the generalizations of the corresponding relations (3.3.3) for the case of the slab in vacuum. However, application of the marching-on-in-time method to the integral equation (3.3.14) clearly results in a solution for which the wave front propagates with a speed of at most $c_1 = c_0/N_1$. As a consequence, the solution obtained will be incorrect if the actual wave speed $c(z) = [\varepsilon(z)\mu_0]^{-\frac{1}{2}} = c_0/N(z)$ locally exceeds c_1, i.e. $\varepsilon_r(z) < \varepsilon_{1r}$ for some z. In fact, numerical experiments for such cases produced results that increased exponentially with increasing t.

The problem indicated above can be circumvented by choosing for the reference medium a lossless, piecewise-homogeneous three-layer medium with $\varepsilon(z) = \varepsilon_1, \varepsilon_0, \varepsilon_3$ for z in $\mathcal{D}_1$, $\mathcal{D}_2$ and $\mathcal{D}_3$, respectively. The relevant integral equation is obtained directly from the frequency-domain integral equation (2.4.20) and the results (2.4.25) - (2.4.29). Expanding $1/\bar{G}(s)$ as in (2.2.45), and recognizing factors of s and exponents as before, we arrive at:

$$\begin{aligned} E(z,t) = {} & \bar{E}(z,t) - \tfrac{1}{2}\int_0^1 J(z',t - |z - z'|)dz' \\ & - \tfrac{1}{2} R_{21}\sum_{k=0}^{\infty}(R_{21}R_{23})^k\int_0^1 J\big(z',t - [z'+z+2k]\big)dz' \\ & - \tfrac{1}{2} R_{23}\sum_{k=0}^{\infty}(R_{21}R_{23})^k\int_0^1 J\big(z',t - [(1-z')+(1-z)+2k]\big)dz' \\ & - \tfrac{1}{2} R_{21}R_{23}\sum_{k=0}^{\infty}(R_{21}R_{23})^k\int_0^1\Big\{J\big(z',t - [(1-z')+z+2k+1]\big) \\ & \qquad + J\big(z', t - [z'+(1-z)+2k+1]\big)\Big\}, \end{aligned} \qquad (3.3.17)$$

where $J(z,t)$ denotes the polarization current density

$$J(z,t) = [\chi(z)\partial_t + \sigma(z)]E(z,t). \tag{3.3.18}$$

According to (2.4.29), we have

$$\begin{aligned} R_{21} &= (1 - N_1)/(1 + N_1), \\ R_{23} &= (1 - N_3)/(1 + N_3), \end{aligned} \tag{3.3.19}$$

where R_{21} and R_{23} denote the plane-wave reflection coefficients for a boundary between vacuum and the corresponding medium. As an illustration, the space-time points that are relevant in the integral equation (3.3.17) are indicated in Figure 3.3.6, along with the relative weighting factors for the fields at those space-time points in terms of the reflection coefficients R_{21} and R_{23}. As specified in Subsection 2.4.4, the auxiliary field $\bar{E}(z,t)$ is the field that would result from the incident field in the reference medium. Expanding $1/\bar{G}(s)$ in (2.4.25) - (2.4.26) as in (2.2.45) directly results in

$$\bar{E}(z,t) = \begin{cases} F(t - N_1 z) + R_{12}F(t + N_1 z) \\ \qquad + T_{12}R_{23}T_{21}G(t + N_1 z - 2) & \text{in } \mathcal{D}_1, \\ T_{12}G(t - z) + T_{12}R_{23}G(t - (1-z) - 1) & \text{in } \mathcal{D}_2, \\ T_{12}T_{23}G(t - N_3(z-1) - 1) & \text{in } \mathcal{D}_3, \end{cases} \tag{3.3.20}$$

with

$$G(t) = \textstyle\sum_{k=0}^{\infty}(R_{21}R_{23})^k F(t - 2k). \tag{3.3.21}$$

In (3.3.20), we have, in addition to (3.3.19), the plane-wave reflection and transmission coefficients:

$$\begin{aligned} R_{12} &= (N_1 - 1)/(N_1 + 1), \\ T_{12} &= 2N_1/(1 + N_1), \\ T_{21} &= 2/(1 + N_1), \text{ and} \\ T_{23} &= 2/(1 + N_3), \end{aligned} \tag{3.3.22}$$

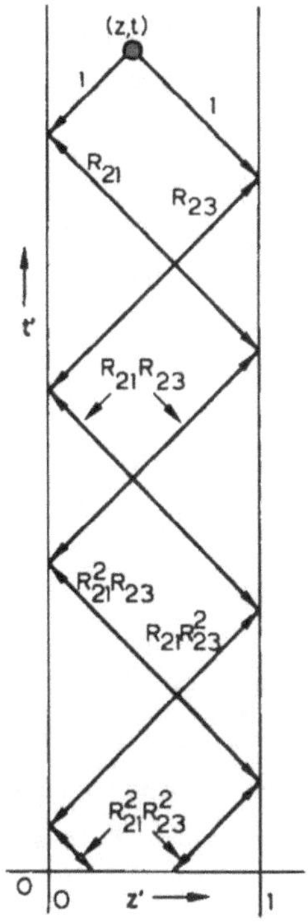

Figure 3.3.6 Space-time points relevant to the integral equation (3.3.17) and relative weighting factors in terms of the reflection coefficients R_{21} and R_{23}. On the zigzag lines, the arrows mark a change in weighting factor.

which also follow directly from (2.4.26) and (2.4.29). The integral equation that is obtained by restricting the space-time domain to $0 \le z \le 1$ and $0 \le t < \infty$ in (3.3.17) does not suffer from the same difficulties as the one obtained from (3.3.14). Hence it does allow the application of the marching-on-in-time method. In the next subsection, we consider this application in more detail.

3.3.5 Slab in between half-spaces: numerical aspects

Although the integral equation (3.3.17) seems complicated, solving it with the aid of the marching-on-in-time method is hardly more involved than solving the integral equation (3.3.2). As Figure 3.3.6

shows, the space-time points that are relevant in (3.3.17) are located on two zigzag lines that remain inside the domain $0 \leq z \leq 1$. Both lines start at the point of observation and have kinks at $z = 0$ and at $z = 1$. With each new kink, the weighting factor for the relevant field values is multiplied by the reflection coefficient at the corresponding slab interface. Redefining S_σ^1, S_σ^2, S_χ^1 and S_χ^2 as the sums along both zigzag lines, we can keep the recurrence relations (3.3.10) as they are. The reflection coefficients R_{21} and R_{23} are taken into account in the initial conditions of the type (3.3.11), which reduce to:

$$\begin{aligned} S_\sigma^1(-1,n-1) &= R_{21}\, S_\sigma^2(-1,n+1), \\ S_\sigma^2(N,n) &= R_{23}\, S_\sigma^1(N,n), \end{aligned} \qquad (3.3.23)$$

and similar conditions for S_χ^1 and S_χ^2. For $0 < m < N$, the unknown field $\tilde{E}(m,n)$ can then still be obtained directly from Equation (3.3.8). For $m = 0$ and $m = N$, a minor adjustment is necessary since, by (3.3.23), $\tilde{E}(m,n)$ also occurs in the sums $S_{\sigma,\chi}^1(-1,n-1)$ and $S_{\sigma,\chi}^2(N,n)$, respectively.

With the scheme outlined above, several numerical results were obtained. As could be expected, the errors in the solutions and the computation times are about the same as the ones found in Subsection 3.3.3 for the slab in vacuum. A representative example is shown in Figure 3.3.7, which shows results of recomputing the field specified in Figure 2.4.7b with the aid of the marching-on-in-time method. Figure 3.3.7a refers to the auxiliary field $\bar{E}(z,t)$, while Figure 3.3.7b presents the computed reflected and transmitted fields. The total computation time required to obtain the fields plotted in Figure 3.3.7b was 5.6 seconds, which compares favorably with the 19 seconds reported in Subsection 2.4.3.

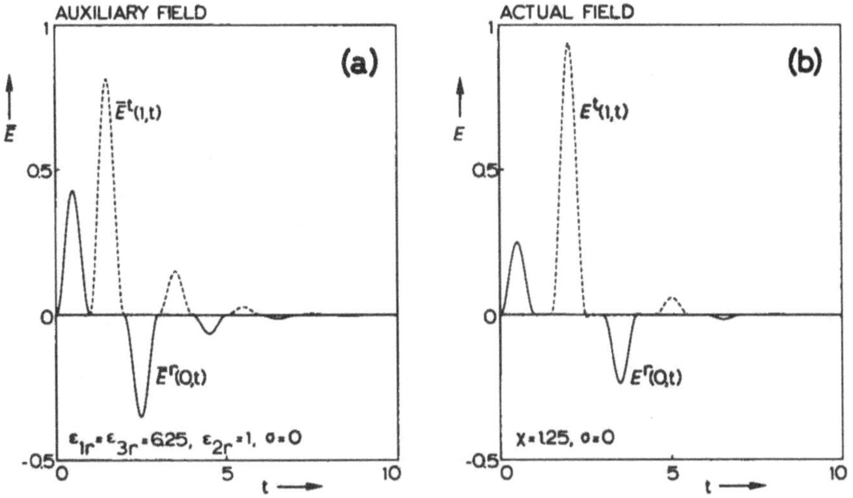

Figure 3.3.7 Results of recomputing the field specified in Figure 2.4.7b by applying the marching-on-in-time method with N = 160 to the integral equation (3.3.17). (a) Auxiliary field as specified in (3.3.20). (b) Reflected and transmitted fields obtained by applying a recursive marching-on-in-time scheme that uses the initial conditions (3.3.23).

3.4 Scattering by a perfectly conducting cylinder

3.4.1 Formulation of the problem

In this and the next section, we apply the marching-on-in-time method to two two-dimensional electromagnetic scattering problems. Both problems can be formulated in terms of a scalar integral equation pertaining to a single field component. In the present section, we consider a pulsed plane wave of finite duration T which is perpendicularly incident on a perfectly conducting cylinder (see Figure 3.4.1). In the next section, we consider the case where the pulse is E-polarized and the cylinder consists of a two-dimensionally inhomogeneous, lossy dielectric. In both these problems, the scattering obstacle is more complicated than the circularly symmetric ones investigated in Sections 2.5 and 2.6. Hence, the derivation of the relevant integral equations requires more effort than in the previous subsection, where the corresponding frequency-domain integral relations were directly available from Subsections 2.2.1 and 2.4.4. This problem is addressed in Appendices 3.4.A and 3.4.B, where space-time integral relations are derived.for two-dimensional electromagnetic fields. With the aid of these relations, integral equations can be formulated for a wide class of two-dimensional, electromagnetic transient scattering problems.

Returning to the problem at hand, let $\mathcal{D}_1$ and $\mathcal{D}_2$ be the exterior and the interior of the perfectly conducting cylinder, respectively. As shown in Figure 3.4.1, the boundary between these domains is a closed contour C, on which we have a normal vector $\underline{n}$ pointing into

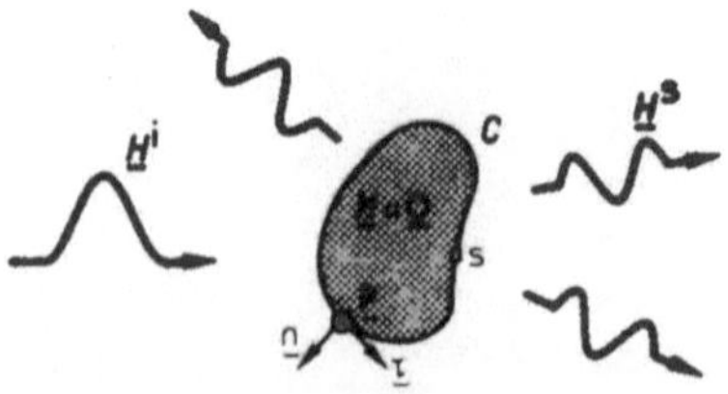

Figure 3.4.1 Scattering of a pulsed electromagnetic plane wave by a perfectly conducting cylinder.

$\mathcal{D}_1$ and a tangential vector $\underline{\tau}$ such that the three unit vectors $\{\underline{n}(\underline{\rho}),\underline{\tau}(\underline{\rho}),\underline{i}_z\}$ form a right-handed system. In addition, we assume that the cylinder has no sharp edges, i.e. $\underline{n}$ and $\underline{\tau}$ are continuous along C. In $\mathcal{D}_1$, we have a homogeneous, lossless dielectric with constant permittivity ε_1 and constant permeability μ_1. As mentioned above, the integral equation for this configuration can be obtained from the integral relations derived in Appendices 3.4.A and 3.4.B.

Let us first consider the scattered field in $\mathcal{D}_1$. The field intensities $\underline{E}^s(\underline{\rho},t)$ and $\underline{H}^s(\underline{\rho},t)$ satisfy the two-dimensional electromagnetic-field equations

$$\nabla_T \times \underline{H}^s(\underline{\rho},t) - \varepsilon_1\partial_t \underline{E}^s(\underline{\rho},t) = \underline{0} \qquad \text{when } \underline{\rho}\in\mathcal{D}_1, \tag{3.4.1a}$$

$$\nabla_T \times \underline{E}^s(\underline{\rho},t) + \mu_1\partial_t \underline{H}^s(\underline{\rho},t) = \underline{0} \qquad \text{when } \underline{\rho}\in\mathcal{D}_1. \tag{3.4.1b}$$

On account of this equation, (3.4.31) leads to

$$\begin{aligned} -\varepsilon_1\partial_t \underline{\Pi}_C^{m,s} + \mu_1^{-1}\int_{-\infty}^{t} dt'\nabla_T[\nabla_T\cdot\underline{\Pi}_C^{m,s}(\underline{\rho},t')] \\ + \nabla_T \times \underline{\Pi}_C^{e,s}(\underline{\rho},t) = \{-1, -\tfrac{1}{2}, 0\}\underline{H}^s(\underline{\rho},t) \end{aligned} \tag{3.4.2}$$

for $\underline{\rho}\in\{\mathcal{D}_1,C,\mathcal{D}_2\}$. In (3.4.2), the potentials $\underline{\Pi}$ are the ones defined in (3.4.32b), while the additional superscript s reflects the fact that these potentials now pertain to the scattered field. Obviously, the wave speed c should be taken equal to $c_1 = (\varepsilon_1\mu_1)^{-\frac{1}{2}}$. Finally, the minus sign in (3.4.2) originates from the fact that $\underline{n}(\underline{\rho})$ points into $\mathcal{D}_1$. In writing down (3.4.2), we have applied the integral relation (3.4.31) to an infinite domain. Strictly speaking, this is not allowed since this relation was derived for a bounded domain. However, we can also arrive at (3.4.2) by considering the bounded domain between C and a circular contour with a large radius ρ_∞. Using the property that the scattered field vanishes when $t \le 0$, we can then make the contribution from the latter contour vanish by choosing a sufficiently large radius ρ_∞.

Next, we consider the incident field in $\mathcal{D}_2$. Since, by definition, the incident field is the field that would be present in the absence of the scatterer, we have

$$\nabla_T \times \underline{H}^i(\underline{\rho},t) - \varepsilon_1 \partial_t \underline{E}^i(\underline{\rho},t) = \underline{0} \qquad \text{when } \underline{\rho}\in\mathcal{D}_2, \tag{3.4.3a}$$

$$\nabla_T \times \underline{E}^i(\underline{\rho},t) + \mu_1 \partial_t \underline{H}^i(\underline{\rho},t) = \underline{0} \qquad \text{when } \underline{\rho}\in\mathcal{D}_2. \tag{3.4.3b}$$

In (3.4.3), the right-hand sides of both equations contain no source terms since the sources of the incident field are located in $\mathcal{D}_1$. Equation (3.4.31) then leads to

$$\begin{aligned} -\varepsilon_1 \partial_t \underline{\Pi}_C^{m,i}(\underline{\rho},t) + \mu_1^{-1} \int_{-\infty}^{t} dt' \; \nabla_T [\nabla_T \cdot \underline{\Pi}_C^{m,i}(\underline{\rho},t')] \\ + \nabla_T \times \underline{\Pi}_C^{e,i}(\underline{\rho},t) = \{0,\tfrac{1}{2},1\}\underline{H}^i(\underline{\rho},t). \end{aligned} \tag{3.4.4}$$

In (3.4.4), the potentials refer to the incident field.

Finally, we make use of the fact that the cylinder is perfectly conducting and, hence, electrically impenetrable. This provides us with the boundary condition

$$\underline{J}_C^{m,i}(\underline{\rho},t) + \underline{J}_C^{m,s}(\underline{\rho},t) = \underline{J}_C^{m}(\underline{\rho},t) = \underline{n}(\underline{\rho}) \times \underline{E}(\underline{\rho},t) = \underline{0} \tag{3.4.5}$$

when $\underline{\rho}\in\mathcal{C}$. This condition shows that we can obtain an integral equation involving only the magnetic-field strength $\underline{H}(\underline{\rho},t)$ by adding up Equations (3.4.2) and (3.4.4). We end up with

$$\underline{H}^i(\underline{\rho},t) - \nabla_T \times \underline{\Pi}_C^{e}(\underline{\rho},t) = \{1,\tfrac{1}{2},0\}\underline{H}(\underline{\rho},t) \tag{3.4.6}$$

for $\underline{\rho}\in\{\mathcal{D}_1,\mathcal{C},\mathcal{D}_2\}$. For $\underline{\rho}\in\mathcal{C}$ and $0 \le t < \infty$, the relation (3.4.6) constitutes an integral equation of the second kind for the unknown field $\underline{H}(\underline{\rho},t)$. Once this field is known, the field outside the cylinder is obtained directly by taking $\underline{\rho}\in\mathcal{D}_1$ in (3.4.6). In principle, the formulation of the problem has therefore been completed.

Nevertheless, we are still faced with the task of breaking (3.4.6) down to a form that is suitable for the marching-on-in-time method. To this end, we first carry out the curl operation. The most direct way to achieve this is to use the equivalence of the integral relations (3.4.27) and (3.4.31). With the aid of the correspondence (3.4.47), we immediately find

$$-\nabla_T \times \underline{\Pi}^e_C(\underline{\rho},t) = \frac{1}{2\pi}\oint_C ds(\underline{\rho}')\,\frac{\underline{i}_R}{R}\times\int_0^{t-R/c_1} dt'\,\frac{t-t'}{[(t-t')^2 - R^2/c_1^2]^{\frac{1}{2}}}\,\partial_{t'}\underline{J}^e(\underline{\rho}',t'), \tag{3.4.7}$$

where, as in Appendix 3.4.A, we have

$$R = |\underline{\rho} - \underline{\rho}'|,$$
$$\underline{i}_R = (\underline{\rho} - \underline{\rho}')/R, \quad \text{and} \tag{3.4.8}$$
$$\underline{J}^e(\underline{\rho},t) = -\underline{n}(\underline{\rho}) \times \underline{H}(\underline{\rho},t).$$

Substitution of (3.4.7) and use of the definition of $\underline{J}^e(\underline{\rho},t)$ in (3.4.6) then results in two scalar integral equations for the tangential components of $\underline{H}(\underline{\rho},t)$:

$$H_z(\underline{\rho},t) = 2H_z^i(\underline{\rho},t) + \frac{1}{\pi}\oint_C ds(\underline{\rho}')\,\frac{(\underline{i}_R\cdot\underline{n}(\underline{\rho}'))}{R}\,I_z(\underline{\rho}';t,R), \tag{3.4.9a}$$

$$H_\tau(\underline{\rho},t) = 2H_\tau^i(\underline{\rho},t) + \frac{1}{\pi}\oint_C ds(\underline{\rho}')\,\frac{(\underline{i}_R\cdot\underline{n}(\underline{\rho}))}{R}\,I_\tau(\underline{\rho}';t,R), \tag{3.4.9b}$$

with

$$I_{z,\tau}(\underline{\rho}';t,R) = \int_0^{t-R/c_1} dt'\,\frac{t-t'}{[(t-t')^2 - R^2/c_1^2]^{\frac{1}{2}}} \times \partial_{t'}H_{z,\tau}(\underline{\rho}',t'), \tag{3.4.10}$$

which hold when $\underline{\rho}\in C$.

It should be remarked that the integral equations (3.4.9) differ from the ones that have been used conventionally in the literature to resolve the present problem (Bennett (1968), Bennett and Weeks (1970)) and similar acoustic scattering problems (Friedman and Shaw (1962), Shaw (1967, 1968, 1975)). These authors use a form that is obtained by carrying out the curl operation in (3.4.6) directly rather than via the correspondence (3.4.47). This results in

$$-\nabla_T \times \underline{\Pi}_C^e(\underline{\rho},t) =$$

$$\frac{1}{2\pi}\oint_C dS(\underline{\rho}')\,\frac{\underline{i}_R}{c_1}\int_0^{t-R/c_1}\frac{dt'}{[(t-t')^2 - R^2/c_1^2]^{\frac{1}{2}}} \times \left\{\frac{\underline{J}(\underline{\rho}',t')}{t - t' + R/c_1} + \partial_{t'}\,\underline{J}(\underline{\rho}',t')\right\}, \tag{3.4.11}$$

which can be handled in a similar manner as the expression in (3.4.7). The equivalence between both forms can be shown by integrating by parts the first term on the right-hand sice of (3.4.11) (see also Tijhuis (1984a,b)). The advantages of (3.4.9) over the corresponding form lie in the numerical implementation. The numerical evaluation of the time integral in (3.4.10) requires only half the effort, while the property

$$\lim_{\underline{\rho}'\to\underline{\rho}} I_{z,\tau}(\underline{\rho}';t,R) = H_{z,\tau}(\underline{\rho},t) \tag{3.4.12}$$

facilitates the subsequent determination of the boundary integrals in (3.4.9). In the next subsection, the numerical implementation will be discussed in more detail.

3.4.2 Numerical implementation

As has been remarked before, it turns out to be favorable to carry out the numerical computations in a dimensionless coordinate system. For the present problem, such a normalization can be achieved by normalizing all space coordinates with respect to some characteristic length a and all times with respect to the corresponding travel time in the outer medium a/c_1. This normalization is similar to the one applied in Subsection 2.5.1, the only difference being the use of the wave speed c_1 instead of the vacuum speed c_0. Its only effect on (3.4.9) - (3.4.10) is that, in these equations, the wave speed c_1 is replaced by its dimensionless equivalent, i.e. one.

Since both Equations (3.4.9a) and (3.4.9b) are solved by the same procedure, we restrict the discussion of the discretization to the case where the magnetic field has only a z-component (H-polarization). Following Bennett (1968), we restrict the space points to the finite set $\{\underline{\rho}_m\}$, m = 1,...,N. The path length between two neighboring points is $h = L/N$, with L the total length of the contour C. The boundary integral along this contour is approximated by a repeated rectangular rule. As a function of the path length $S(\underline{\rho}')$, the integrand of (3.4.9a) is a periodic function. Therefore, it follows from discrete Fourier theory that the approximation error is proportional to the N'th Fourier coefficient in an expansion of the type (2.5.4). Moreover, the integrand of (3.4.9a) has an integrable derivative. Consequently, this Fourier coefficient will be of $O(N^{-2})$ or, equivalently, of $O(h^2)$. This leaves us with the determination of the integrand to the same order of accuracy. For a specific obstacle, we assume the factor $(\underline{i}_R \cdot \underline{n}(\underline{\rho}'))/R$ to be known for $R > 0$, while for $R \to 0$, we have

$$(\underline{i}_R \cdot \underline{n}(\underline{\rho}'))/R \to -[2a(\underline{\rho})]^{-1}, \tag{3.4.13}$$

with $a(\underline{\rho})$ being the radius of curvature at the point of observation. (see Bennett (1968)). To determine the time integral for $R > 0$, we restrict the dimensionless distance R to multiples of the dimensionless time step Δt (note that this step is smaller than the dimensionless arc-length step h). For $t = n\Delta t$ and $R = k\Delta t$, the dimensionless version of (3.4.10) reduces to

$$I_z(\underline{\rho}';n\Delta t,k\Delta t) = \sum_{n'=0}^{n-k-1} \int_{n'\Delta t}^{(n'+1)\Delta t} dt' \frac{n\Delta t - t'}{[(n\Delta t-t')^2 - k^2\Delta t^2]^{\frac{1}{2}}} \times \partial_{t'} H_z(\underline{\rho}',t'). \tag{3.4.14}$$

The integrals over the subintervals $n'\Delta t < t' < (n'+1)\Delta t$ are obtained by approximating $H_z(\underline{\rho}',t')$ by a quadratic interpolation polynomial in t', also using the field value at $t' = (n'-1)\Delta t$. The time differentiation and the subsequent integration are carried out analytically. By considering the interpolation error (see Hildebrand (1956), Stoer (1972)), it can be shown that this procedure leads to an approximation of order h^2 of the time integral $I_z(\underline{\rho}_m;n\Delta t,k\Delta t)$. The integral $I_z(\underline{\rho}_{m'};n\Delta t,R_{mm'})$, which is required in the discretized contour integral, is obtained from the values at $R = k\Delta t$ by linear interpolation in R. For small R, this may not seem justified since the approximation error is proportional to $h^2\partial_R^2 I_z$, with $\partial_R^2 I_z$ logarithmically singular at R = 0. However, the interpolation only needs to be carried out for $R > \Delta t$, and in that region the interpolation error is of order $h^2\ln(h)$. Hence, the total error in the discretization is also of order $h^2\ln(h)$. Since the analysis presented in Subsection 3.2.2 can directly be generalized to include the factor of ln(h) in the discretization error, we may consider criterion I to be met.

Furthermore, we need not have problems in the inversion of the weighting matrix as described in (3.2.3) - (3.2.6). From the description of the discretization given above, it follows that the weighting matrices are of the form given in (3.2.7). Moreover, with (3.4.12) and (3.4.13), we find

$$\tilde{K}(m,m;0) = -h/2\pi a(\underline{\rho}_m), \tag{3.4.15}$$

where $\tilde{K}$ is the discretized operator specified in Subsection 3.2.1. Hence, we can always choose h sufficiently small for the inverse weighting matrix in (3.2.7b) to be well defined.

As in the case of the dielectric slab, the complicated forms of the integrals in (3.4.9) - (3.4.10) make it hard to show that criterion II holds for the corresponding discretized form. Clearly, the regularity of the integrand as $R\to0$ rules out the first interpretation of (3.2.18). This leaves us again with the second interpretation, in which a number of errors in the marching-on-in-time solution obtained add up systematically. Hence, we can revert to the intuitive argument given in Subsection 3.3.2, which predicts instabilities only when the exact field $H_z(\underline{\rho},t)$ exhibits a systematic behavior in some nonvanishing region of the space-time domain.

A special feature of the present discretization is that all the weighting factors $\tilde{K}(m,m';n - n')$ in the discretized integral equation are of $O(1)$ as $h\to0$. This can be seen as follows. The time differentiation in (3.4.10) provides a factor of $O(h^{-1})$ and the space integration in (3.4.9a) a factor of $O(h)$. As the resulting product is of $O(1)$, the factors K(m,m';n - n') will be of the same order of magnitude as the integral of the function $(t - t')/[(t - t')^2 - R^2/c_1^2]^{\frac{1}{2}}$ in (3.4.10) over a subinterval of length Δt. This leads to two different estimates: Near the end of the integration interval we find

$$\tilde{K}(m,m';n - n') = \mathcal{O}(h^{\frac{1}{2}})\mathcal{O}(R_{mm'}^{\frac{1}{2}}) \tag{3.4.16a}$$

because of the root-like singularity at $t' = t - R$. In the remainder of the integration interval, we have

$$\tilde{K}(m,m';n - n') = \mathcal{O}(h). \tag{3.4.16b}$$

For space-time points neighboring the point of observation, both estimates coincide, since for those points $R_{mm'} = \mathcal{O}(h)$. In view of these estimates, it seems plausible that, even for a systematic error, the parameters w_0 and w_1 in the corresponding difference model (3.2.18) decrease when the mesh size h is reduced. As discussed in the second part of Subsection 3.2.3, this would make it possible to eliminate the instability by refining the discretization.

3.4.3 Numerical results

To investigate how the considerations of the previous subsection work out in a practical scattering problem, we have applied the marching-on-in-time method to the transient scattering by a perfectly conducting circular cylinder. In Figure 3.4.2, results are presented for illumination by an H-polarized, sine-squared incident pulse of duration $T = 2$ for 32 and 64 points on the integration contour. Apparently, the field $H_z(\underline{\rho},t)$ on the cylinder boundary exhibits no systematic behavior and, consequently, an accurate, stable solution is obtained when the mesh size h is taken small enough. The accuracy of the results shown in Figure 3.4.2b could be verified since, for this special configuration, the frequency-domain solution can be obtained by carrying out the same decomposition as in Subsections 2.5.1 and 2.5.2 (see Jones (1964), pp. 450-452). The results of applying a direct Fourier transformation to that solution agreed

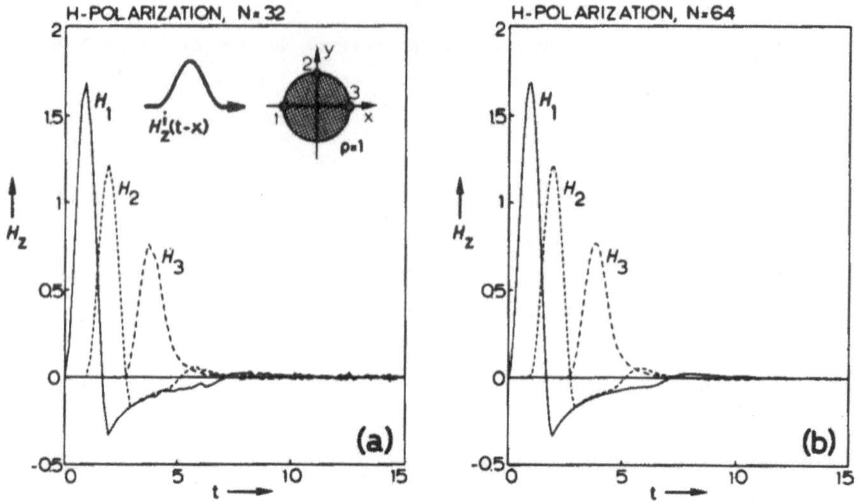

Figure 3.4.2 H_z as a function of normalized time for a perfectly conducting cylinder illuminated by an H-polarized incident pulse with $H_z^i(\underline{\rho},t) = F(t - x - 1)$, where $F(t) = \sin^2(\pi t/T)\mathrm{rect}(t - T/2;T)$ with $T = 2$. Figures (a) and (b) show marching-on-in-time results computed with N = 32 and N = 64, respectively.

well with the ones displayed in Figure 3.4.2b.

Results for an E-polarized incident pulse of the same duration are shown in Figure 3.4.3. In the discretization, the same approximations were used as for the case of H-polarization, the only difference being that replacing $\underline{n}(\underline{\rho}')$ with $\underline{n}(\underline{\rho})$ in (3.4.13) cancels the minus sign on the right-hand side. Now, the tangential field $H_\tau(\underline{\rho},t)$ does show a systematic behavior since, for large t, it turns out to be constant around the contour. Obviously, the incident field excites a "stationary" current propagating in the z-direction and a corresponding "stationary" magnetic field around the cylinder. This current may be considered as the limiting case of the stationary behavior for the penetrable cylinder, as observed in Figures 2.5.6, 2.5.7 and 2.5.10 and interpreted in the second part of Subsection 2.5.5. Because of this behavior, the marching-on-in-time solution shown in Figure 3.4.3 exhibits a systematic instability whose start coincides in time with the onset of the "stationary" field. Compar-

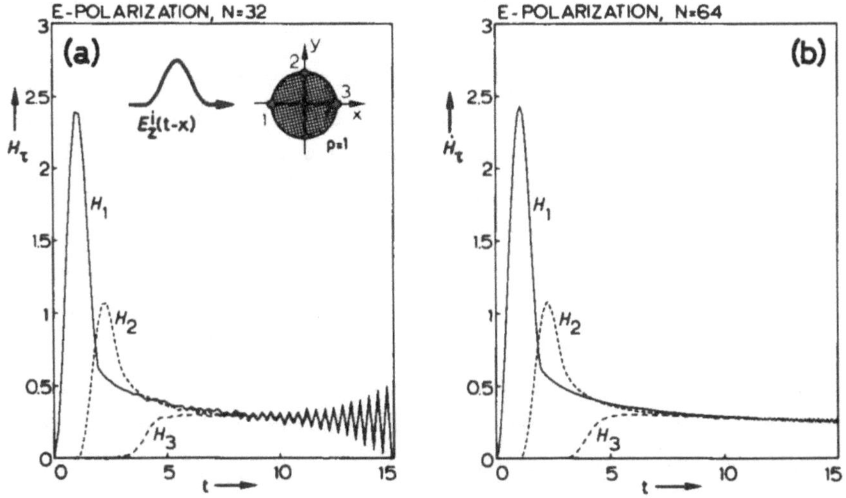

Figure 3.4.3 H_τ as a function of time for a perfectly conducting cylinder illuminated by the pulse specified in (2.5.1) for $F(t) = \sin^2(\pi t/T)\mathrm{rect}(t - T/2;T)$ with $T = 2$. Figures (a) and (b) show marching-on-in-time results of computing H_τ with $N = 32$ and $N = 64$, respectively.

ing Figures 3.4.3a and 3.4.3b confirms the conjecture made at the end of Subsection 3.4.2 that even systematic instabilities may be controllable via the mesh size h. The solution to the special instability problem that crops up in Figure 3.4.3 is obvious: the problem formulation should be revised such that the quantity to be computed by the marching-on-in-time method does not exhibit the systematic behavior. This can be effectuated by computing $\partial_t H_\tau(\underline{\rho},t)$ rather than $H_\tau(\underline{\rho},t)$ itself. The relevant integral equation is obtained from (3.4.9b) when both the incident field $H_\tau^i(\underline{\rho},t)$ and the total field $H_\tau(\underline{\rho},t)$ are replaced by their time derivatives. This follows directly from the correspondence between the last terms on the right-hand sides of (3.4.27) and (3.4.31). The field $H_\tau(\underline{\rho},t)$ can then be computed from the $\partial_t \tilde{H}_\tau(m,n)$ obtained via the repeated trapezoidal rule

$$\tilde{H}_\tau(m,n) \triangleq \sum_{n'=0}^{n-1} \Delta t[\partial_t \tilde{H}_\tau(m,n') + \partial_t \tilde{H}_\tau(m,n'+1)]/2. \qquad (3.4.17)$$

This has the additional advantage that, in (3.4.17), any alternating behavior that may be present in the numerically obtained $\partial_t\tilde{H}_\tau(m,n)$ is smoothed out by the repeated combination of two values at subsequent instants as well as reduced by a factor of Δt. The effectiveness of this procedure is illustrated in Figure 3.4.4, which shows the results of recomputing the signals specified in Figure 3.4.3 in the manner explained above. Now we only have a "global" error in the level of $H_\tau(\underline{\rho},t)$ for large t, which can be handled by increasing N.

A possible disadvantage of the marching-on-in-time method is the sharp increase in computation time upon reduction of h. This may make it very time-consuming to compute the field excited by a short incident pulse. For a computation as described in this and the previous section, we must evaluate, at $\mathcal{O}(N)$ instants, at N space points, $\mathcal{O}(N)$ different time integrals, each of which requires a weighted summation of $\mathcal{O}(N)$ field values. As a result, the number of opera-

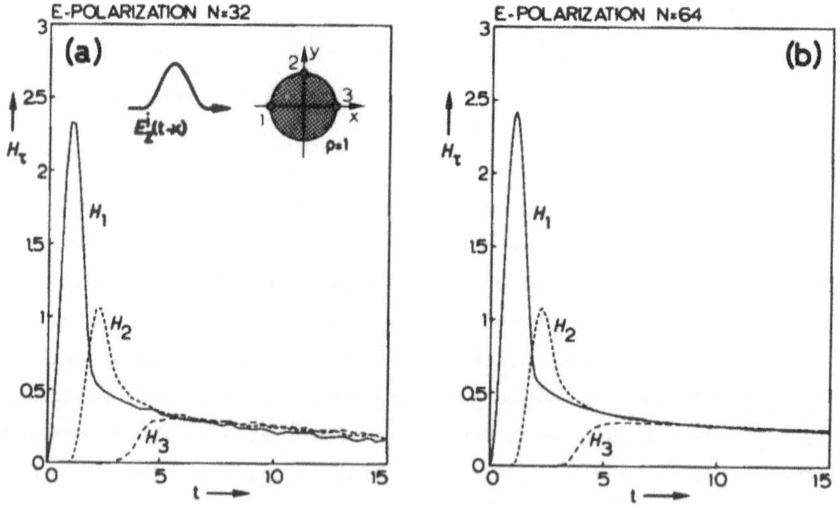

Figure 3.4.4 Results of recomputing the field specified in Figure 3.4.3 by applying the marching-on-in-time method to determine $\partial_t H_\tau(\underline{\rho},t)$ and, subsequently, a numerical time integration to obtain $H_\tau(\underline{\rho},t)$.

tions is approximately proportional to N^4 as N becomes large. For the fields shown in Figures 3.4.2 - 3.4.4, the computation times were 0.38, 4.4 and 61 seconds for N = 16, 32 and 64, respectively.

Appendix 3.4.A Source representations for two-dimensional electromagnetic-field quantities

The basic tool in the integral-equation formulation of scattering problems is a certain integral relation that, for points of observation located in a certain domain in space, leads to a source-type integral representation for the electromagnetic-field quantities. De Hoop (1977) discussed integral relations of this kind for three-dimensional frequency-domain problems. From his results, the corresponding time-domain relations are obtained by inspection. In this appendix, we carry De Hoop's derivation over to two-dimensional time-domain problems involving electromagnetic fields.

We start with the situation where the electromagnetic-field quantities are defined in a bounded subdomain $\mathcal{D}_2$ of $\mathbb{R}^2$. The boundary of $\mathcal{D}_2$ is the closed contour C (Figure 3.4.5). C is assumed to be sufficiently regular, i.e. the unit vector $\underline{n}$ along its outward normal is a piecewise-continuous vector function of position. The unbounded domain exterior to C is called $\mathcal{D}_1$. In $\mathcal{D}_2$, we assume an electric-current density $\underline{J}^e_D(\underline{\rho},t)$ and a magnetic-current density $\underline{J}^m_D(\underline{\rho},t)$ to be

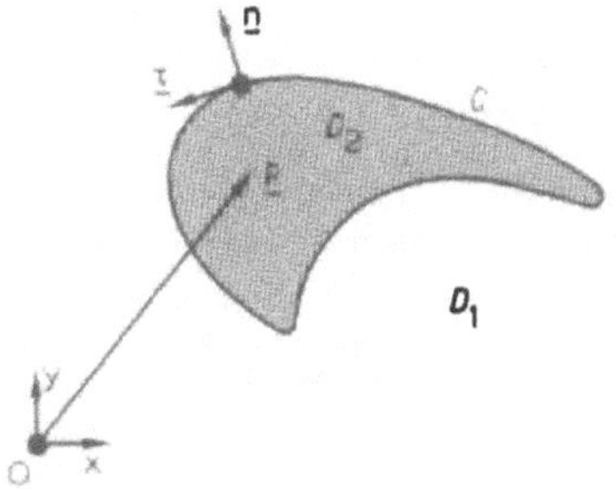

Figure 3.4.5 The bounded domain $\mathcal{D}_2$ interior to the closed contour C, for which source-type integral relations for two-dimensional electromagnetic fields are derived, and an arbitrary position vector $\underline{\rho}$.

present in a homogeneous, lossless background medium with constant permittivity ε and constant permeability μ. The current densities appear as source terms in the electromagnetic-field equations, which are written as

$$\nabla_T \times \underline{H}(\underline{\rho},t) - \varepsilon\partial_t \underline{E}(\underline{\rho},t) = \underline{J}_D^e(\underline{\rho},t), \tag{3.4.18a}$$

$$\nabla_T \times \underline{E}(\underline{\rho},t) + \mu\partial_t \underline{H}(\underline{\rho},t) = -\underline{J}_D^m(\underline{\rho},t), \tag{3.4.18b}$$

with

$$\begin{aligned} \underline{\rho} &= \underline{i}_x x + \underline{i}_y y, \\ \nabla_T &= \underline{i}_x \partial_x + \underline{i}_y \partial_y. \end{aligned} \tag{3.4.19}$$

Next, we transform the system of equations (3.4.19) in two steps into an equivalent system of algebraic equations. First, a temporal Laplace transformation reduces (3.4.18) to

$$\nabla_T \times \underline{H}(\underline{\rho},s) - s\varepsilon\, \underline{E}(\underline{\rho},s) = \underline{J}_D^e(\underline{\rho},s), \tag{3.4.20a}$$

$$\nabla_T \times \underline{H}(\underline{\rho},s) + s\mu\, \underline{E}(\underline{\rho},s) = -\underline{J}_D^m(\underline{\rho},s). \tag{3.4.20b}$$

Next, we define the spatial Fourier transform $\underline{\tilde{V}}(\underline{\kappa})$ of a vector field $\underline{V}(\underline{\rho})$ over the domain $\mathcal{D}_2$ as

$$\underline{\tilde{V}}(\underline{\kappa}) \triangleq \iint_{\mathcal{D}_2} dA(\underline{\rho}) \exp(-i\underline{\kappa}\cdot\underline{\rho})\underline{V}(\underline{\rho}), \tag{3.4.21}$$

where $\underline{\kappa} = k_x\underline{i}_x + k_y\underline{i}_y$. In order to apply this transformation to (3.4.20), we further need the corresponding transform of $\nabla_T \times \underline{V}(\underline{\rho})$. With the aid of Gauss' theorem, we find

$$(\nabla_T \tilde{\times} \underline{V}) = i\underline{\kappa} \times \underline{\tilde{V}}(\underline{\kappa}) + \oint_C dS(\underline{\rho}) \exp(-i\underline{\kappa}\cdot\underline{\rho})[\underline{n}(\underline{\rho}) \times \underline{V}(\underline{\rho})]. \tag{3.4.22}$$

The second term on the right-hand side of (3.4.22) is just the spatial Fourier transform of the vector function $\underline{n}(\underline{\rho}) \times \underline{V}(\underline{\rho})$ over its domain of definition $\mathcal{C}$. Using (3.4.20) and (3.4.22), we obtain the following equations for $\underline{\tilde{H}}(\underline{\kappa},s)$ and $\underline{\tilde{E}}(\underline{\kappa},s)$:

$$i\underline{\kappa} \times \underline{\tilde{H}} - s\varepsilon\, \underline{\tilde{E}} = \underline{\tilde{J}}^e_D + \underline{\tilde{J}}^e_C, \tag{3.4.23a}$$

$$i\underline{\kappa} \times \underline{\tilde{E}} + s\mu\, \underline{\tilde{H}} = -\underline{\tilde{J}}^m_D - \underline{\tilde{J}}^m_C, \tag{3.4.23b}$$

in which $\underline{\tilde{J}}^e_C$ and $\underline{\tilde{J}}^m_C$ are the spatial Fourier transforms over $\mathcal{C}$ of the quantities

$$\underline{J}^e_C(\underline{\rho},s) \stackrel{\Delta}{=} -\underline{n}(\underline{\rho}) \times \underline{H}(\underline{\rho},s) \quad \text{when } \underline{\rho}\in\mathcal{C}, \tag{3.4.24}$$

and

$$\underline{J}^m_C(\underline{\rho},s) \stackrel{\Delta}{=} \underline{n}(\underline{\rho}) \times \underline{E}(\underline{\rho},s) \quad \text{when } \underline{\rho}\in\mathcal{C}, \tag{3.4.25}$$

respectively. Note that in the right-hand sides of (3.4.24) and (3.4.25) the limiting values of the quantities upon approaching $\mathcal{C}$ via $\mathcal{D}_2$ are to be taken. The structure of (3.4.23) leads to the interpretation that $\underline{J}^e_C$ and $\underline{J}^m_C$ are the Laplace-transformed source densities of the electric and the magnetic surface-currents. With (3.4.23), we now have arrived at the desired system of algebraic equations. Solving them yields

$$\begin{aligned}\underline{\tilde{E}} = (\underline{\kappa}\cdot\underline{\kappa} + s^2\varepsilon\mu)^{-1}\{&-s\mu(\underline{\tilde{J}}^e_D + \underline{\tilde{J}}^e_C) \\ &+ (s\varepsilon)^{-1}i\underline{\kappa}[i\underline{\kappa}\cdot(\underline{\tilde{J}}^e_D + \underline{\tilde{J}}^e_C)]-i\underline{\kappa}\times(\underline{\tilde{J}}^m_D + \underline{\tilde{J}}^m_C)\},\end{aligned} \tag{3.4.26}$$

and

$$\begin{aligned}\underline{\tilde{H}} = (\underline{\kappa}\cdot\underline{\kappa} + s^2\varepsilon\mu)^{-1}\{&-s\varepsilon(\underline{\tilde{J}}^m_D + \underline{\tilde{J}}^m_C) \\ &+ (s\mu)^{-1}i\underline{\kappa}[i\underline{\kappa}\cdot(\underline{\tilde{J}}^m_D + \underline{\tilde{J}}^m_C)] + i\underline{\kappa}\times(\underline{\tilde{J}}^e_D + \underline{\tilde{J}}^e_C)\}.\end{aligned} \tag{3.4.27}$$

In Appendix 3.4.B it will be shown that the factor

$$\tilde{G}(\underline{\kappa},s) \triangleq (\underline{\kappa}\cdot\underline{\kappa} + s^2/c^2)^{-1}, \tag{3.4.28}$$

with $c = (\varepsilon\mu)^{-\frac{1}{2}}$ being the wave speed in the background medium, corresponds to the two-dimensional space-time domain Green's function

$$G(\underline{\rho},t) \triangleq (2\pi)^{-1}(t^2 - \rho^2/c^2)^{-\frac{1}{2}}\mathcal{U}(t - \rho/c). \tag{3.4.29}$$

In (3.4.29), $\mathcal{U}(t)$ denotes the unit time-step function. This correspondence allows us to identify the right-hand sides of (3.4.26) and (3.4.27) in terms of space-time convolution integrals. In the case of the space integrals, we have to account for the domain over which the relevant forward transformations were defined. We end up with

$$\begin{aligned}\{1,\tfrac{1}{2},0\}\underline{E}(\underline{\rho},t) = &-\mu\partial_t[\underline{\Pi}_D^e(\underline{\rho},t) + \underline{\Pi}_C^e(\underline{\rho},t)] \\ &+ \varepsilon^{-1}\int_{-\infty}^{t}dt'\ \nabla_T\{\nabla_T\cdot[\underline{\Pi}_D^e(\underline{\rho},t') + \underline{\Pi}_C^e(\underline{\rho},t')]\} \\ &- \nabla_T \times [\underline{\Pi}_D^m(\underline{\rho},t) + \underline{\Pi}_C^m(\underline{\rho},t)],\end{aligned} \tag{3.4.30}$$

and

$$\begin{aligned}\{1,\tfrac{1}{2},0\}\underline{H}(\underline{\rho},t) = &-\varepsilon\partial_t[\underline{\Pi}_D^m(\underline{\rho},t) + \underline{\Pi}_C^m(\underline{\rho},t)] \\ &+ \mu^{-1}\int_{-\infty}^{t}dt'\ \nabla_T\{\nabla_T\cdot[\underline{\Pi}_D^m(\underline{\rho},t') + \underline{\Pi}_C^m(\underline{\rho},t')]\} \\ &+ \nabla_T \times [\underline{\Pi}_D^e(\underline{\rho},t) + \underline{\Pi}_C^e(\underline{\rho},t)],\end{aligned} \tag{3.4.31}$$

for $\underline{\rho}\in\{\mathcal{D}_2,C,\mathcal{D}_1\}$. In (3.4.30) and (3.4.31), we have implicitly introduced the potentials

$$\underline{\Pi}_D^{e,m}(\underline{\rho},t) = \frac{1}{2\pi}\iint_{\mathcal{D}_2}dA(\underline{\rho}')\int_0^{t-R/c}dt'\ \frac{\underline{J}_D^{e,m}(\underline{\rho}',t')}{[(t-t')^2 - R^2/c^2]^{\frac{1}{2}}}, \tag{3.4.32a}$$

$$\underline{\Pi}_C^{e,m}(\underline{\rho},t) = \frac{1}{2\pi}\oint_C ds(\underline{\rho}')\int_0^{t-R/c}dt'\ \frac{\underline{J}_C^{e,m}(\underline{\rho}',t')}{[(t-t')^2 - R^2/c^2]^{\frac{1}{2}}}, \tag{3.4.32b}$$

with $R = |\underline{\rho} - \underline{\rho}'|$.

It should be pointed out that, in writing down (3.4.30) and (3.4.31), we have systematically translated the factors $i\underline{\kappa}$ and s occurring in (3.4.26) and (3.4.27) into space and time differentiations acting on the complete space-time integrals. For the time differentiations and integrations, an alternative translation is obtained by attributing the factors of s in (3.4.26) and (3.4.27) to the corresponding operations acting on $\underline{J}^{e,m}_{D,C}(\underline{\rho}',t')$. For the space differentiations, such an ambiguity is not available because of the finite extent of the domains of definition of the spatial Fourier transforms. However, their elaboration can readily be incorporated in the Fourier inversion procedure for the Green's function discussed in Appendix 3.4.B. From (3.4.47) and (3.4.48), it is observed that each space differentiation produces a factor of s. These factors, too, can be attributed to time differentiations acting on either the complete space-time integrals or the current distributions $\underline{J}^{e,m}_{D,C}(\underline{\rho}',t')$.

With (3.4.30) and (3.4.31), we now have at our disposal the basic integral relations for the integral-equation formulation of two-dimensional time-domain electromagnetic scattering problems.

Appendix 3.4.B Determination of $G(\underline{\rho},t)$ and its space derivatives

The remaining issue from the analysis of Appendix 3.4.A is the determination of the two-dimensional space-time domain Green's function $G(\underline{\rho},t)$. Conventionally, $G(\underline{\rho},t)$ is the causal solution of

$$[\nabla_T^2 - \partial_t^2/c^2]G(\underline{\rho},t) = -\delta(\underline{\rho})\delta(t), \tag{3.4.33}$$

with c being the wave speed in some homogeneous background medium. The frequency domain equivalent $G(\underline{\rho},s)$ of $G(\underline{\rho},t)$ is obtained directly by subjecting (3.4.33) to a temporal Laplace transformation. This reduces (3.4.33) to

$$[\nabla_T^2 - s^2/c^2]G(\underline{\rho},s) = -\delta(\underline{\rho}), \tag{3.4.34}$$

which has the solution

$$G(\underline{\rho},s) = (2\pi)^{-1}K_0(s\rho/c) \tag{3.4.35}$$

(see e.g. Morse and Feshbach (1953), Jones (1964), pp. 92 - 94). However, transforming (3.4.35) to the time domain is rather complex. Hence, we follow the same procedure as in Appendix 3.4.A and subject (3.4.34) to a spatial Fourier transformation over $\mathbb{R}^2$. This results in

$$\tilde{G}(\underline{\kappa},s) = (\underline{\kappa}\cdot\underline{\kappa} + s^2/c^2)^{-1}, \tag{3.4.36}$$

which is exactly the function specified in (3.4.28). In this appendix, we present one way to transform $\tilde{G}(\underline{\kappa},s)$ and related functions containing factors $(i\underline{\kappa})$ to the space-time domain.

We start by carrying out the spatial Fourier inversion:

$$G(\underline{\rho},s) = \frac{1}{4\pi^2}\int_{-\infty}^{\infty}\int_{-\infty}^{\infty}dk_x dk_y \frac{\exp(ik_x x + ik_y y)}{k_x^2 + k_y^2 + s^2/c^2}. \tag{3.4.37}$$

In (3.4.37), the k_x-integral can be evaluated directly by decomposing the fraction in the integrand, closing the contour in the upper or lower half of the complex k_x-plane for $x > 0$ and $x < 0$, respectively, and applying Jordan's lemma and Cauchy's theorem. This leads to

$$G(\underline{\rho},s) = \frac{1}{4\pi}\int_{-\infty}^{\infty}dk_y \frac{\exp[ik_y y - (k_y^2 + s^2/c^2)^{\frac{1}{2}}|x|]}{(k_y^2 + s^2/c^2)^{\frac{1}{2}}}, \tag{3.4.38}$$

where $\mathrm{Re}(k_y^2 + s^2/c^2)^{\frac{1}{2}} \geq 0$. The integral in (3.4.38) is now in a form amenable to the Cagniard-De Hoop method (De Hoop (1960), Achenbach (1973)). In this method, s is taken real and positive, which allows the substitution of $k_y = s\lambda/c$, with λ being a real-valued, dimensionless parameter. We find

$$G(\underline{\rho},s) = \frac{1}{4\pi}\int_{-\infty}^{\infty}d\lambda \frac{\exp[is\lambda y/c - s(\lambda^2 + 1)^{\frac{1}{2}}|x|/c]}{(\lambda^2 + 1)^{\frac{1}{2}}}. \tag{3.4.39}$$

Next, the contour of integration is shifted in the complex λ-plane to the Cagniard-De Hoop contour, given by

$$\lambda = [ic\tau y \pm |x|(c^2\tau^2 - \rho^2)^{1/2}]/\rho^2, \qquad \rho/c < \tau < \infty, \tag{3.4.40}$$

along which the argument of the exponent in (3.4.39) equals $-s\tau$. For λ on this contour, we have

$$(\lambda^2 + 1)^{-1/2}d\lambda = \pm(\tau^2 - \rho^2/c^2)^{-1/2}d\tau, \tag{3.4.41}$$

where the $\pm$ signs correspond to the ones in (3.4.40). The contributions from the connecting segments at infinity vanish by Jordan's lemma. Taking into account the direction of the integration, we finally arrive at

$$G(\underline{\rho},s) = \frac{1}{2\pi}\int_{\rho/c}^{\infty}d\tau\,\frac{\exp(-s\tau)}{(\tau^2 - \rho^2/c^2)^{1/2}} \tag{3.4.42}$$

for s real and positive. Now we invoke Lerch's theorem (Widder (1946)), which prescribes that a causal function is uniquely determined by its known Laplace transform on the positive real s-axis. With this theorem and (3.4.42), we obtain by inspection

$$G(\underline{\rho},t) = \frac{\mathcal{U}(t - \rho/c)}{2\pi(t^2 - \rho^2/c^2)^{1/2}}, \tag{3.4.43}$$

in which $\mathcal{U}(t)$ is the unit time-step function.

The space-time-domain functions corresponding to products of $\tilde{G}(\underline{\kappa},s)$ and factors of $(i\underline{\kappa})$, as occurring in (3.4.26) and (3.4.27), can be obtained in a similar way. For example, the integrals in the Fourier transformation

$$\partial_y G(\underline{\rho},s) = \frac{1}{4\pi}\int_{-\infty}^{\infty}\int_{-\infty}^{\infty}dk_x dk_y\,\frac{ik_y\exp(ik_x x + ik_y y)}{k_x^2 + k_y^2 + s^2/c^2} \tag{3.4.44}$$

reduce to

$$\partial_y G(\underline{\rho},s) = \frac{is}{4\pi c}\int_{-\infty}^{\infty}d\lambda\,\frac{\lambda\exp[is\lambda y/c - s(\lambda^2 + 1)^{1/2}|x|/c]}{(\lambda^2 + 1)^{1/2}}. \tag{3.4.45}$$

Shifting the path of integration to the Cagniard-De Hoop contour specified in (3.4.40), and substituting the intermediate results

(3.4.40) and (3.4.41) yields

$$\partial_y G(\underline{\rho},s) = \frac{-sy}{2\pi\rho^2}\int_{\rho/c}^{\infty} d\tau \, \frac{\tau \exp(-s\tau)}{(\tau^2 - \rho^2/c^2)^{\frac{1}{2}}} \, . \tag{3.4.46}$$

The factor of s in the right-hand side can be translated into a time-differentiation acting on the current distribution with which the Green's function is convolved. The remaining integral can again be converted to the time domain with the aid of Lerch's theorem. Moreover, the corresponding derivative with respect to x is found immediately by interchanging x and y in the result. Hence, we have

$$\frac{i\underline{\kappa}}{s}\tilde{G}(\underline{\kappa},s) \leftrightarrow -\underline{i}_\rho \, \frac{t\, U(t-\rho/c)}{2\pi\rho(t^2 - \rho^2/c^2)^{\frac{1}{2}}} \, . \tag{3.4.47}$$

In a similar manner, we arrive at

$$[(ik_\ell)(ik_m)/s^2 - \varepsilon\mu\delta_{\ell m}]\tilde{G}(\underline{\kappa},s) \leftrightarrow$$

$$\frac{[(2t^2 - \rho^2)x_\ell x_m - \rho^2 t^2 \delta_{\ell m}]U(t-\rho/c)}{2\pi\rho^4(t^2 - \rho^2/c^2)^{\frac{1}{2}}} \, , \tag{3.4.48}$$

where $\ell = 1,2$ and $m = 1,2$. The latter correspondence allows the identification of the transverse components of the first two terms on the right-hand sides of both (3.4.26) and (3.4.27) in terms of a single space-time convolution integral. In particular, such a combined form may be useful in analyzing the singularity of the resulting integrand near the point of observation. For the z-components, the first two terms need not be combined since the second terms have only transverse components.

3.5 Scattering by an inhomogeneous, lossy dielectric cylinder

3.5.1 Formulation of the problem

In this section, we turn our attention to a generalized version of the problem investigated in Sections 2.5 and 2.6, namely the scattering of the pulse specified in (2.5.1) by a lossy dielectric cylinder with an arbitrary shape and inhomogeneity, embedded in vacuo (see Figure 3.5.1). For this configuration, the electromagnetic-field strengths $\underline{E}(\underline{\rho},t)$ and $\underline{H}(\underline{\rho},t)$ satisfy the two-dimensional version of Maxwell's equations

$$\nabla_T \times \underline{H}(\underline{\rho},t) - \varepsilon(\underline{\rho})\partial_t \underline{E}(\underline{\rho},t) = \sigma(\underline{\rho})\underline{E}(\underline{\rho},t), \tag{3.5.1a}$$

$$\nabla_T \times \underline{E}(\underline{\rho},t) + \mu_0\partial_t \underline{H}(\underline{\rho},t) = \underline{0}, \tag{3.5.1b}$$

for all $\underline{\rho}$. In (3.5.1a), we have $\varepsilon(\underline{\rho}) = [1 + \chi(\underline{\rho})]\varepsilon_0$, with ε_0 and μ_0 being the permittivity and permeability in vacuo, respectively. As indicated in Figure 3.5.1, we have $\chi(\underline{\rho}) \geq 0$ and $\sigma(\underline{\rho}) \geq 0$ in a finite interior domain $\mathcal{D}_2$, and $\chi(\underline{\rho}) = 0$ and $\sigma(\underline{\rho}) = 0$ in an infinite exterior domain $\mathcal{D}_1$. For the plane-wave incident field specified in (2.5.1), we have

$$\nabla_T \times \underline{H}^i(\underline{\rho},t) - \varepsilon_0\partial_t\underline{E}^i(\underline{\rho},t) = \underline{0}, \tag{3.5.2a}$$

$$\nabla_T \times \underline{E}^i(\underline{\rho},t) + \mu_0\partial_t\underline{H}^i(\underline{\rho},t) = \underline{0}, \tag{3.5.2b}$$

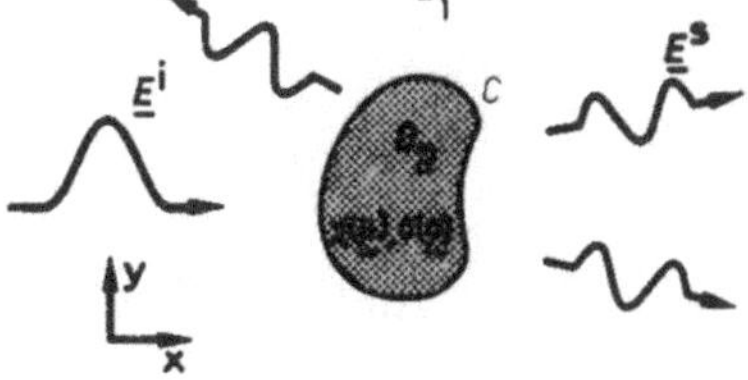

Figure 3.5.1 Scattering of a pulsed electromagnetic plane wave by an inhomogeneous, lossy dielectric cylinder.

for all $\underline{\rho}$. As before, we define the scattered field by

$$\underline{E}^{s}(\underline{\rho},t) = \underline{E}(\underline{\rho},t) - \underline{E}^{i}(\underline{\rho},t), \qquad \underline{H}^{s}(\underline{\rho},t) = \underline{H}(\underline{\rho},t) - \underline{H}^{i}(\underline{\rho},t). \tag{3.5.3}$$

The two-dimensional electromagnetic-field equations for this scattered field can be obtained directly by substracting (3.5.2) from (3.5.1). This yields

$$\nabla_T \times \underline{H}^{s}(\underline{\rho},t) - \varepsilon_0\partial_t \underline{E}^{s}(\underline{\rho},t) = \underline{J}_D^{e}(\underline{\rho},t), \tag{3.5.4a}$$

$$\nabla_T \times \underline{E}^{s}(\underline{\rho},t) + \mu_0\partial_t \underline{H}^{s}(\underline{\rho},t) = \underline{0}, \tag{3.5.4b}$$

with

$$\underline{J}_D^{e}(\underline{\rho},t) = [\varepsilon_0\chi(\underline{\rho})\partial_t + \sigma(\underline{\rho})]\underline{E}(\underline{\rho},t) \tag{3.5.5}$$

being the polarization current density in $\mathcal{D}_2$. Now we can choose the characteristic length a in (2.5.1) such that the front of the incident field does not reach the scattering obstacle before $t = 0$. Then, causality ensures that the scattered field in $\mathcal{D}_1$ and $\mathcal{D}_2$ as well as the total field in $\mathcal{D}_2$ vanish in the time interval $-\infty < t < 0$. By virtue of the same argument that we used in writing down (3.4.2) we can therefore apply the relation (3.4.30) to the entire two-dimensional space $\mathbb{R}^2$. Writing the electromagnetic field as in (2.5.2), evaluating the relevant vector products, and substituting (3.4.32a), we end up with the integral relation

$$E(\underline{\rho},t) = E^{i}(\underline{\rho},t) - \frac{\mu_0}{2\pi}\iint_{\mathcal{D}_2} dA(\underline{\rho}')\int_0^{t-R/c_0} dt' \, \frac{\partial_{t'} J_D^{e}(\underline{\rho}',t')}{[(t-t')^2 - R^2/c_0^2]^{\frac{1}{2}}}, \tag{3.5.6}$$

where $J_D^{e}(\underline{\rho}',t')$ denotes the z-component of the polarization current density introduced in (3.5.5). When $\underline{\rho}\in\mathcal{D}_2$, (3.5.6) constitutes an integral equation of the second kind in $E(\underline{\rho},t)$.

As in the two-dimensional scattering problems discussed in Sections 2.5, 2.6 and 3.4, it is convenient to normalize all space coordinates with respect to the characteristic length a and all times with respect to the corresponding vacuum travel time a/c_0. In addition, we introduce the dimensionless conductivity $\bar{\sigma}(\bar{\underline{\rho}}) = Z_0\sigma(\underline{\rho})a$. The formulation of the integral equation (3.5.6) in terms of these dimensionless quantities is obtained directly by setting $\bar{\varepsilon}_0 = 1$, $\bar{\mu}_0 = 1$, $\bar{c}_0 = 1$ in (3.5.5) and (3.5.6). From now on, we will use this normalized version instead of (3.5.6). As before, we will omit the bars.

Compared to the integral equations discussed in the previous sections, we now have the additional difficulty of a logarithmic singularity in the space integration. This singularity is observed immediately from the frequency-domain Green's function given in (3.4.35). Using the correspondence between (3.4.35) and (3.4.43) and substituting formulas (9.6.12) and (9.6.13) of Abramowitz and Stegun (1965) results in the estimate

$$\int_0^{t-R} dt' \frac{f(t')}{[(t-t')^2 - R^2]^{\frac{1}{2}}} = -f(t)\ln(R) + \text{constant} + O[R^2 \ln(R)] \tag{3.5.7}$$

as $R \to 0$. In (3.5.7), $f(t)$ denotes a sufficiently differentiable real-valued time signal. The estimate (3.5.7) can also be obtained by carrying out a repeated partial integration with respect to t' in the integral on its left-hand side (see Tijhuis (1984a,b)). The logarithmic singularity can be isolated from the remainder of the integrand in (3.5.6) by rewriting the dimensionless version of that equation as

$$\begin{aligned} E(\underline{\rho},t) = {} & E^i(\underline{\rho},t) + \frac{1}{2\pi}\iint_{D_2} dA(\underline{\rho}')\left\{\ln(R)\,\partial_t\, J_D^e(\underline{\rho}',t)\right\} \\ & - \frac{1}{2\pi}\iint_{D_2} dA(\underline{\rho}')\left\{\int_0^{t-R} dt' \frac{\partial_{t'} J_D^e(\underline{\rho}',t')}{[(t-t')^2 - R^2]^{\frac{1}{2}}} \right. \\ & \left. + \ln(R)\,\partial_t\, J_D^e(\underline{\rho}',t)\right\}. \end{aligned} \tag{3.5.8}$$

In the second integral, both the integrand and its first-order space derivatives remain bounded as $R \to 0$, while the second-order space-derivatives are logarithmically singular.

Both Equations (3.5.6) and (3.5.8) are in a form which is suitable for the application of the marching-on-in-time scheme. In the remainder of this section, we will consider two different methods to implement this scheme.

3.5.2 Method I: numerical implementation and results

In both methods, the space-time points were limited to the simple cubic grid $x_\ell = \ell h$, $y_m = mh$, $t_n = n\Delta t$, with the dimensionless time step Δt being equal to the dimensionless space step $h = 1/N$. In method I, the boundary of the domain of integration $\mathcal{D}_2$ was piecewise approximated by straight lines within each quarter square subdomain of width $h/2$. Then, the conventional approach was followed and $J_D^e(x',y',t')$ was approximated by a piecewise-constant function

$$J_D^e(x,y,t) = [\tilde{\chi}(\ell,m)\partial_t + \tilde{\sigma}(\ell,m)]E(x_\ell,y_m,t) \qquad (3.5.9)$$

for $\underline{\rho}\in\mathcal{D}_2 \cap \mathcal{D}_{\ell m}$, where $\mathcal{D}_{\ell m}$ denotes the domain $\max(|x - x_\ell|, |y - y_m|) < h/2$, and where $\tilde{\chi}(\ell,m)$ and $\tilde{\sigma}(\ell,m)$ are taken at some point inside $\mathcal{D}_2 \cap \mathcal{D}_{\ell m}$. Substitution of (3.5.9) in (3.5.6) results, for each combination of an observation point $\underline{\rho}_{\ell m} = x_\ell \underline{i}_x + y_m \underline{i}_y$ and a square $\mathcal{D}_{\ell' m'}$ in an integral over $\mathcal{D}_{\ell' m'}$ of a function of $R_{\ell m} \stackrel{\Delta}{=} |\underline{\rho}' - \underline{\rho}_{\ell m}|$ only. The numerical evaluation of this function requires the determination of a time integral of the type (3.5.7). This time integral was handled by restricting, as in (3.4.14), the radius R to multiples of Δt, and approximating in each subinterval of length Δt, the field by a cubic interpolation polynomial. The resulting integrals over each subinterval were integrated analytically. Where possible, the interpolation over a subinterval $n'\Delta t < t' < (n' + 1)\Delta t$ was carried out using the field values at the boundaries of that interval and at $t' = (n' - 1)\Delta t$ and $t' = (n' + 2)\Delta t$. When the latter value was not available, the value at $t' = (n' - 2)\Delta t$ was used. Subsequently, the integral over $\mathcal{D}_{\ell' m'}$ was obtained by analytical integration of a linear approximation in $R_{\ell m} = |\underline{\rho}' - \underline{\rho}_{\ell m}|$. An exception was made in the self patch $\mathcal{D}_{\ell m}$, where the exact upper boundary $n\Delta t - R_{\ell m}$ was used and the resulting singular function of $R_{\ell m}$ was integrated analytically.

From the description given above, it follows that we need not encounter problems in the inversion of the weighting matrix. The discretization has been organized such that this matrix is of the form (3.2.7a). Moreover, it follows from the estimate (3.5.7) that (3.2.8a) holds for a sufficiently small mesh size h. On the other hand, the logarithmic singularity may cause an instability problem. From (3.5.5) and (3.5.8), it is observed that, for a nonvanishing susceptibility $\chi(\underline{\rho})$, the field values $\tilde{E}(\ell',m',n')$ obtained are multiplied by a factor of $\mathcal{O}(h^{-2})$ due to the double time differentiation. In the singular part of the integrand, the product of the resulting approximation and $\ln(R_{\ell m})$ is integrated over a square with sides of length h. Thus, we end up with a weighting factor that is of the same order of magnitude as the average of $\ln(R_{\ell m})$ over $\mathcal{D}_{\ell' m'}$. For the point of observation and neighboring space-time points, where the logarithmic term dominates the integrand of (3.5.6), this implies that the relevant weighting factors are of the order of ln(h). Hence, they will even increase when h is reduced. In view of this estimate, we must expect an unstable marching-on-in-time solution, regardless of the size of the discretization step. A different situation arises when $\chi(\underline{\rho})$ vanishes in $\mathcal{D}_2$. In that case, we need only carry out a single time differentiation. Consequently, the weighting factors are at most of $\mathcal{O}[h\ \ln(h)]$, which suggests a stability behavior similar to that predicted at the end of Subsection 3.4.2.

It is noted that the discretization method discussed in this subsection does not meet criterion I. This criterion has already been violated by the basic approximation (3.5.9). In fact, the difference in behavior between the results obtained with this discretization and the slab results discussed in Subsection 3.3.3 was one of the motives for starting the stability analysis written up in Tijhuis (1984a,b) and refined in Section 3.2 of the present chapter. In accordance with the discussion given above, we concentrated on obtaining results for a susceptibility contrast only, i.e. $\sigma(x,y) = 0$. The results obtained with the discretization above exhibit a short-term instability. As argued above, this instability is caused by interpolation errors at the space points next to the point of observation, where at $t' = t - R$ the time derivative must be determined by backward time interpolation. It can be removed by using, after each time step, the field obtained to determine a new approximation to those time derivatives by central time interpolation instead of backward time interpolation. The computed field values are subsequently improved by correcting the discretized version of the integral in (3.5.6) for half the difference of the two approximations of the relevant time derivatives. With this correction procedure, the computational scheme yields stable results for a fairly wide range of incident-pulse durations and contrasts. An example is shown in Figure 3.5.2. The increase in computation time is even more drastic than for the impenetrable cylinder. Now, we must evaluate, at $\mathcal{O}(N)$ instants, at $\mathcal{O}(N^2)$ space points, $\mathcal{O}(N)$ different time integrals, each of which requires a weighted summation of $\mathcal{O}(N)$ field values. Once these time integrals have been determined, $\mathcal{O}(N^2)$ space integrals must be computed, each of which requires, in turn, a weighted summation of $\mathcal{O}(N^2)$ time integrals. This implies that the computation time is approximately proportional to N^5. This count also demonstrates the advantage of restricting R to multiples of Δt and interpolating in between: if we had used the $\mathcal{O}(N^2)$ exact values $|\rho_{\ell m} - \rho_{\ell',m'}|$ for all space points, the number of operations would have run up to $\mathcal{O}(N^6)$. For not too small a h, the proportionality found above is not yet noticed and the computation times are of the same order of magnitude as in the previous subsection. For a homogeneous circular cylinder, the results coincide with a reference solution obtained by applying a direct Fourier inversion to the frequency-domain result given by (2.5.16), (2.5.17), (2.5.19), and (2.5.20). For pulse lengths which are too short or contrasts which are too high, long-term instabilities are observed at late times. Since the discretization violates criterion I, these instabilities may not be regarded as "global" errors that can be removed by reducing the discretiza-

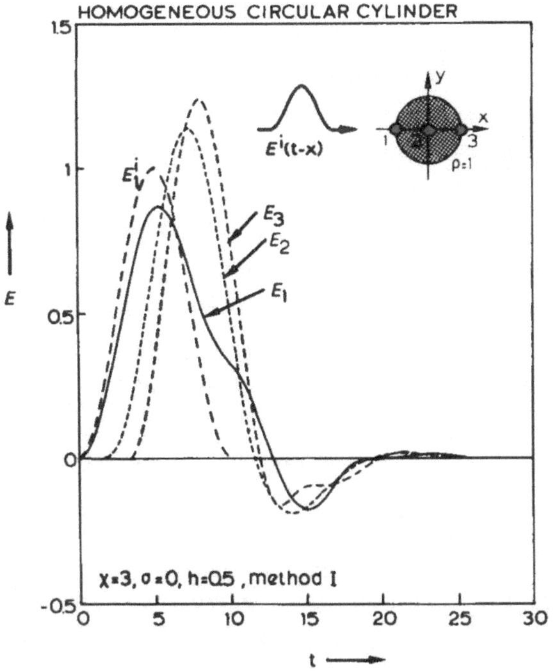

Figure 3.5.2 E as a function of normalized time for a homogeneous, circular cylinder illuminated by the pulse specified in (2.5.1) for $F(t) = \sin^2(\pi t/T)\text{rect}(t - T/2;T)$ with T = 10 at the points indicated in the inset.

tion step. In fact, such a reduction may even raise the short-term instabilities again. In line with the argument given above, this phenomenon can be understood from the increased values of the dominant weighting factors in the discretized integral equation.

3.5.3 Method II: numerical implementation

In view of the imperfections in the discretization of method I, the numerical experimentation with this method was terminated and a second method was devised. Until now, this method has only been implemented for a square cylinder with dimensionless diameter 2. Moreover, the constitutive coefficients $\chi(\underline{\rho})$ and $\sigma(\underline{\rho})$ were assumed to be twice differentiable with respect to x and y. However, this restriction is not really necessary; a boundary line along which $\chi(\underline{\rho})$, $\sigma(\underline{\rho})$ or one of their derivatives is discontinuous can be included, as in method I, by approximating it piecewise by straight lines. Since this second discretization method was developed simultaneously with the stability analysis of Section 3.2, it was put together such that a number of the observations made in Subsection 3.2.3 could be verified. It is a modification of the second method discussed in Tijhuis (1984a,b), which almost met criterion I.

The approximation underlying the discretization is that, in each square $x_\ell < x < x_{\ell+1}$, $y_m < y < y_{m+1}$, a function $g(x,y,t)$ may be replaced by the bilinear interpolation formula

$$\begin{aligned} g'_{\ell m}(x,y,t) &= a_{\ell m}(t) + b_{\ell m}(t)(x - x_\ell) + c_{\ell m}(t)\ (y - y_m) \\ &\quad + d_{\ell m}(t)(x - x_\ell)(y - y_m). \end{aligned} \tag{3.5.10}$$

In (3.5.10), the time-dependent coefficients a, b, c and d are chosen such that $g'(x_\ell,y_m,t) = g(x_\ell,y_m,t)$ for all relevant ℓ and m. This results in

$$\begin{aligned} a_{\ell m}(t) &= g(x_\ell,y_m,t) \\ b_{\ell m}(t) &= [g(x_{\ell+1},y_m,t) - g(x_\ell,y_m,t)]/h \\ c_{\ell m}(t) &= [g(x_\ell,y_{m+1},t) - g(x_\ell,y_m,t)]/h \\ d_{\ell m}(t) &= [g(x_{\ell+1},y_{m+1},t) - g(x_{\ell+1},y_m,t) \\ &\quad - g(x_\ell,y_{m+1},t) + g(x_\ell,y_m,t)]/h^2. \end{aligned} \tag{3.5.11}$$

The approximations $g'_{\ell m}(x,y,t)$ have the advantage that their combination $g'(x,y,t)$ is continuous across the boundaries of adjoining squares. Furthermore, $g'(x,y,t)$ is piecewise linear along each line of constant x or y. An error estimate for the bilinear interpolation can be found by regarding it, in each square, as the result of a linear interpolation in the x-direction, followed by one in the y-direction or vice versa. Keeping track of the errors made in each interpolation step (see Hildebrand (1956), Stoer (1972)), we end up with

$$|g'_{\ell m}(x,y,t) - g(x,y,t)| < \\ < \frac{3}{8} h^2 \max\left\{|\partial^2_{x'} g(x',y',t)|, |\partial^2_{y'} g(x',y',t)|\right\}, \tag{3.5.12}$$

where the point (x',y') varies over the range of the interpolation as specified in connection with (3.5.10).

Next, we have to consider which parts of the integrands in the integrals on the right-hand sides of (3.5.6) and (3.5.8) are eligible for an approximation of the type (3.5.10). In view of the restrictions imposed on $\chi(\underline{\rho})$ and $\sigma(\underline{\rho})$ at the beginning of this subsection, this approximation applies to the polarization current density $J^e_D(\underline{\rho},t)$ defined in (3.5.5). However, replacing $J^e_D(\underline{\rho},t)$ in (3.5.6) by its bilinear approximation still leaves us with an implicit logarithmic singularity. Hence, we resort to the alternative form (3.5.8), in which this singularity shows up explicitly. In the first integral on the right-hand side we carry out the substitution mentioned above for $J^e_D(\underline{\rho},t)$. This leaves us with the spatial integration of the product of a bilinear form of the type (3.5.10) with $\ln(R)$, which can be carried out analytically over any polygon. In the second integral on the right-hand side of (3.5.8), we distinguish between two cases. In squares that do not contain the point of observation, we approximate the whole integrand

$$I(\underline{\rho}',t) = \int_0^{t-R} dt' \frac{\partial_{t'} J^e_D(\underline{\rho}',t')}{[(t-t')^2 - R^2]^{\frac{1}{2}}} + \ln(R)\partial_t J^e_D(\underline{\rho}',t) \tag{3.5.13}$$

by the bilinear interpolation formula (3.5.10). From (3.5.12) and (3.5.7), we observe that this results in an error of at most $O[h^2 \ln(h)]$, which is just acceptable. Subsequently we can then use the two-dimensional trapezoidal rule

$$\int_{x_\ell}^{x_{\ell+1}} dx \int_{y_m}^{y_{m+1}} dy\, g'_{\ell m}(x,y,t) = \frac{h^2}{4} \sum_{\ell'=\ell}^{\ell+1} \sum_{m'=m}^{m+1} g(x_{\ell'},y_{m'},t), \tag{3.5.14}$$

which attributes an equal weighting factor to the spatially sampled function values at each corner. For squares with one corner coinciding with the point of observation, the logarithmic singularity in the second space derivatives prevents this approximation. However, in that region (3.5.7) produces the estimate

$$\int_0^{t-R} dt' \frac{g(x',y',t')}{[(t-t')^2 - R^2]^{\frac{1}{2}}} + g(x',y',t)\ln(R) =$$
$$\int_0^{t-\Delta t} dt' \frac{g(x',y',t')}{[(t-t')^2 - \Delta t^2]^{\frac{1}{2}}} + g(x',y',t)\ln(\Delta t) + O[h^2 \ln(h)] \tag{3.5.15}$$

when $R = O(h)$, which is consistent with the approximation made in the remaining squares. Substituting (3.5.15) in the second integrand in (3.5.8) limits its spatial dependence to that in $\partial_{t'} J_D^e(\underline{\rho}',t')$, which can be approximated by a bilinear form of the type (3.5.10). Hence, the resulting space integral can be approximated by applying the two-dimensional trapezoidal rule (3.5.14) to that bilinear form. Note that, with this discretization technique, we have attained the same result as for the perfectly conducting cylinder, namely that the time integrals need only be determined for $R \geq \Delta t$.

For the time differentiations in $\partial_t J_D^e(\underline{\rho},t)$, backward interpolation formulas were used similar to those applied in the case of the dielectric slab (see Subsection 3.3.2). At each space-time point, we defined

$$\begin{aligned}\partial_t \mathcal{J}_D^e(\ell,m,n) = \tilde{\chi}(\ell,m)\{&2\tilde{\mathcal{E}}(\ell,m,n) - 5\tilde{\mathcal{E}}(\ell,m,n-1) \\ &+ 4\tilde{\mathcal{E}}(\ell,m,n-2) - \tilde{\mathcal{E}}(\ell,m,n-3)\}/h^2 \\ + \tilde{\sigma}(\ell,m)\{&\tfrac{3}{2}\tilde{\mathcal{E}}(\ell,m,n) - 2\tilde{\mathcal{E}}(\ell,m,n-1) + \tfrac{1}{2}\tilde{\mathcal{E}}(\ell,m,n-2)\}/h,\end{aligned} \tag{3.5.16}$$

where $\tilde{\chi}(\ell,m) = \chi(\underline{\rho}_{\ell m})$ and $\tilde{\sigma}(\ell,m) = \sigma(\underline{\rho}_{\ell m})$. In the first integral of (3.5.8), this approximation was substituted directly. In the second integral, the time integral was handled in the same manner as the one for the perfectly conducting cylinder (see (3.4.10) and (3.4.14)). The only difference was that, in each subinterval $n'\Delta t < t' < (n' + 1)\Delta t$, the quantity $\partial_{t'} J_D^e(\underline{\rho}_{\ell m},t')$ was approximated by linear interpolation between the values $\partial_t \mathcal{J}_D^e(\ell,m,n')$ and $\partial_t \mathcal{J}_D^e(\ell,m,n' + 1)$ as specified in (3.5.16). This approach seemed preferable to the one used in method I, i.e. using a cubic time-interpolation polynomial for $E(\underline{\rho}_{\ell m},t')$, for two reasons. In the first place, it seemed more consistent to use the same approximation for

$\partial_t\, J_D^e(\underline{\rho}_{\ell m},t)$ in both integrals in (3.5.8), particularly since these integrals originated from taking out the logarithmic singularity in (3.5.6). In the second place, a systematic use of (3.5.16) allows us to experiment with doubling the time step in the difference formulas for the time derivatives, as proposed in the second part of Subsection 3.2.3.

The principal advantages of the discretization outlined above are this flexibility and the fact that it does meet criterion I. In addition, the discretized time integral in (3.5.8) depends only on field values at previous instants. The major disadvantage is that we have lost the property (3.2.7) for the weighting matrices. However, the occurrence of the off-diagonal terms is caused by treating the space integral of $\ln(R)\partial_t\, J_D^e(\underline{\rho}',t)$ in two different ways, namely by including or excluding the factor of ln(R) in the bilinear interpolation employed. Hence these terms will become small as ln(R) becomes regular, i.e. as R increases. Since the diagonal terms behave similar to those in method I, this implies that we need not encounter problems in the determination of the inverse weighting matrix. Nevertheless, special attention was given to the necessary matrix inversion in order to avoid any possible doubt with regard to the origin of the instabilities observed in the results. To this end, the inverse weighting matrix was determined up to machine precision (a 16 digit accuracy) by Crout's factorization method, followed by repeated back substitution (NAG (1984), subroutine F04AEF).

As far as the stability is concerned, the present discretization will behave similar to the one used in method I: For $\chi(\underline{\rho}) = 0$, we only expect an instability when the excited field exhibits some systematic behavior. Possibly, this instability may be controlled by reducing the discretization step. For an increasing susceptibility contrast, however, we must expect an instability that cannot be controlled by any of the methods proposed in the second part of Subsection 3.2.3. In that case, we will have to modify the method of solution as proposed in the third part of that subsection. The correspondence with method I also holds for the increase in computation time upon reduction of the mesh size. The estimate given in Subsection 3.5.2 only needs to be augmented with the amount of work required in the multiplication with the system matrix as given in abstract form in (3.2.6). At $\mathcal{O}(N)$ instants, this multiplication continuously requires $\mathcal{O}(N^4)$ operations. Thus, we still have a computation time that is proportional to N^5.

3.5.4 Method II: numerical results

With the descretization outlined in Subsection 3.5.3, several numerical computations were performed. Because of the different stability behavior in the presence or absence of a susceptibility contrast, we concentrated on configurations with either a vanishing susceptibility ($\chi = 0$) or a vanishing conductivity ($\sigma = 0$). Specifically, we considered the scattering of a Gaussian incident pulse as specified in (2.5.1) with

$$F(t) = \exp[-16(t - \tfrac{1}{2}T)^2/T^2], \qquad (3.5.17)$$

scattered by a homogeneous square cylinder of dimensionless diameter 2. Figure 3.5.3 shows results of applying the explicit marching-on-in-time method for the case of conductivity contrast only. Figure 3.5.3a pertains to the case where $T = 10$, $\sigma = 5$ and $N = 2$, and was obtained with the single-step difference rules used in (3.5.16). The value of σ was chosen such that the marching-on-in-time results are just unstable. Note that the alternating, unstable component of the solution that shows up for $t > 15$ moves in the same direction across the entire cylinder. Moreover, the start of the unstable behavior coincides in time with the onset of a "stationary" behavior of the actual field, which is precisely the behavior discussed in the second part of Subsection 2.5.5. These observations are in agreement with the conclusion reached above that, for zero susceptibility, only a systematic behavior of the excited field can cause an instability. From Figure 3.5.3a, it is also observed that the alternating component of the solution obtained has a larger amplitude at the center of the cylinder than at its boundary. This can be explained as a bulk effect due to the larger influence of field values at grid points surrounding the point of observation. Figure 3.5.3b shows the

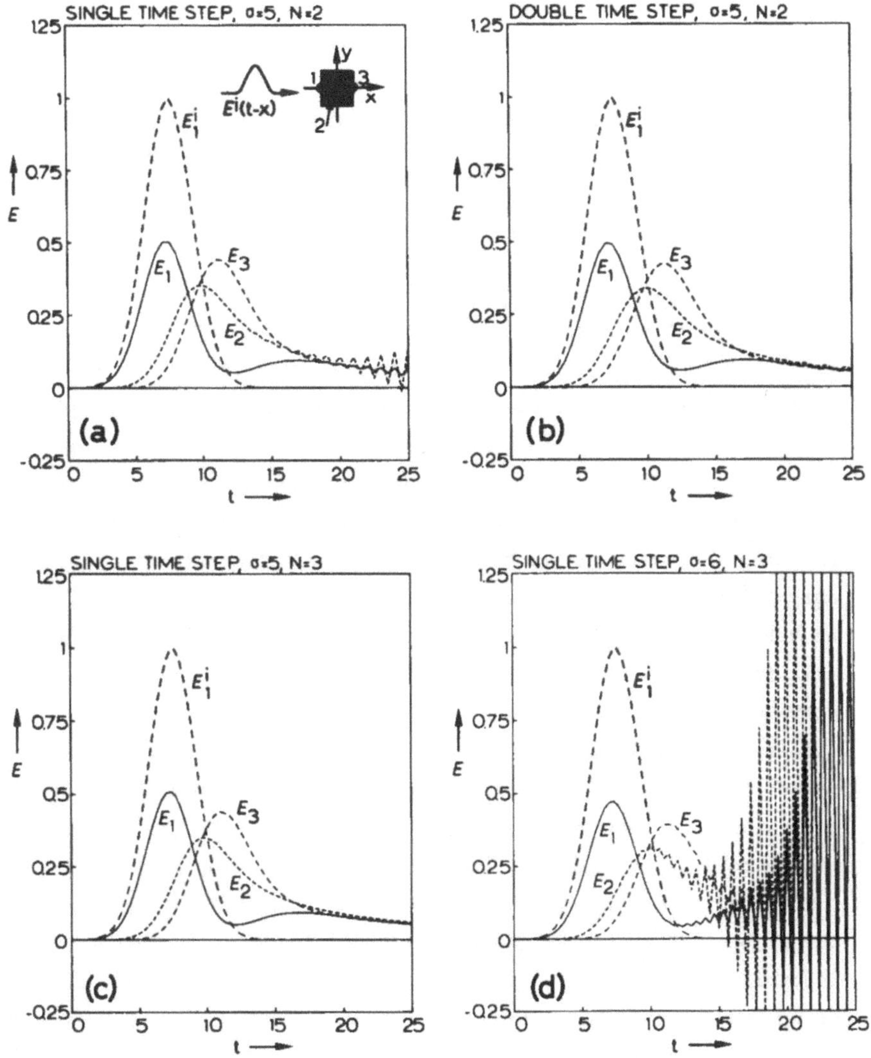

Figure 3.5.3 Results of applying the explicit marching-on-in-time method to obtain the field excited in a homogeneous square cylinder with $\chi = 0$ by a Gaussian pulse as specified in (3.5.17) with $T = 10$. Conductivity and discretization as indicated in the figure headings.

result of repeating the computation with the double time step in the difference rules, i.e. with (3.5.16) replaced by

$$\begin{aligned}\partial_t \mathcal{J}^e_D(\ell,m,n) = \tilde{\chi}(\ell,m)\Big\{&2\tilde{E}(\ell,m,n) - 5\tilde{E}(\ell,m,n-2) \\ &+ 4\tilde{E}(\ell,m,n-4) - \tilde{E}(\ell,m,n-6)\Big\}/4h^2 \\ + \tilde{\sigma}(\ell,m)\Big\{&\tfrac{3}{2}\tilde{E}(\ell,m,n) - 2\tilde{E}(\ell,m,n-2) + \tfrac{1}{2}\tilde{E}(\ell,m,n-4)\Big\}/2h.\end{aligned} \tag{3.5.18}$$

The results appear to be on the brink of instability. This was verified by repeating the computation for varying values of the conductivity. This result confirms the interpretation of the instability as a systematic one: doubling the time step will mainly take effect when the discretized integral is dominated by the contributions from a few space-time points near the point of observation. Figure 3.5.3c shows the result of repeating the original computation with N = 3. Reducing the mesh size in this manner clearly leads to a stable marching-on-in-time solution. This corroborates the conjecture made at the end of Subsection 3.4.2 that, to some extent, even a systematic instability may be eliminated by refining the discretization, provided that the weighting factors for the field values in the discretized integrals are of $O(1)$ as $h \to 0$. However, as Figure 3.5.3d shows, the extension of the stability range is limited. Increasing the value of the conductivity from $\sigma = 5$ to $\sigma = 6$ in the computation of Figure 3.5.3c causes the instabilities to show up again.

The results of applying the explicit marching-on-in-time method to the problem of transient scattering by a cylinder with only a susceptibility contrast are even more discouraging. For the discretization with a single time step in the backward difference rules for the time derivatives, the stability range is not worth mentioning. When (3.5.18) is used, and the discretization is not too fine, stable results can be obtained until at most $\chi = 2$. Refining the discretization, however, may stir up the instabilities again. This is due to the fact that, for nonvanishing susceptibility contrast, reducing the discretization step enhances the dominance in the discretized integrals of the contributions involving field values at "nearby" space-time points. An example is given in Figure 3.5.4,

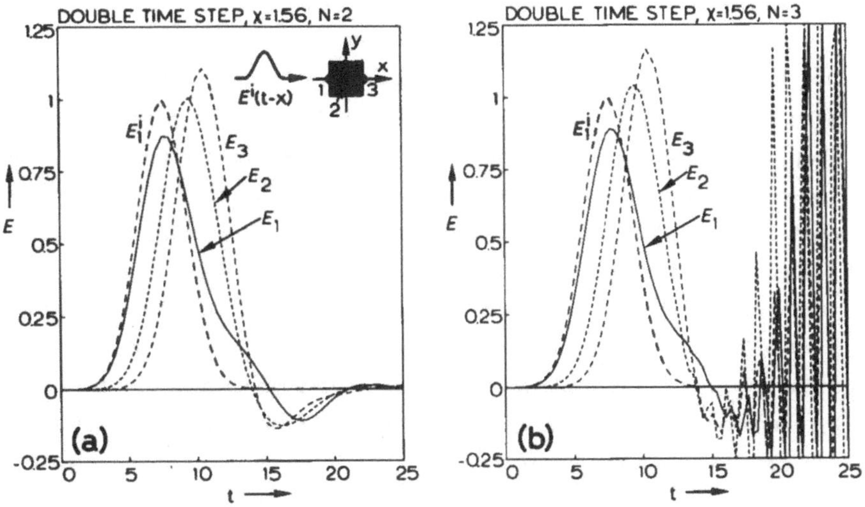

Figure 3.5.4 Results of applying the explicit marching-on-in-time method to a homogeneous, lossless square cylinder with $\chi = 1.56$, illuminated by the Gaussian pulse specified in (3.5.17) with $T = 10$. Discretization as indicated in the figure headings.

which shows, for one and the same configuration, a stable result obtained for $N = 2$ and an unstable result obtained for $N = 3$. In Figure 3.5.4b it is observed that, due to the choice of the double time step, the unstable part of the marching-on-in-time result no longer changes sign at each new instant. Rather, we have the interplay of a number of different instabilities that alternate with the double time step.

The computation times for the numerical results displayed in Figures 3.5.3 and 3.5.4 were quite acceptable. A single marching-on-in-time computation took about 1.5 and 8.5 seconds for $N = 2$ and $N = 3$, respectively. However, the relation between both computation times already indicates that increasing N may considerably enlarge the numerical effort required. This was checked further by repeating the computations of Figures 3.5.3a and 3.5.3c with $N = 4$, which took about 35 seconds. As could be expected, no instabilities showed up in the solution obtained. The relation between the computation times for $N = 3$ and $N = 4$, in particular, is in good agreement with the estimate that the computation time is of $O(N^5)$.

With the experiments above, we have exhausted the supply of possible modifications to the discretization listed in the second part of Subsection 3.2.3. Since neither of the modifications proposed is completely effective in eliminating the instabilities in the marching-on-in-time solutions obtained, it seems justified to employ the modified method of solution formulated in the third part of that subsection. With this method, several numerical experiments were carried out for the two types of scattering problem described above. In view of the computation time consumed in carrying out a single iteration step, the progression of the iterative procedure was monitored by computing, after each iteration step, a discretized version of the root-mean-square error

$$err \triangleq \left\{ \frac{\int_{\mathcal{D}_2} dA(\underline{\rho}) \int_0^{T_{max}} dt \; dev(\underline{\rho},t)^2}{\int_{\mathcal{D}_2} dA(\underline{\rho}) \int_0^{T_{max}} dt \; E^i(\underline{\rho},t)^2} \right\}^{\frac{1}{2}} . \tag{3.5.19}$$

In (3.5.19), $dev(\underline{\rho},t)$ is defined as the error that results in the equality sign of the integral equation (3.5.8) when a given approximation to $E(\underline{\rho},t)$ is substituted. Further, T_{max} is the termination instant of the time interval under consideration. The normalization in (3.5.19) can be understood from the fact that $E^i(\underline{\rho},t)$ is just the deviation in the equality sign of (3.5.8) that results upon substitution of $E(\underline{\rho},t) = 0$. The error defined in (3.5.19) can be envisaged as a global measure of the relative accuracy of a given approximate solution of (3.5.8). In addition, it is the same error that is being used in the application of gradient-type methods to measure the quality of the results obtained (see e.g. Van den Berg (1984)).

In all computations, the iterative procedure was started from the initial value $E^{(0)}(\underline{\rho},t) = 0$. Since this initial value has zero components along all eigenvectors of the system matrix of the discretized integral equation, it certainly fulfils the requirement that it should have zero components along the problematic ones. Moreover, this choice has the advantage that the corresponding root-mean-square error defined in (3.5.19) need not be computed. We directly have

$$err^{(0)} = 1 \tag{3.5.20}$$

for both the exact and the discretized version of that error. In particular, this property is useful in the selection of the proper value of p. It turns out that a sufficient condition for convergence to a smooth solution is that the error after the first iteration step should meet the criterion

$$err^{(1)} < err^{(0)}. \tag{3.5.21}$$

This can be explained from the fact that the instabilities that show up in solving, for too small a $p > 0$, the modified marching-on-in-time equation (3.2.29) differ from the ones obtained in solving the corresponding explicit one. For the initial estimate $E^{(0)}(\underline{\rho},t) = 0$, the verification of the criterion (3.5.21) requires a computational effort equivalent to performing a single marching-on-in-time computation. For any other initial estimate, twice that effort is required to check (3.5.21). The increased efficiency in verifying (3.5.21) is important since it allows a more accurate determination of the most appropriate value for the parameter p. In this sense, it also improves the overall efficiency of the iterative procedure since choosing too large a p may unnecessarily slow down the convergence of the results.

Some illustrative results obtained with the modified method of solution are shown in Figures 3.5.5 and 3.5.6. Figure 3.5.5 shows results for the same situation as specified in Figures 3.5.3c,d. Now, the conductivity has been increased up to $\sigma = 10$, which exceeds the upper limit of the stability range for the explicit marching-on-in-time method by a factor of 2. Figure 3.5.5a shows the result obtained after 15 iteration steps with $p = 1$ and Figure 3.5.5b illustrates the root-mean-square error $err^{(n)}$ found after each iteration step. From the latter, it is observed that, after ten iteration steps, the iterative result meets the equality sign in the discretized integral equation with a global accuracy of three significant

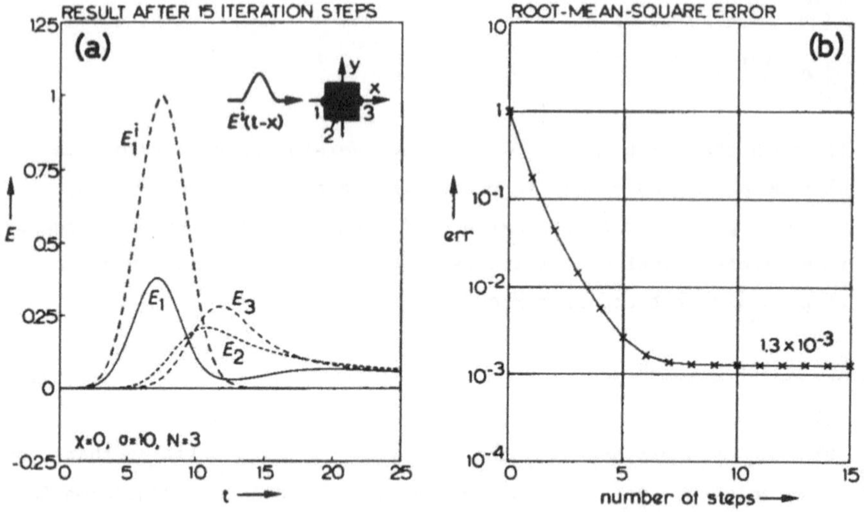

Figure 3.5.5 Results of applying the relaxation method with p = 1 to obtain the field excited in a homogeneous, square cylinder with $\chi = 0$ and $\sigma = 10$ by a Gaussian pulse as specified in (3.5.17) with T = 10. (a): Result after 15 iteration steps. (b): Root-mean-square error as a function of the number of iteration steps. Discretization: single time step, N = 3.

digits. Note that performing the 15 iteration steps takes about half the computational effort required in solving the explicit marching-on-in-time equation with a twice-refined discretization, which is the least we would have to do to extend the stability range of that technique by a factor of two. The results shown in Figure 3.5.5 may also serve to demonstrate the poor condition of the discretized integral equation. As remarked above, the smooth solution displayed in Figure 3.5.5a meets the equality sign in that equation with an average accuracy of about 10^{-3}. On the other hand, the exact solution to the discretized equation is dominated by a large, alternating unstable component. This is in complete agreement with the conclusion of Subsection 3.2.2 that this unstable component can be associated with an eigenvector of the system matrix of the discretized equation with a small eigenvalue.

In case of a nonvanishing susceptibility contrast, both refining the discretization and increasing the susceptibility contrast intensify the unstable behavior in the results of the explicit marching-on-in-time method. Nevertheless, stable results can be obtained by raising the regularization parameter p to a sufficiently large value. An extreme case is presented in Figure 3.5.6, which shows results for T = 10, χ = 5, N = 3 and a discretization with the single-step difference rules used in (3.5.16). As Figure 3.5.6 indicates, a stable solution can be obtained by applying the relaxation method with p = 4. However, the computation becomes rather time consuming since it takes at least 80 iteration steps, each of which lasts 12.5 seconds, to arrive at the residual value of the root-mean-square error. The larger value of this error is due to the more rapid variation of the exact field with $\underline{\rho}$ and t, which results in a larger discretization error. The increase in the computation time per iteration step compared with the computations in Figures 3.5.3c,d, 3.5.4b

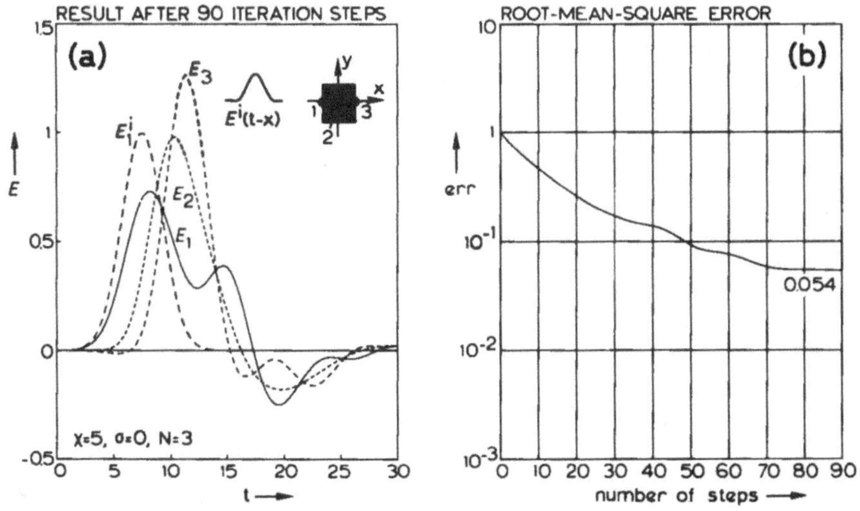

Figure 3.5.6 Results of applying the relaxation method with p = 4 to obtain the field in a homogeneous, square cylinder with χ = 5 and σ = 0, excited by a Gaussian pulse as specified in (3.5.17) with T = 10. (a): Result after 90 iteration steps. (b): Root-mean-square error as a function of the number of iteration steps. Discretization: single time step, N = 3.

and 3.5.5 is due to the extension of the time interval under consideration. Finally, the numerical experiment summarized in Figure 3.5.6 demonstrates that, even when one aims at applying the relaxation method, it is recommended to modify the discretization such that instabilities are suppressed as much as possible. This may allow a smaller value of the parameter p to be selected and, thus, it may lead to a faster convergence of the iterative procedure. For example, replacing the single-step difference formula (3.5.16) by the double-step formula (3.5.18) in the discretization employed in Figure 3.5.6 allows a reduction of the regularization parameter to $p = 1$. As a consequence, it takes only 20 iteration steps for the relaxation method to come up with the eventual solution.

3.6 Conclusions

In this chapter, a consistent analysis was given of the general properties of the marching-on-in-time method and of its application to a number of one- and two-dimensional electromagnetic problems. In the solutions obtained, two types of errors can be observed. In the first place, we observe instabilities that can be regarded as "local" errors that are located in some subspace of a vast vectorial space of possible solutions. The occurrence of these instabilities depends on the structure of the time-domain integral equation and on the space-time behavior of the exact field rather than on the accuracy of the discretization. In the second place, we observe a gradually increasing error due to the recursive method of solution that can be regarded as a "global" error in the remainder of the solution space.

Several methods were indicated to resolve the instability problem by modifying either the discretization, the problem formulation or the method of solution. When applicable, either of the first two seems preferable because of the saving in computation time compared with the last one. However, their applicability is limited, whereas the modified solution method seems generally applicable provided that the approximate behavior of the instabilities is known. Our modification of the solution method amounts to a regularized solution of the discretized integral equation in matrix form. In that respect, it is similar to applying a gradient-type method to minimize a squared error. However, it seems more suitable for the determination of a large solution vector since it operates on all components of that vector simultaneously. Moreover, the strict convergence of our method has been shown.

The global error can only be controlled by the method described in Subsection 3.2.2. This requires that the integrals in the time-domain integral equation be discretized with an error of the order of the square of the space-time discretization step. In such a discretization, the "global" error in a given time interval reduces linearly with this discretization step.

The improved stability analysis was also applied to analyze the solution of a number of specific one- and two-dimensional transient scattering problems. It turns out that the present analysis allows a more consistent explanation of this stability behavior than was given in Tijhuis (1984a,b). In addition, the results obtained by Rynne (1985), which deviate from the latter analysis, can be fitted in. The most striking result, however, is that the modified solution method derived with its aid can master the hitherto unmanageable instabilities that show up in the problem of plane-wave scattering by an inhomogeneous, lossy dielectric cylinder.

Finally, it should be remarked that the iterative nature of the modified solution procedure should not be considered a true objection to its repeated application in solving an inverse-scattering problem. In the iterative procedures that will be discussed in Chapters 5 and 6, the approximate field obtained in an iteration step provides an excellent starting value for the determination of the transient field in the next one.

References

Abramowitz, M., and Stegun, I.A. (1965), *Handbook of mathematical functions*, Dover Publications, New York.

Achenbach, J.D. (1973), *Wave propagation in elastic solids*, North-Holland Publishing Company, Amsterdam, p. 298.

Bennett, C.L. (1968), *A technique for computing approximate impulse response of conducting bodies*, Ph.D. Thesis, Purdue University, Lafayette, Indiana.

Bennett, C.L. (1978), The numerical solution of transient electromagnetic scattering problems. In: *Electromagnetic Scattering*, Uslenghi, P.L.E. (Ed.), Academic Press, New York, Chap. 11.

Bennett, C.L., and Mieras, H. (1981a), Time domain integral equation solution for acoustic scattering from fluid targets, *J. Acoust. Soc. Am. 69*, 1261-1265.

Bennett, C.L., and Mieras, H. (1981b), Time domain scattering from open thin conducting surfaces, *Radio Sci. 16*, 1231-1239.

Bennett, C.L., and Mieras, H. (1983), Space-time integral equation solution for hard or soft targets in the presence of a hard or soft half space, *Wave Motion 5*, 399-411.

Bennett, C.L., and Ross, G.F. (1978), Time-domain electromagnetics and its applications, *Proc. IEEE 66*, 299-318.

Bennett, C.L., and Weeks, W.L. (1970), Transient scattering from conducting cylinders, *IEEE Trans. Antennas Propagat. 18*, 627-633.

Bolomey, J.Ch., Durix, Ch., and Lesselier, D. (1978), Time domain integral equation approach for inhomogeneous and dispersive slab problems, *IEEE Trans. Antennas Propagat. 26*, 658-667.

Damaskos, N.J., Brown, R.T., Jameson, J.R., and Uslenghi, P.L.E. (1985), Transient scattering by resistive cylinders, *IEEE Trans. Antennas Propagat. 33*, 21-25.

De Hoop, A.T. (1960), A modification of Cagniard's method for solving seismic pulse problems, *Appl. Sci. Res. 3*, 179-188.

De Hoop, A.T. (1977), General considerations on the integral-equation formulation of diffraction problems. In: *Modern topics in electromagnetics and antennas*, Maanders, E.J., and Mittra, R. (Eds.), PPL Conference Publications 13, Peter Peregrinus, Stevenage, U.K., Chap. 6.

Friedman, M.B., and Shaw, R.P. (1962), Diffraction of pulses by cylindrical obstacles of arbitrary cross section, *J. Appl. Mech. 29*, 40-46.

Herman, G.C. (1981a), Scattering of transient acoustic waves by an inhomogeneous obstacle, *J. Acoust. Soc. Am. 69*, 909-915.

Herman, G.C. (1981b), *Scattering of transient acoustic waves in fluids and solids*, Ph.D. Thesis, Delft University of Technology, Delft, The Netherlands.

Herman, G.C. (1982), Scattering of transient elastic waves by an inhomogeneous obstacle: Contrast in volume density of mass, *J. Acoust. Soc. Am. 71*, 264-272.

Herman, G.C., and Van den Berg, P.M. (1982), A least-square iterative technique for solving time-domain scattering problems, *J. Acoust. Soc. Am. 72*, 1947-1953.

Hildebrand, F.B. (1956), *Introduction to numerical analysis*, McGraw-Hill, New York, Chap. 3.

Jones, D.S. (1964), *The theory of electromagnetism*, Pergamon Press, Oxford.

Lesselier, D. (1982), *Diagnostic optimal de la lame inhomogène en régime temporel. Applications à l'électromagnétisme et à l'acoustique*, Ph.D. Thesis, l'Université Pierre et Marie Curie, Paris.

Lesselier, D. (1983), P waves transient scattering by 2-D penetrable targets: A direct solution, *J. Acoust. Soc. Am. 74*, 1274-1278.

Marx, E. (1984), Integral equation for scattering by a dielectric, *IEEE Trans. Antennas Propagat. 32*, 166-172.

Marx, E. (1985a), Self-patch integrals in transient electromagnetic scattering, *IEEE Trans. Antennas Propagat. 33*, 763-767.

Marx, E. (1985b), Neighboring-patch integrals in transient electromagnetic scattering, *IEEE Trans. Antennas Propagat. 33*, 768-773.

Mieras, H., and Bennett, C.L. (1982), Space-time integral approach to dielectric targets, *IEEE Trans. Antennas Propagat. 30*, 2-9.

Miller, E.K., and Landt, J.A. (1980), Direct time-domain techniques for transient radiation and scattering from wires, *Proc. IEEE 68*, 1396-1423.

Mittra, R. (1976), Integral equation methods for transient scattering. In: *Transient electromagnetic fields*, Felsen, L.B. (Ed.), Springer Verlag, Berlin, Chap. 2.

Mitzner, K.M. (1967), Numerical solution for transient scattering from a hard surface of arbitrary shape-Retarded potential technique, *J. Acoust. Soc. Am. 42*, 391-397.

Morse, P.H., and Feshbach, H. (1953), *Methods of theoretical physics*, McGraw-Hill, New York, pp. 803-828.

NAG Fortran Library Mark 11 (1984), Numerical Algorithms Group, Oxford.

Neilson, H.C., Lu, Y.P., and Wang, Y.F. (1978), Transient scattering by arbitrary axisymmetric surfaces, *J. Acoust. Soc. Am. 63*, 1719-1726.

Poggio, A.J., and Miller, E.K. (1973), Integral equation solutions of three-dimensional scattering problems. In: *Computer techniques for electromagnetics*, Mittra, R. (Ed.), Pergamon Press, Oxford, Chap. 4.

Rao, S.M., Sarkar, T.K., and Dianat, S.A. (1984), The application of the conjugate gradient method to the solution of transient electromagnetic scattering from thin wires, *Radio Sci. 19*, 1319-1326.

Rynne, B.P. (1985), Stability and convergence of time marching methods in scattering problems, *IMA J. Appl. Math. 35*, 297-310.

Sarkar, T.K. (1984), The application of the conjugate gradient method to the solution of operator equations arising in the electromagnetic scattering from wire antennas, *Radio Sci. 19*, 1156-1172.

Sarkar, T.K., and Rao, S.M. (1984), The application of the conjugate gradient method for the solution of electromagnetic scattering from arbitrarily oriented wire antennas, *IEEE Trans. Antennas Propagat. 32*, 398-403.

Sarkar, T.K., Weiner, D.D., and Jain, V.K. (1981), Some mathematical considerations in dealing with the inverse problem, *IEEE Trans. Antennas Propagat. 29*, 373-379.

Shaw, R.P. (1967), Diffraction of acoustic pulses by obstacles of arbitrary shape with a Robin boundary condition, *J. Acoust. Soc. Am. 41*, 855-859.

Shaw, R.P. (1968), Diffraction of plane acoustic pulses by obstacles of arbitrary cross section with an impedance boundary condition, *J. Acoust. Soc. Am. 44*, 1062-1068.

Shaw, R.P. (1975), Transient scattering by a circular cylinder, *J. Sound Vibr. 42*, 295-304.

Stoer, J. (1972), *Einführung in die numerische Mathematik I*, Springer Verlag, Berlin, pp. 41-44.

Tijhuis, A.G. (1981), Iterative determination of permittivity and conductivity profiles of a dielectric slab in the time domain, *IEEE Trans. Antennas Propagat. 29*, 239-245.

Tijhuis, A.G. (1984a), Toward a stable marching-on-in-time method for two-dimensional transient electromagnetic scattering problems, *Radio Sci. 19*, 1311-1317.

Tijhuis, A.G. (1984b), Stability analysis of the marching-on-in-time method for one- and two-dimensional scattering problems. In: *Colloquium topics in applied numerical analysis*, Verwer, J.G. (Ed.), CWI Syllabus 5, Centre for Mathematics and Computer Science, Amsterdam, pp. 363-386.

Tikhonov, A.N., and Arsenin, V.Y. (1977), *Solutions of ill-posed problems*, V.H. Winston & Sons, Washington D.C., Chapters II, III.

Van den Berg, P.M. (1984), Iterative computational techniques in scattering based upon the integrated square error criterion, *IEEE Trans. Antennas Propagat. 32*, 1063-1071.

Widder, D.V. (1946), *The Laplace Transform*, Princeton University Press, Princeton, pp. 61-63.

IDENTIFICATION

4. A PRONY-TYPE METHOD

4.1 Introduction

In this chapter, we consider the identification of a scattering obstacle in terms of natural-mode parameters. To this aim, we try to determine the relevant parameters from the transient field excited by an incident pulse of finite duration. As we saw in Chapter 2, at each point of observation $\underline{x} = \underline{x}_\beta$, there exists a finite initial instant after which either the total or the scattered field can be written in the form

$$\underline{E}(\underline{x}_\beta,t) = \sum_\alpha F(s_\alpha)D_\alpha \underline{E}_\alpha(\underline{x}_\beta)\exp(s_\alpha t). \tag{4.1.1}$$

In (4.1.1), the complex natural frequencies $\{s_\alpha\}$ and the natural-mode distributions $\{\underline{E}_\alpha(\underline{x})\}$ depend on the configuration only. The coupling coefficients $\{D_\alpha\}$ depend on the configuration as well as on the spatial distribution of the incident field, while $F(s)$ is the Laplace-transformed time distribution of that field. The special form of (4.1.1) suggests the following identification procedure. First, the field $\underline{E}(\underline{x}_\beta,t)$ should be decomposed into exponentials with frequencies $\{s_\alpha\}$ and amplitudes $\{F(s_\alpha)D_\alpha \underline{E}_\alpha(x_\beta)\}$ by the application of a Prony-type method. For a known incident field, the time distribution of the incident field can then be factored out by dividing by $F(s_\alpha)$. This provides us with the set of configuration-dependent parameters $\{s_\alpha, D_\alpha \underline{E}_\alpha(\underline{x}_\beta)\}$, which may be regarded as the "signature" of the scattering obstacle. Next, the scatterer should be identified from that signature. In this identification, a priori information is possibly needed. The procedure outlined above is illustrated in Figure 4.1.1, which shows the role of a Prony-type method in the process of identifying an unknown object.

In the present chapter, we only concern ourselves with the steps indicated by the solid lines, i.e. with recovering the characteristic parameters of the exponentials constituting a transient signal. Most of the methods employed to effectuate such a reconstruction are based on an algorithm devised by Prony (1795). Although this algorithm was well known in the literature (Hildebrand (1956), Weiss and

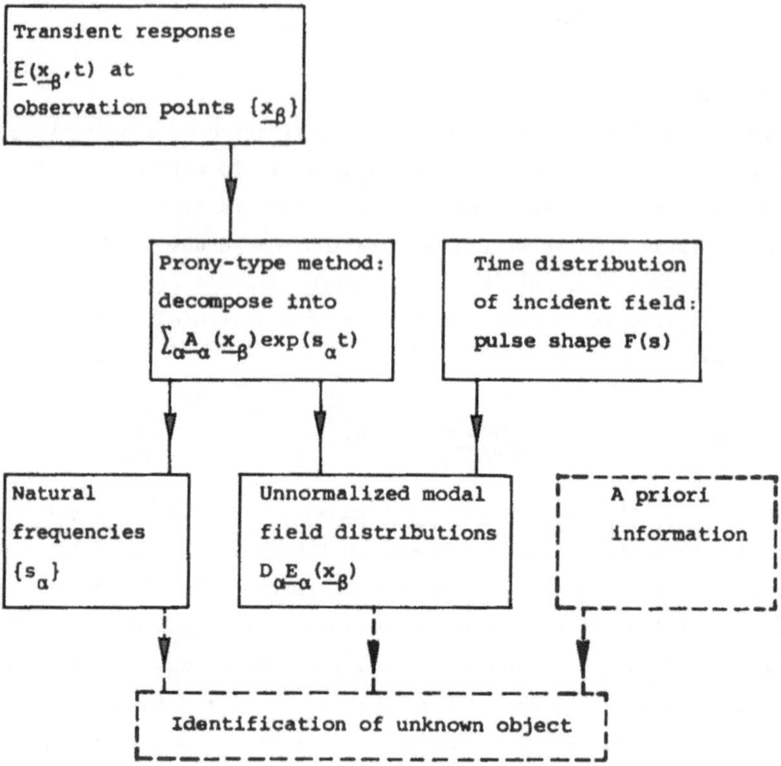

Figure 4.1.1 The role of a Prony-type method in the identification of an unknown obstacle. $\{\underline{x}_\beta\}$ denotes a set of observation points; other symbols as defined in Figure 2.1.1. Only the steps indicated by solid lines are treated in the present discussion.

McDonough (1963)), it was not until 1975 that it was first applied to a transient electromagnetic field (Van Blaricum and Mittra (1975)). Since then, a number of papers dealing with various aspects of Prony's method have appeared. The practical application of the technique as such to a realistic signal was treated in Poggio et al. (1978), Van Blaricum and Mittra (1978), Miller (1981) and Kulp (1981). A generalization to pole extraction from real-frequency information was described in Brittingham et al. (1980). The connection with current methods from system theory and from speech processing was studied by Dudley (1979,1983), Sarkar et al. (1982,1984), and Knockaert (1984). Within the restrictions imposed by the conditions for the validity of the SEM representation, most of these methods turn out to be equivalent. An exception is the pencil-of-functions method (Jain et al. (1983)), which provides a more systematic way of handling noise at the cost of more involved computations. Comparatively, only a few successful applications to actual signals, either measured or synthesized by some time-domain code, have been reported. Measured data were used in Pearson and Roberson (1980), and synthetic ones in Lin and Cordaro (1983).

As for the identification of the scatterer from the natural frequencies obtained, even less is known. The present state of affairs is that such an identification cannot be carried out without some *a priori* information. Such information can appear in two forms. In the first place, additional information may be available on the geometry of the configuration. The most relevant examples are a one-dimensional, lossless dielectric medium consisting of a specified number of layers (Lytle and Lager (1976), Miller and Lager (1982)), and a radially distributed, lossless dielectric sphere (Hoenders (1982), Eftimiu and Huddleston (1984)). Only the last reference actually includes a detailed description of a reconstruction procedure. In the second place, a stored library of known pole sets may be available. In that case, the unknown obstacle may be identified with the aid of some target classification algorithm (Miller et al. (1977), Bennett and Toomey (1981)).

The discussion in this chapter is organized as follows. In Section 4.2 we formulate Prony's method and describe all of those aspects which are relevant in a practical application. The material contained in that section is not new. Rather, it is a systematic compilation of results that have been dispersed over the literature referred to above. The argumentation is illustrated with the aid of results for a judiciously chosen test signal. In Section 4.3, we apply Prony's method to the reflected and transmitted fields caused by a sine-squared pulse, incident on a dielectric slab as discussed in Sections 2.3 and 3.3. A preprocessing procedure will be proposed that adapts these signals to the limitations of Prony's algorithm. When this preprocessing is included in the procedure, the resulting Prony-type scheme can indeed recover the natural frequencies of a given slab. By repeating the computation several times, even noisy signals can be handled. Section 4.4 speculates on the possibility of applying Prony's algorithm to a cylinder signal. Finally, the conclusions are stated in Section 4.5.

4.2 Prony's algorithm

4.2.1 Formulation of Prony's method

In this section, we consider the formal as well as the numerical aspects of Prony's method. To this end we first apply it in an idealized situation that is not likely to occur in practice. In this situation, we assume that we have a signal $f(t)$ consisting of M linearly independent exponentials. Since we aim to analyze time-domain electromagnetic fields, we will, in addition, assume $f(t)$ to be real-valued. However, this restriction is not really necessary. The application of Prony's method in more practical situations will be discussed further on.

Let us first discuss the principles of the algorithm. A signal $f(t)$ as specified above can be written in the form

$$f(t) = \sum_{m=1}^{M} A_m \exp(s_m t), \tag{4.2.1}$$

where A_m denotes a complex amplitude and s_m a complex frequency. The parameters $\{A_m\}$ and $\{s_m\}$ completely determine the signal and, hence, can be designated as the characteristic parameters. The problem we want to resolve is the reconstruction of these parameters from N + 1 samples $f[n]$ of $f(t)$ taken with sample period Δt at the equally spaced instants $t = n\Delta t$, $n = 0,1,\ldots,N$. At these instants (4.2.1) reduces to

$$f[n] \triangleq f(n\Delta t) = \sum_{m=1}^{M} A_m z_m^n, \quad n = 0,1,\ldots,N, \tag{4.2.2}$$

with

$$z_m \triangleq \exp(s_m \Delta t), \quad m = 1,2,\ldots,M. \tag{4.2.3}$$

The relation (4.2.2) constitutes a system of N + 1 nonlinear equations for the 2M unknown variables $\{A_m\}$ and $\{z_m\}$. Since such nonlinear equations are hard to solve, it seems favorable to convert (4.2.2) into two systems of linear equations for M unknowns.

The essence of Prony's method (Prony (1795)) is that it provides such a conversion. This conversion is achieved by defining the polynomial

$$P(z) = \prod_{m=1}^{M}(z - z_m) = -\sum_{\ell=0}^{M} p_\ell z^{M-\ell}, \tag{4.2.4}$$

whose zeros are $z = z_m$, $m = 1,2,\ldots,M$. From (4.2.4), we have $p_0 = -1$. The definition (4.2.4) prescribes an equivalence relation between the unknown zeros $\{z_m\}$ and the polynomial coefficients $\{p_\ell\}$. Consequently, it also suffices to determine the latter. From (4.2.2) and (4.2.4), it follows that

$$\begin{aligned}\sum_{\ell=0}^{M}\delta[n-\ell]p_\ell &= \sum_{\ell=0}^{M}p_\ell\sum_{m=1}^{M}A_m z_m^{n-\ell}\\ &= \sum_{m=1}^{M}A_m z_m^{n-M}\sum_{\ell=0}^{M}p_\ell z_m^{M-\ell}\\ &= \sum_{m=1}^{M}A_m z_m^{n-M}P(z_m) = 0,\end{aligned} \tag{4.2.5}$$

for $M \le n \le N$. With $p_0 = -1$, (4.2.5) can be rewritten as

$$\sum_{\ell=1}^{M}\delta[n-\ell]p_\ell = \delta[n], \quad M \le n \le N, \tag{4.2.6}$$

which constitutes a system of $N + 1 - M$ linear equations for these unknown p_ℓ's. For $N + 1 \ge 2M$, we can therefore solve them from (4.2.6) with the aid of some matrix-inversion procedure.

With (4.2.6), we now have available the basis of the procedure to reconstruct the characteristic parameters $\{A_m\}$ and $\{s_m\}$. The total procedure should progress as follows:

i) Determine the polynomial coefficients $\{p_\ell\}$ from the system (4.2.6). Since the elements of the time sequence $\delta[n]$ are real-valued, this will result in a set of real-valued coefficients $\{p_\ell\}$.

ii) Determine the zeros $\{z_m\}$ of the polynomial $P(z)$ defined according to (4.2.4). Since this polynomial has real-valued coefficients, these zeros must either be real-valued or occur in complex-conjugate pairs.

iii) The complex frequencies $\{s_m\}$ are then obtained directly by inverting the relation (4.2.3). Like the zeros $\{z_m\}$, these frequencies either are real-valued or occur in complex-conjugate pairs.

iv) Once the zeros $\{z_m\}$ are known, Equation (4.2.2) can be regarded as a system of N + 1 linear equations for the unknown amplitudes $\{A_m\}$. Since the samples $\mathfrak{f}[n]$ are real-valued, these amplitudes exhibit the same symmetry behavior as the corresponding $\{z_m\}$ and $\{s_m\}$.

It should be remarked that the symmetry behavior of $\{A_m\}$, $\{s_m\}$ and $\{z_m\}$ observed above is in agreement with similar observations regarding the natural modes made in Chapter 2.

In the reconstruction procedure formulated above, it is implicitly assumed that the total number of exponentials M is known. As we shall see in Section 4.3, this is indeed what happens in a practical situation, where the desired number of exponentials is an input parameter for the low-pass filtering that must precede the application of Prony's algorithm. Nevertheless, let us, for the sake of completeness, consider the determination of M from the sampled signal. This number can be found according to a procedure due to Van Blaricum and Mittra (1978). In this procedure, it is supposed that the number of samples is considerably larger than required for a unique determination of $\{A_m\}$ and $\{s_m\}$, i.e. $N >> 2M$. Furthermore, it is assumed that, from a priori information, a conservative estimate K of M is available such that $M \le K$ and $M < N - K$. In that case, the problem can be resolved by introducing the $N - K + 1$ vectors $\underline{\mathfrak{f}}^{(i)}$, whose components are defined by

$$\mathfrak{f}_k^{(i)} \triangleq \mathfrak{f}[i + k], \quad 0 \le i \le N - K, \; 0 \le k \le K. \tag{4.2.7}$$

By applying the relation

$$\sum_{i=0}^{M} \mathfrak{f}[i + k] p_{M-i} = 0, \quad 0 \le k \le N - M, \tag{4.2.8}$$

which is obtained from (4.2.5) by taking $n = k + M$ and $\ell = M - i$, it follows that these vectors satisfy the equation

$$\sum_{i=0}^{M} p_{M-i} \underline{f}^{(i)} = \underline{0}. \tag{4.2.9}$$

This equation states that the vectors $\{\underline{f}^{(i)}\}$ with $0 \le i \le M$ are linearly dependent. On the other hand, each set consisting of less than $M + 1$ vectors must be linearly independent since the exponentials in (4.2.1) are. It is this property that can be used to determine M with the aid of a Gram-Schmidt orthogonalization procedure. Starting from $\underline{f}^{(i)}$, we define the orthogonal vectors $\underline{g}^{(i)}$ as

$$\underline{g}^{(i)} = \underline{f}^{(i)} - \sum_{j=0}^{i-1} \frac{(\underline{g}^{(j)} \cdot \underline{f}^{(i)})}{(\underline{g}^{(j)} \cdot \underline{g}^{(j)})} \underline{g}^{(j)}. \tag{4.2.10}$$

As soon as $(\underline{g}^{(i)} \cdot \underline{g}^{(i)})$ vanishes for the first time, we have reached the value $i = M$.

4.2.2 Numerical implementation

Next, we consider the numerical implementation of Prony's algorithm. We restrict ourselves to the practical situation where the number of exponentials M is known. This leaves us with the implementation of steps i), ii) and iv) as outlined in Subsection 4.2.1. In the implementation of steps i) and iv), we have some freedom in the number of equations that we take into consideration. One variant is to use exactly M equations, i.e. the minimum number required. In step i), this amounts to choosing $N = 2M - 1$; in step iv), it can be achieved by selecting the equations (4.2.2) with $0 \le n \le M - 1$. In both steps, we thus obtain a square matrix equation that can, for example, be resolved by Gaussian elimination. It is this variant that was essentially proposed by Prony (1795).

In our implementation, we have chosen to use *all* the equations available. In doing so, we follow, among others, Miller (1981). In step i) as well as in step iv), we end up with an overdetermined system of equations. These two systems are solved with the aid of a linear-least-squares inversion procedure based on Householder transformations, which is available from the NAG library (NAG (1984), subroutine F04AMF). Ideally, there should be no difference between both variants, since the linear dependence for systems of more than M equations is exact. In practice, however, even slightly overdetermined systems turn out to be considerably better conditioned numerically than the corresponding square matrix equations.

In both variants, we have to determine the zeros of the polynomial P(z). For this problem, we again invoke a standard subroutine from the NAG library (loc. cit., subroutine C02AEF). This subroutine finds all the roots of a real polynomial equation within limiting machine precision, using the method of Grant and Hitchins.

As an illustration, we present results of the reconstruction of the characteristic parameters of a test signal. This test signal was chosen such that it should resemble the residual contribution from a subset of slab modes to the transmitted field excited by a delta-function incident pulse. Moreover, the specific choice of the test signal was inspired by a similar time signal used by Miller (1981). The amplitudes and frequencies of our signal are given by

$$\begin{aligned}
s_0 &= 0, & A_0 &= -1,\\
s_{\pm 1} &= -0.1 \pm i, & A_{\pm 1} &= 1,\\
s_{\pm 2} &= -0.25 \pm i2.5, & A_{\pm 2} &= -1,\\
s_{\pm 3} &= -0.5 \pm i5.0, & A_{\pm 3} &= 1,\\
s_{\pm 4} &= -0.75 \pm i7.5, & A_{\pm 4} &= -1,\\
s_{\pm 5} &= -1.0 \pm i10.0, & A_{\pm 5} &= 1.
\end{aligned} \tag{4.2.11}$$

Note that the numbering of the exponentials in (4.2.11) differs from the one used in (4.2.1). The aim of this renumbering was to bring out the symmetry in the characteristic parameters and to match their ordering with the one used in Sections 2.2 and 2.3 for the slab modes. The range of the complex frequencies $\{s_m\}$ has been derived from the numerical results obtained in Subsections 2.3.5 and 2.5.5. Apart from an irrelevant multiplication factor, the amplitudes were taken equal to the asymptotic coupling coefficients for a slab (see e.g. Table 2.3.3).

The way Prony's algorithm works is illustrated in Figure 4.2.1,

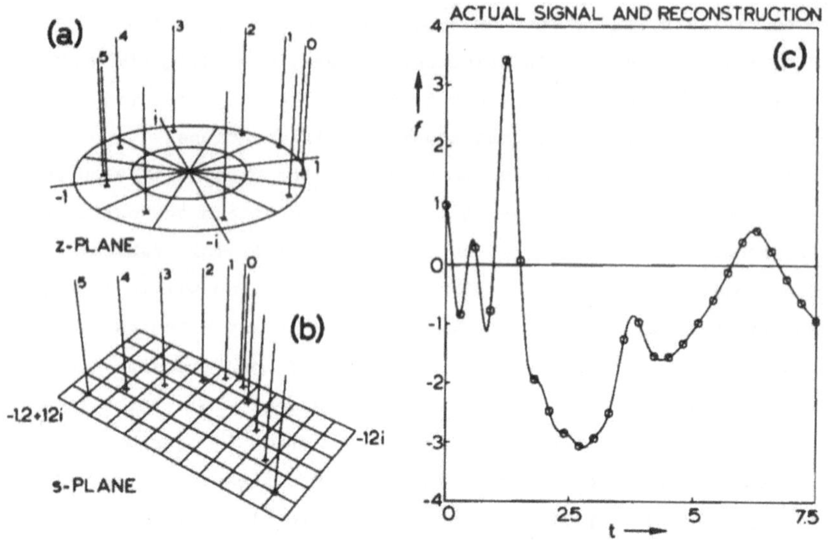

Figure 4.2.1 Results obtained from applying Prony's method to the test signal specified in (4.2.11), sampled with $\Delta t = 0.3$ and $N = 25$. (a): $\{z_m\}$; (b): $\{s_m\}$. In both figures, the lengths of the vertical lines indicate $|A_m|$; the values with $m \geq 0$ have been marked for comparison with Figure 4.2.2. (c): Solid line: actual signal; dashed line: reconstructed signal (indistinguishable); o: reconstructed signal at the sample points.

which shows results obtained from a sampling with sampling period $\Delta t = 0.3$ and $N = 25$ (this accounts for four samples more than necessary). In Figures 4.2.1a,b, the recovered frequency parameters $\{z_m\}$ and $\{s_m\}$ have been plotted in the complex z- and s-plane, respectively, along with the magnitudes of the corresponding amplitudes $\{A_m\}$. In Figure 4.2.1c, the actual signal $f(t)$ is compared with the reconstruction according to (4.2.1) at and in between the sample points. The reconstructed values of the characteristic parameters turned out to be accurate up to machine precision, i.e. up to a 16 digit accuracy. This example illustrates that, in principle, Prony's algorithm indeed allows an accurate numerical computation of such parameters.

4.2.3 Problems associated with Prony's method

The accuracy of the results obtained in the numerical experiment described above was only achieved because we applied the Prony scheme to an "ideal" sampling of an "ideal" signal. Generally, the recovered data will be considerably less accurate. In the present subsection, we will briefly discuss some of the problems that may lead to this loss of accuracy. We will demonstrate these problems by repeating the numerical experiment outlined above in less ideal conditions.

Aliasing effect (undersampling)

As we suggested above, particularly the choice of the sampling period Δt influences the accuracy of the reconstruction. This influence can be understood from the behavior of the $\{z_m\}$ defined in (4.2.3) in the complex z-plane. A necessary condition for the $\{s_m\}$ to be uniquely represented by

$$s_m = \ln(z_m)/\Delta t, \tag{4.2.12}$$

with $-\pi < \mathrm{Im}[\ln(z)] \le \pi$, is that

$$-\pi < \mathrm{Im}(s_m \Delta t) \le \pi. \tag{4.2.13}$$

For a given sampling period Δt, (4.2.13) implies that we can only handle complex frequencies whose imaginary part is smaller than the *Nyquist frequency* $\pi/\Delta t$. Frequencies with larger imaginary parts are only recovered from (4.2.12) modulo $2\pi i/\Delta t$. This phenomenon is known as the *aliasing effect*. For a given set of frequencies, (4.2.13) implies that a unique reconstruction can only be obtained if the sampling period meets the criterion

$$\Delta t < \pi/\omega_{max}, \tag{4.2.14}$$

where ω_{max} denotes the maximum magnitude of $\mathrm{Im}(s_m)$. Hence, an a priori estimate of ω_{max} is needed for a successful application of Prony's algorithm.

The effect of choosing the sampling period Δt too large is illustrated in Figure 4.2.2, which presents results from a computation with $\Delta t = 0.6$ and $N = 25$. This time step is considerable larger than

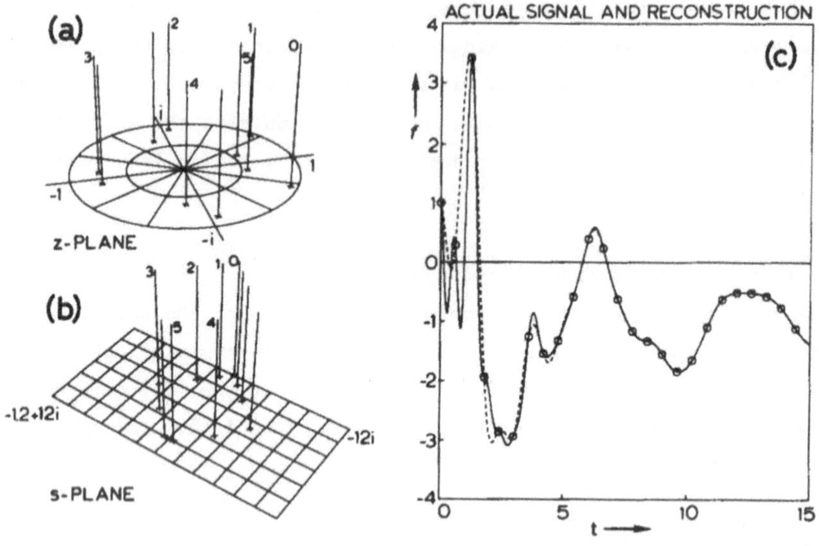

Figure 4.2.2 Results obtained from applying Prony's method to the test signal specified in (4.2.11), sampled with $\Delta t = 0.6$ and $N = 25$. Symbols and ordering as given in Figure 4.2.1.

the maximum prescribed by (4.2.13), i.e. $\pi/10$ (see (4.2.11)). It is observed that the values of $s_{\pm 4}$ and $s_{\pm 5}$ are indeed recovered with a shift of $\mp 2\pi/\Delta t$ along the imaginary s-axis. As a result, the reconstructed signal differs considerably from the actual one in between the sample points.

For real-valued signals, some caution should be excercised in applying criterion (4.2.14). When Δt is chosen such that a particular z_m gets close to the negative real z-axis, its complex conjugate will as well. In that case, it may be difficult to numerically distinguish between the pair of them, which will result in a loss of accuracy in the reconstruction of *all* characteristic parameters. This observation, too, was confirmed by numerical experiments for the test signal specified in (4.2.11).

Oversampling

The aliasing effect explained above can be avoided by choosing a sample period smaller than π/ω_{max}. In theory, this can be achieved by choosing Δt very small, i.e. $\Delta t << \pi/\omega_{max}$. However, for too small a value of Δt, we may have $z_m = \exp(s_m t) \approx 1$ for some or all of the $\{z_m\}$. This would lead to similar numerical ambiguities as outlined above for the case of $\Delta t \approx \pi/\omega_{max}$. Since these difficulties may concern more than a single pair of $\{z_m\}$, oversampling the signal in this manner will have quite a severe effect on the accuracy of the reconstructed characteristic parameters. An example is shown in Figure 4.2.3, which shows results for $\Delta t = 0.03$ and $N = 25$, i.e. a ten times finer sampling than in the experiment of Figure 4.2.1. Clearly, only the parameters with $|m| > 3$ are reconstructed with some accuracy. In addition, the error in the reconstruction causes a sizable deviation at late times between the actual signal and the obtained representation of the form (4.2.1). This deviation suggests that a better reconstruction could be obtained by sampling over a longer time interval. To verify this, we repeated the computation with $N = 250$, which means sampling the signal over the same time range as shown in Figure 4.2.1c, with a ten times finer sampling.

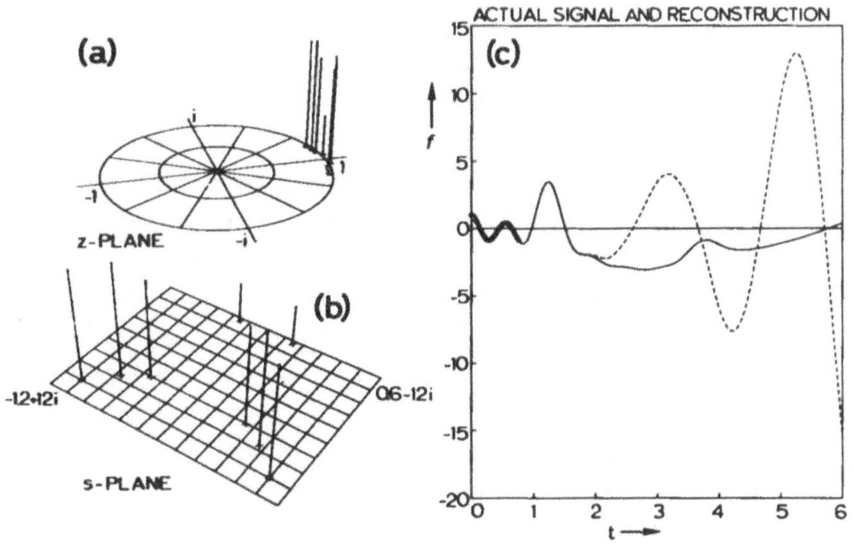

Figure 4.2.3 Results obtained from applying Prony's method to the test signal specified in (4.2.11), sampled with $\Delta t = 0.03$ and $N = 25$. Symbols as given in Figure 4.2.1.

Note that this sampling contains all the data used in the computation leading to Figure 4.2.1, along with nine times as many other data values. Nevertheless, the frequencies were only reconstructed with an accuracy varying from 10^{-3} (for $m = 0$) to 10^{-6} (for $m = \pm 5$), while the error in all recovered amplitudes was about 10^{-2}.

From the numerical experiments outlined above, it seems fair to conclude that we can only hope to apply Prony's algorithm with some success when the signal is sampled such that the $\{z_m\}$ are more or less evenly dispersed in the entire region $|z| < 1$, as shown in Figure 4.2.1a. An additional confirmation of our conclusion is obtained from the numerical experiments performed by Kulp (1981), who proposed a condition number by which the optimum sampling for a given, discretized signal can be established when a priori information is not available.

Noise

As mentioned in the introduction, the presence of even a small amount of noise on the data may deteriorate the accuracy of the parameter reconstruction considerably. One way of understanding this instability in Prony's method is to realize that, in decomposing (4.2.2) into two linear systems of equations, we have forfeited the possibility to compute $\{A_m\}$ and $\{z_m\}$ such that they "best fit" the sampled data $\mathfrak{f}[n]$ with respect to some error criterion. Because of this property, solving (4.2.6) in least-squares sense, as described in Subsection 4.2.2, is known in the literature as "semi-least-squares inversion" (Miller (1981)). The only motivation of this approach seems to be that it attaches an equal value to each individual equation of the overdetermined system.

Alternatively, the instability mentioned above can be understood from the occurrence of values of $\mathfrak{f}[n]$ in the system matrix of (4.2.6), the elements of which are inaccurate when the elements of the sequence $\mathfrak{f}[n]$ are. As demonstrated by Sarkar et al. (1981), such an inaccuracy may result in considerable errors in the solution vector $\{p_\ell\}$. Moreover, Dudley (1979) has shown both theoretically and with the help of specific examples that even small errors in these polynomial coefficients $\{p_\ell\}$ can produce large errors in the zeros

$\{z_m\}$ of the polynomial P(z) constructed from them.

To obtain an impression of the influence of noise in our application, we have used two different procedures to generate a "noisy" signal. Both these methods start from a set of "exact" sampled data $f[n]$ and a set of random numbers $\xi[n]$, distributed uniformly over the interval $-1 < \xi < 1$. In the case of *multiplicative noise*, we combine these sets into $f[n](1 + r\xi[n])$, where the ratio r relates the amplitude of the noise to the local magnitude of the signal. One way to envisage this type of noise is to consider it as being caused by inaccuracies in measuring the sampled signal. Such inaccuracies involve, for example, timing errors and gain errors due to the detection of $f[n]$ up to a limited number of significant digits. An alternative interpretation is obtained from the property that multiplicative noise decays in time with the same rate as the actual signal. In that sense, it resembles the residual contribution from higher-order poles, neglected in truncating the summation in (4.2.1) at $m = M$. In the case of *additive noise*, the noisy signal is given by $f[n] + r f_{max}\xi[n]$, where f_{max} denotes the maximum magnitude of $f[n]$ for $0 \le n \le N$. This type of noise can be understood as originating from a constant-amplitude background that interferes with the actual signal.

As a first example, we present in Table 4.2.1 and in Figure 4.2.4 the result of applying Prony's method to three different realiza-

Table 4.2.1 Results of applying Prony's algorithm to three different realizations of the sampled signal defined in Figure 4.2.1, contaminated with 1% multiplicative noise.

Reconstructed frequencies s_m			
m=0	-0.015	0.019	0.040
m=1	-0.110+i1.009	-0.099+i0.985	-0.095+i0.968
m=2	-0.243+i2.501	-0.249+i2.507	-0.259+i2.512
m=3	-0.503+i5.003	-0.506+i4.988	-0.499+i4.993
m=4	-0.721+i7.507	-0.745+i7.518	-0.791+i7.506
m=5	-1.091+i9.822	-0.901+i10.040	-0.975+i10.033
Reconstructed amplitudes A_m			
m=0	-1.070	-0.913	-0.823
m=1	1.027-i0.046	1.014+i0.066	1.015+i0.149
m=2	-0.975+i0.005	-0.992+i0.033	-1.032+i0.052
m=3	1.025-i0.002	1.008+i0.042	0.981+i0.027
m=4	-0.931+i0.032	-1.004+i0.066	-1.051+i0.015
m=5	0.893+i0.500	0.932-i0.148	1.002-i0.138

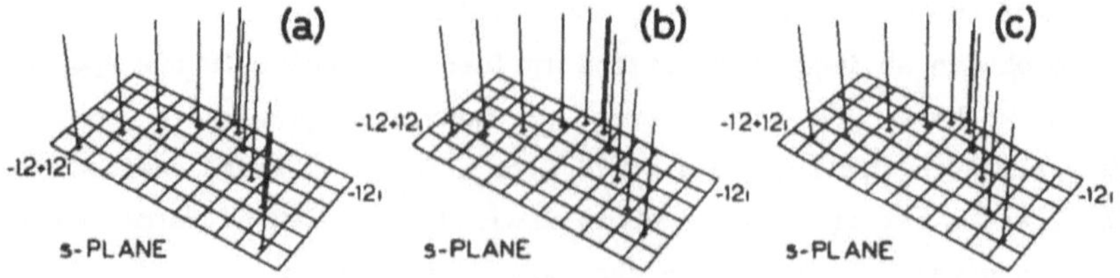

Figure 4.2.4 Three-dimensional representation of the results obtained from noisy data as specified in Table 4.2.1.

tions of the sampled signal used in Figure 4.2.1, contaminated with one percent multiplicative noise ($r = 0.01$). It is observed that even this small amount of noise causes deviations up to 10% in the reconstructed frequencies and up to 50% in the corresponding amplitudes. This indicates that noise should be handled with extreme care. On the other hand, the reconstructed values of $\{s_m\}$ and $\{A_m\}$ seem to spread around the actual ones. This indicates that a more accurate estimate of the latter could be obtained by repeating the Prony calculation for subsequent measurements. In Section 4.3, results of such a repetition will be presented for both types of noise.

4.2.4 Connection with other methods

In this last subsection, we discuss briefly the connection between Prony's method as formulated above and other methods that are currently being applied in the processing of transient signals. For a more detailed study of the connections listed here and related ones, the reader is referred to Sarkar et al. (1982,1984), Dudley (1979, 1983), and Knockaert (1984). The first connection follows from (4.2.6). From that equation, it is observed immediately that the polynomial coefficients $\{p_\ell\}$ are nothing but the predictor coefficients of a *linear prediction filter*. In fact, apart from the possible effects of numerical round-off errors, solving (4.2.6) by least-squares inversion is equivalent to solving the square matrix equation

$$\sum_{\ell=1}^{M}\left\{\sum_{n=M}^{N} f[n-k]f[n-\ell]\right\}p_\ell = \sum_{n=M}^{N} f[n-k]f[n], \tag{4.2.15}$$

with $k = 1,2,\ldots,M$. Equation (4.2.14) is precisely the result in the *covariance method* in linear prediction.

A further connection with system theory is obtained when the sampled signal $f[n]$ with $0 \le n \le N$ is envisaged as a subset of the causal time sequence $f[n]$ with $n\in\mathbb{Z}$, where $f[n]$ is given by (4.2.2) for $n \ge 0$, and where $f[n] = 0$ for $n < 0$. The sequence $f[n]$ can then be regarded as the linear response to an input sequence $g[n]$ passing before $t = 0$, i.e. with $g[n] = 0$ for $n \ge 0$. In that context, (4.2.6) can be viewed upon as being a corollary of the linear difference equation

$$f[n] = \sum_{\ell=1}^{M} p_\ell f[n-\ell] + \sum_{\ell=0}^{L} q_\ell g[n-\ell]. \tag{4.2.16}$$

Moreover, for the equation (4.2.16) to be consistent with (4.2.6) for $n \ge M$ and with the specifications of $f[n]$ and $g[n]$ given above, we have to choose $q_0 = 0$, $L = M$ and $g[n] = G\delta[n+1]$, where G denotes a real-valued amplitude and where $\delta[n]$ denotes the delta sequence

$$\delta[n] \triangleq \begin{cases} 1, & \text{for } n = 0, \\ 0, & \text{for } n \ne 0. \end{cases} \tag{4.2.17}$$

Because of the linearity of (4.2.16), we can, without loss of generality, take $G = 1$.

In order to obtain the system function of the difference equation (4.2.16), we introduce the bilateral z-transform of a time sequence $f[n]$ as

$$F(z) \triangleq \sum_{n=-\infty}^{\infty} f[n]z^{-n}. \tag{4.2.18}$$

Taking the z-transform of (4.2.16) and substituting $q_0 = 0$ and $L = M$ as obtained above results in

$$H(z) \triangleq \frac{F(z)}{G(z)} = \frac{\sum_{\ell=1}^{M} q_\ell z^{-\ell}}{1 - \sum_{\ell=1}^{M} p_\ell z^{-\ell}} = \sum_{\ell=1}^{M} q_\ell z^{M-\ell} / P(z), \tag{4.2.19}$$

with P(z) being the characteristic polynomial defined in (4.2.4). H(z) represents the quotient between the output F(z) and the input G(z) of the linear system and, hence, can be regarded as its system function. From (4.2.19), it is observed that the poles of this system function are exactly the $\{z_m\}$ defined in 4.2.3.

The relation between the amplitudes $\{A_m\}$ and the coefficients occurring in (4.2.16) and (4.2.19) is obtained by subjecting the sequence $\delta[n]$ specified above to the z-transformation (4.2.18). This yields

$$F(z) = z\sum_{m=1}^{M} A_m/(z - z_m) = z\sum_{m=1}^{M} \left\{A_m \Pi_{m' \neq m}(z - z_{m'})\right\}/P(z). \tag{4.2.20}$$

With G(z) = z, we directly identify

$$Q(z) \triangleq \sum_{\ell=1}^{M} q_\ell z^{M-\ell} = \sum_{m=1}^{M} \left\{A_m \Pi_{m' \neq m}(z - z_{m'})\right\}, \tag{4.2.21}$$

which, for given $\{z_m\}$, prescribes an equivalence relation between the $\{q_\ell\}$ and the $\{A_m\}$.

The first part of (4.2.20) represents $H(z) = z^{-1}F(z)$ in terms of an *all-pole model*. In view of the connections made above, Prony's method is equivalent to all techniques that are based on such a representation, such as linear prediction filtering, and the maximum entropy method used in speech processing (see also Miller (1981)). With (4.2.19) and the second part of (4.2.20), we also have available the *pole-zero model* corresponding with Prony's method.

Finally, it should be pointed out that the equivalences listed above refer to the case of *exact data*. In the presence of noise, however, the results of various procedures may differ. This does not refer to the sampling problems; these will show up in all solution techniques based on the difference equation, the all-pole model, or the pole-zero model specified above. However, it makes a considerable difference whether one chooses to minimize the actual reconstruction error in (4.2.2) or the error in the equality sign of either of the difference equations (4.2.6) and (4.2.16). In the former case, the reconstructed sampled signal will aways resemble the actual one. In the latter case, it may happen that the reconstructed signal does not resemble the actual one at all. In Subsection 4.3, an example of a test signal that exhibits this effect will be presented and discussed.

In order to weigh the consequences of minimizing the error in the equality sign of (4.2.2) against those of minimizing the error in (4.2.6) as in Prony's method, let us consider how the former can be achieved. From Parseval's theorem, it follows that we should minimize the real-frequency integral

$$\frac{\Delta t}{2\pi}\int_{-\pi/\Delta t}^{\pi/\Delta t} \left| H(\exp(i\omega\Delta t)) - \exp(-i\omega\Delta t)F(\exp(i\omega\Delta t)) \right|^2 d\omega, \qquad (4.2.22)$$

where $H(z)$ is given by (4.2.19), with respect to the real-valued polynomial coefficients $\{p_\ell\}$ and $\{q_\ell\}$. Once these coefficients have been determined, the characteristic parameters $\{z_m\}$ and $\{A_m\}$ are obtained directly with the aid of the correspondences derived above. The principal problem in the numerical implementation of this alternative scheme is the determination of the coefficients $\{p_\ell\}$ and $\{q_\ell\}$, which requires the use of a nonlinear optimization procedure like the conjugate-gradient method or the quasi-Newton method. A discussion of such methods will be given in Section 6.3, where they are applied in a different context. A well known disadvantage is that their application is hindered by convergence problems and by the occurrence of local minima (see also Jain et al. (1983)).

Because of these problems, it seems fair to conclude that any reconstruction of characteristic parameters from a sampled time signal should begin with an attempt to apply Prony's algorithm. As discussed in Subsection 4.2.2, the numerical implementation of the algorithm is straightforward and can be achieved with the aid of commonly available standard software. The accuracy of the reconstructed parameters can then be measured from the error in the equality sign of (4.2.2). Moreover, as Table 4.2.1 and Figure 4.2.4 indicate, the results thus obtained can be used to generate suitable starting values for a nonlinear optimization procedure that minimizes the global squared error in that equality sign. In this manner, the convergence and ambiguity problems associated with the application of such a method can possibly be avoided.

4.3 Application to a slab signal

4.3.1 Formulation of the problem

In this section, we discuss the application of Prony's method to transient signals in a slab configuration as investigated in Sections 2.3 and 3.3. With the marching-on-in-time procedure outlined in the latter section, we are able to determine such signals directly in the time domain. Moreover, by choosing the space-time step small enough, we can attain a specified accuracy. On the other hand, the analysis of Section 2.3 has provided us with an accurate procedure to determine the natural frequencies, and with initial instants for the validity of the SEM representation at each space point. Therefore, we are in the unique position to test Prony's algorithm for an actual, albeit simple scattering configuration.

4.3.2 Filtering procedure (preprocessing)

The main obstacle to a successful application of Prony's algorithm to a slab signal seems to be the total residual contribution from higher-order poles. Whether this contribution is regarded as a high-frequency term subject to the aliasing effect, or as multiplicative noise, the experiments performed in the previous section show that it may considerably disturb the reconstruction. As the experiment in Figure 4.2.3 shows, this problem cannot be resolved by increasing the sampling rate. This would destroy the accuracy in the reconstruction of the lower-order poles, which are most characteristic for the slab configuration (see Subsections 2.3.3 and 2.3.5).

As an illustration of the effect of higher-order pole contributions, let us first consider a simple synthetic signal that contains such contributions. We define

$$\mathcal{G}(t) \triangleq \sum_{n=0}^{\infty} D^n F(t - n\tau_D), \qquad (4.3.1)$$

where $F(t)$ denotes the sine-squared pulse

$$F(t) = \sin^2(\pi t/T)\,\mathrm{rect}(t - T/2;T). \qquad (4.3.2)$$

The expression (4.3.1) can be envisaged as a simplified relation for the high-frequency approximations to the reflected and transmitted fields as introduced in (2.2.46) and (2.3.35). In (4.3.1) and (4.3.2), T is a pulse duration, D a damping factor and τ_D a delay time. The characteristic parameters of $\delta(t)$ are

$$s_m = \ln(D)/t_D + im2\pi/\tau_D,\ A_m = F(s_m)/\tau_D, \tag{4.3.3}$$

with $m \in \mathbb{Z}$. The signal vanishes in the time intervals $T + n\tau_D \leq t \leq (n + 1)\tau_D$, with $n = 0,1,2,\ldots,\infty$. For a small enough T and a sufficiently fine sampling, the system (4.2.6) may contain equations involving only samples from these intervals. Such equations, with zero coefficients, will not affect the obtained $\{s_m\}$ at all. In that case, these frequencies are not constructed from the whole signal and, hence, cannot be correct. This observation was verified by the following numerical experiment. First Prony's algorithm was applied with $M = 25$, $\Delta t = 0.04$ and $N = 54$ to the signal given by (4.3.1) and (4.3.2) with $T = 0.4$, $\tau_D = 1.5$ and $D = 0.9$. The representation in terms of exponentials thus obtained did not even resemble the actual signal. Next, the computation was repeated for a signal synthesized from exponentials as specified in (4.3.3) with $|m| \leq 12$. Now, the characteristic parameters were obtained within machine precision. Nevertheless, the deviation between both sampled signals was of the order of 10^{-3}.

This experiment demonstrates that Prony's algorithm can only be applied successfully to a sampled signal from which the contributions of the higher-order poles have been removed. This need not prevent the possible identification of the unknown slab configuration, for which primarily the lower-order poles are important (see (2.3.19) - (2.3.21)). The conclusion stated above also has a bearing on the interpretation of a one-dimensional scattering problem within the framework of system theory. As shown in Section 2.3, the coupling coefficients for the slab configuration are all of the same order of magnitude. This implies that the resulting pole-zero system function for the scattering problem itself is of infinite order. Moreover, that system function applies only after a finite initial instant. To arrive at a system of finite order that holds at all sample points, we have to adapt our picture of the scattering problem as shown in Figure 4.3.1. The interpretation of the input sequence $\delta[n + 1]$ as a discrete version of a delta-function incident pulse that hits the slab immediately before $t = 0$ can remain as it is. However, this pulse must pass through a low-pass frequency filter before it actually hits the slab. This filter should be chosen such that its output signal excites only the slab modes with $-M \leq m \leq M$ for some specified M and yet is contained as much as possible within a time interval of known length T. Subsequently, the thus generated excited field must pass through a time delay filter that accounts for the validity conditions for the SEM representation summarized in Table 2.3.2. This filter transmits the signal only after $t = t_0(z)$, with $t_0(z)$ being the first instant where a nonvanishing field is represented solely by natural-mode contributions. The signal originating from the latter filter can then be subjected to Prony's method.

We will now describe one way to simulate a composite system as specified above for the slab configuration. The procedure is based on work of Poggio et al. (1978), and on some of the observations made in Subsections 2.4.3 and 4.2.3. We consider the determination

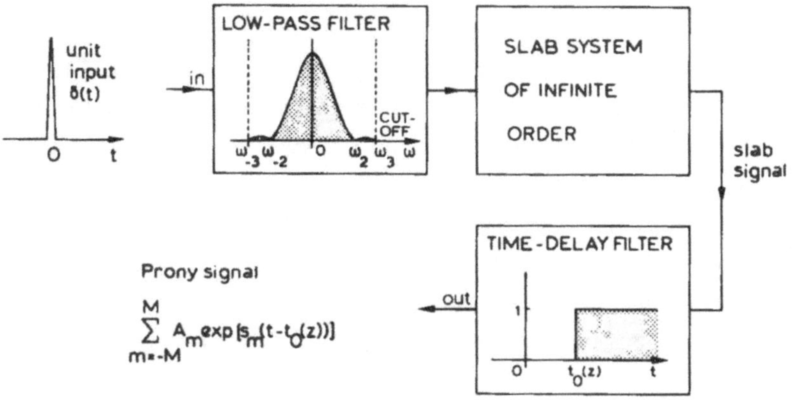

Figure 4.3.1 Interpretation of the scattering by a dielectric slab in terms of a system of finite order.

of the modal parameters with $-M \le m \le M$. We suppose that the signal is sampled considerably finer than prescribed by (4.2.13), and that the incident field is a sine-squared pulse whose duration T can be adjusted. In that case, the following steps lead to a sampled signal tailored to the application of Prony's algorithm. First, we determine the spectrum of the sampled signal with the aid of an FFT algorithm. For a sine-squared incident pulse, this spectrum has zeros at the known values of $\omega_m = 2\pi m/T$, with $m = \pm 2, \pm 3, \ldots, \pm\infty$. In between these zeros, local minima and maxima that are associated with the natural frequencies are observed. This frequency behavior makes it possible to adjust T such that ω_2, or possibly ω_3, lies in between $\mathrm{Im}(s_M)$ and $\mathrm{Im}(s_{M+1})$. From the frequency-domain response of the thus adjusted signal, we compute a new, synthetic signal by setting all components with $|\omega| > \omega_{2,3}$ to zero, and carrying out the inverse Fourier transform. Thus, we have accomplished the low-pass filtering. This leaves us with the time-delay filtering, and the resampling of the new synthetic signal such that the $z_m = \exp(s_m \Delta t)$ with $-M \le m \le M$ will be evenly dispersed across the unit disk $|z| < 1$. To this end, we utilize the property of the low-pass filtering procedure that each sample of its output signal depends on *all* input data. Therefore, we can thin these samples out by selecting every j'th one with j being the minimum integer value such that $j\Delta t \ge \pi/\omega_{2,3}$, i.e. such that the cut-off exceeds the Nyquist frequency $\pi/j\Delta t$. The time-delay filtering is effectuated by starting this operation at the first sample point where the condition for closure to the right from Table 2.3.2 does not hold and that for closure to the left does (see Table 2.3.2). This limits the resampled data $\mathfrak{f}[n]$ to nonvanishing values that can be generated by a truncated version of the conventional SEM representation. Hence, these data are suitable as input for Prony's algorithm. Obviously, the number of unknown exponentials should be set to 2M + 1. Note that, when the original sampling is not fine enough, the recovered frequency pair $s_{\pm M}$ may be aliased. This is not a true objection, because the proximity of the cut-off causes the reconstruction to be poor anyway. Moreover, choosing the Nyquist frequency above the cut-off produced inferior results.

4.3.3 Numerical results

Using the preprocessing procedure explained above and the computational scheme described in Section 4.2, we have obtained numerical results for several slab configurations. In this subsection we present some illustrative examples. The course of a typical computation is shown in Figure 4.3.2, which displays results for the "standard" configuration considered in Figures 2.3.4 - 2.3.7c and 2.4.4 - 2.4.7a, excited by a sine-squared pulse with T = 1.2. Figure 4.3.2a shows part of the reflected field obtained from carrying out 2000 marching-on-in-time steps with h = 1/160. In Figure 4.3.2b, the corresponding frequency spectrum has been plotted. The dashed line marks the cut-off at $\omega = \pm\omega_3$. As argued in Subsection 2.4.3, the ab-

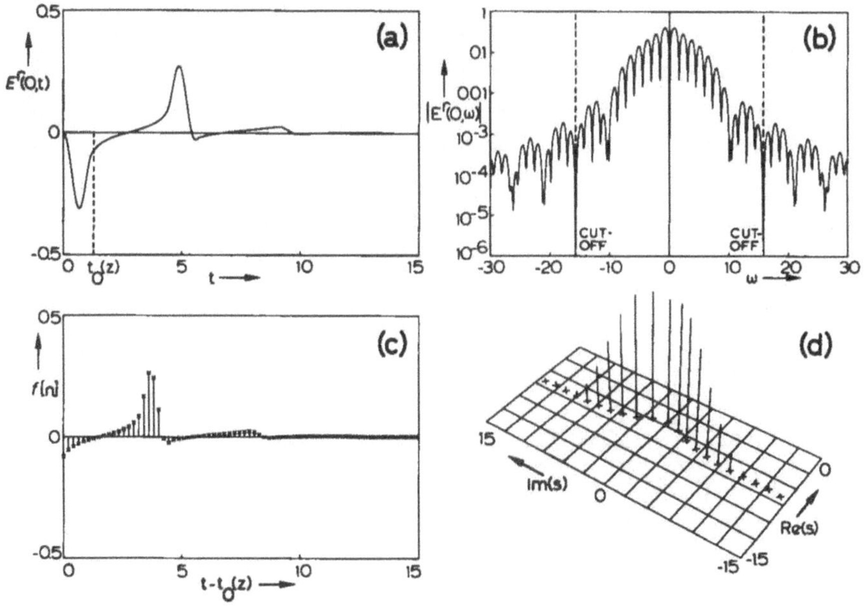

Figure 4.3.2 Results of reconstructing the SEM parameters of the lossless slab specified in Figure 2.3.4. (a): reflected field computed by the marching-on-in-time method with h = 1/160; (b): frequency spectrum and cut-off; (c): vertical lines: resampled data; x: reconstruction; (d): three-dimensional illustration of recovered SEM parameters.

sorption peaks in this reflection spectrum correspond to the natural frequencies. Taking into account the zeros at $\omega = \pm\omega_2$, we then observe that the filtered signal contains the modal contributions with $-10 \le m \le 10$. The resampled data values that are put into Prony's algorithm are indicated by the vertical lines in Figure 4.3.2c. Note that the resampling starts at $t = T$, i.e. when the incident pulse has completely emerged into the slab. Figure 4.3.2d presents a three-dimensional picture of the results obtained. Compared with Figure 2.3.6a, only the magnitudes of the weighting factors differ slightly due to the different incident-pulse duration. The accuracy of the reconstructed natural frequencies varied from 5×10^{-5} for $m = 0$ to 4×10^{-2} for $m = 10$. Furthermore, the reconstructed s_m with $|\mathrm{Im}(s_m)| < \omega_2$ were significantly better than the remaining ones. Finally, the x's in Figure 4.3.2c denote the values at the sample points of the exponential series obtained. Similar computations were also carried out for the transmitted field, with the resampling starting at $t = \tau$, and for both fields with the cut-off at $|\omega| = \omega_2$. These experiments led to similar results and similar accuracies.

Results for the discontinuous slab defined in Figure 2.3.11b are presented in Figure 4.3.3 and in Table 4.3.1. In this case, the incident-pulse duration was shortened to $T = 0.65$, which put the poles

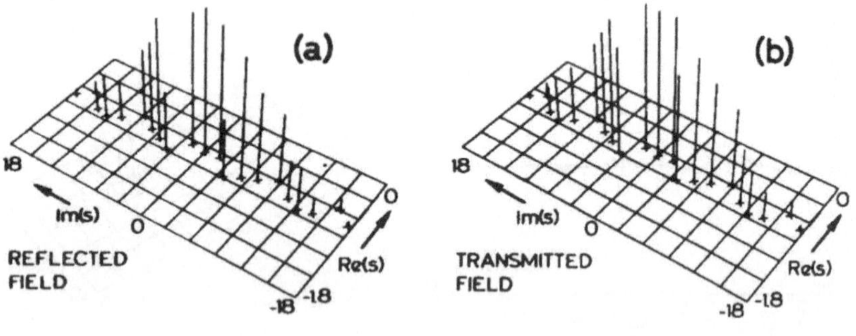

Figure 4.3.3 Final results of applying the preprocessing procedure and Prony's algorithm to the reflected and transmitted field excited by a sine-squared incident pulse of duration $T = 0.65$, incident on the piecewise-homogeneous four-layer slab specified in Figure 2.3.11b. The time-domain fields were computed by the marching-on-in-time method with $h = 1/320$.

Table 4.3.1 Comparison of the actual natural frequencies displayed in Figure 2.3.11b with the reconstructed ones shown in Figure 4.3.3.

Actual natural frequency	Reconstruction from reflected field	Reconstruction from transmitted field
-0.6970	-0.6969	-0.6973
-0.6703+i1.7688	-0.6703+i1.7686	-0.6705+i1.7687
-0.8899+i3.6560	-0.8918+i3.6552	-0.8904+i3.6550
-0.7945+i5.2723	-0.7952+i5.2710	-0.7935+i5.2712
-0.7210+i6.8686	-0.7217+i6.8665	-0.7201+i6.8675
-0.6047+i8.8195	-0.6043+i8.8170	-0.6037+i8.8162
-0.7757+i10.8264	-0.7676+i10.8214	-0.7724+i10.8202
-0.8541+i12.3558	-0.8410+i12.3558	-0.8553+i12.3505
-0.8143+i13.8820	-0.8045+i13.8980	-0.8159+i13.8624
-0.6177+i15.8791	-0.5998+i15.8506	-0.6226+i15.5854
-0.7070+i17.7771	-0.7500+i17.5923	-0.7400+i17.5958

with $|m| \leq 10$ within the main lobe of the frequency spectrum. The shorter pulse length was handled by executing the marching-on-in-time method with $h = 1/320$ in 4000 steps. In the discretization, appropriate measures were taken to deal with the discontinuities in the permittivity. The results of carrying out the entire reconstruction scheme for the reflected and the transmitted field are represented pictorially in Figure 4.3.3. In Table 4.3.1, the recovered natural frequencies are compared with the actual ones shown in Figure 2.3.11b. The accuracy is about the same as for the computation of Figure 4.3.2. It should be pointed out that the theory of Section 2.3 is not decisive with respect to the validity of the SEM representation for discontinuous profiles. The good agreement observed in Table 4.3.1 seems to justify the conjecture that the closing conditions listed in Table 2.3.2 also apply to such profiles.

The influence of noise is illustrated in Figure 4.3.4. Figure 4.3.4a contains the results of the whole reconstruction scheme, including preprocessing, to ten different realizations of contaminating the reflected-field data used in Figure 4.3.3a with 10% multiplicative noise. Figure 4.3.4b contains the results obtained from ten realizations with 1% additive noise. The first observation to be made is that the definitions of the noise ratio introduced in Subsection 4.2.3 are clearly inconsistent, since 1% additive noise has more effect than 10% multiplicative noise. This was confirmed by comparing the frequency spectra of noisy signals of both types. The constant background level of the former was slightly higher than

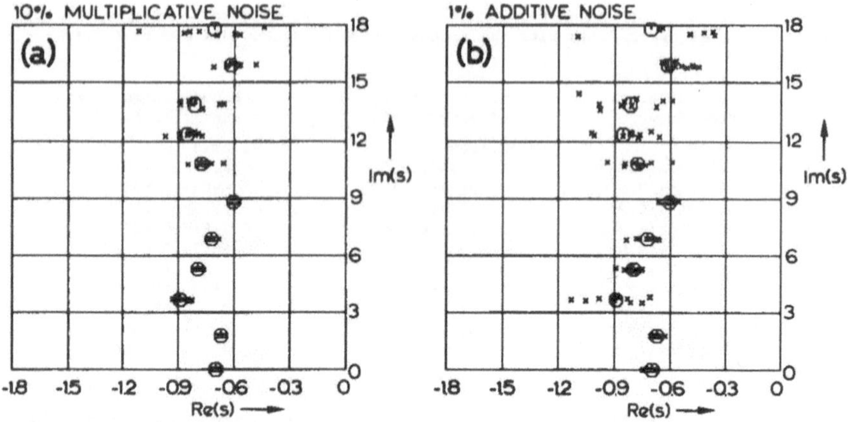

Figure 4.3.4 Reconstructed natural frequencies obtained by repeatedly applying the preprocessing procedure to ten different noisy signals generated from the input data used in Figure 4.3.3a and Table 4.3.1. (a): 10% multiplicative noise; (b): 1% additive noise; O: actual value; x: reconstructed value.

that of the latter. Secondly, we observe that, although individual results may deviate considerably from their actual counterparts, the overall picture shows a clustering around the proper values. The only possible exception is the pole of highest order, which is inaccurate anyway because of its proximity to the cut-off frequency. Even the poles with m = 7,8,9 in Figure 4.3.4b, whose contribution is barely above the noise level, are reconstructed fairly. Finally, it is observed that those residual contributions that have less influence on the sampled data, either because they decay more rapidly in time or because they have smaller amplitudes, indeed suffer the most from the presence of noise.

From a computational point of view, the reconstruction procedure applied in this subsection turns out to be highly efficient. In the longest computation, i.e. the one leading to Figures 4.3.3 and 4.3.4, the preprocessing of a single sampled signal and the subsequent application of Prony's algorithm took 4 and 2 seconds, respectively.

4.4 On the application to a cylinder signal

To complete this chapter, we briefly assess the possibilities of applying a Prony-type method to the cylindrical scattering problems described in Sections 2.5, 2.6, 3.4 and 3.5. The results obtained in Sections 4.2 and 4.3 and in Section 2.5 indicate that, for cylindrical scattering problems, there will be two major difficulties in applying such an algorithm.

In the first place, we now have a doubly denumerable set of poles $\{s_{mn}\}$, with both indices m and n having an infinite range. As a consequence, we have to recover infinitely more poles than in the case of the slab discussed in Section 4.3. This problem can possibly be overcome by taking a priori information into account. If, for example, we restrict ourselves, as in Section 2.5, to a radially inhomogeneous, lossy dielectric cylinder with $\varepsilon(\rho)$ and $\sigma(\rho)$ continuous inside it, the following procedure might be followed. First, the infinity in the angular order m can be removed by measuring the field at $M + 1$ equally spaced points on a semicircular arc of constant radius. Because of symmetry, we then have the field at 2M angles $\phi = \frac{2\pi k}{M}$ with $k = 0,1,\ldots,2M - 1$. By applying a discrete version of the Fourier transformation (2.5.4) to these field values, one can then obtain the Fourier coefficients $E_m(\rho,t)$ with $|m| \leq M$ with an error of at most $O(E_M)$ (see Hamming (1973)). Subsequently, for each angular order m, the convergence of the actual modal parameters to the asymptotic equivalents can be exploited to truncate the infinite range of n. Since these approximations only depend on some global properties of the cylinder, most of the information about the configuration is contained in the modes with low n. Hence, a high-frequency filtering of the transient signal can be employed as in Subsection 4.3.2 to generate a signal consisting of contributions from these lower-order modes only. Furthermore, the creeping-wave modes depend, in a first approximation, on $\varepsilon_r(1)$ only. In an optimization-type approach as proposed in connection with (4.2.22), the relevant modal parameters can therefore be handled as known functions of this single parameter. The procedure proposed above is in agreement with Dudley (1985), who argued that the natural frequencies of a multidimensional scattering problem should be reconstructed from multiple-output data, i.e. from the excited fields at several observation points.

In the second place, the application of a Prony-type algorithm is hindered by the presence of a branch-cut contribution. In view of the number of exponentials needed to approximate this contribution in its numerical evaluation (see the results in Subsection 2.5.5), it does not seem likely that it can be described accurately by a limited number of exponentials. This also applies to the branch-cut contribution that occurs in the SEM representation for a single Fourier coefficient $E_m(\rho,t)$. Therefore, the presence of this integral in particular seems to prevent a successful numerical recovery of the modal parameters of two-dimensional scatterers from time-domain data.

4.5 Conclusions

In this chapter, we have studied the limitations of Prony's method for the decomposition of a transient signal into a series of exponentials of given order. In addition, we have investigated the consequences of these limitations for the application to actual transient electromagnetic fields. Two important conclusions seem to protrude from our analysis. In the first place, it is essential to use an optimum sampling rate, even when exact data are available. In the second place, the transient field must consist of no more than the residual contributions from a specified number of poles. One way to achieve this is to use a judiciously chosen incident pulse and to process the excited field such that small, unwanted contributions from higher-order poles are eliminated. This was implemented successfully for the one-dimensional problem of scattering by a slab. For two- and three-dimensional problems, such a preprocessing will probably involve combining transient fields at multiple observation points. For two-dimensional problems, the presence of a branch-cut contribution may even then prevent a successful recovery. Future work should include a better handling of noise, for example by invoking the pencil-of-functions method (Jain et al. (1983)) or nonlinear optimization techniques. In addition, a cluster analysis of the results obtained from repeated scattering experiments producing noisy data seems imminent.

References

Bennett, C.L., and Toomey, J.P. (1981), Target classification with multiple frequency illumination, *IEEE Trans Antennas Propagat. 29*, 352-358.

Brittingham, J.N., Miller, E.K., and Willows, J.L. (1980), Pole extraction from real-frequency information, *Proc. IEEE 68*, 263-273.

Dudley, D.G. (1979), Parametric modeling of transient electromagnetic systems, *Radio Sci. 14*, 387-395.

Dudley, D.G. (1983), Parametric identification of transient electromagnetic systems, *Wave Motion 5*, 369-384.

Dudley, D.G. (1985), A state-space formulation of transient electromagnetic scattering, *IEEE Trans. Antennas Propagat. 33*, 1127-1130.

Eftimiu, C., and Huddleston, P.L. (1984), Reconstruction of a spherical scatterer from its natural frequencies, *IEEE Trans. Antennas Propagat. 32*, 694-698.

Hamming, R.W. (1973), *Numerical methods for scientists and engineers*, McGraw-Hill, New York, second edition, pp. 503-515.

Hildebrand, F.B. (1956), *Introduction to numerical analysis*, McGraw-Hill, New York, pp. 378-382.

Hoenders, B.J. (1982), The unique determination of an object with a radially dependent index of refraction by its natural frequencies, *Optica Acta 29*, 55-62.

Jain, V.K., Sarkar, T.K., and Weiner, D.D. (1983), Rational modeling by the pencil-of-functions method, *IEEE Trans. Acoustics, Speech, and Signal Processing 31*, 564-573.

Knockaert, L. (1984), Parametric modeling of electromagnetic waveforms, *Electromagnetics 4*, 415-430.

Kulp, R.W. (1981), An optimum sampling procedure for use with the Prony method, *IEEE Trans. Electromagn. Compat. 23*, 67-71.

Lin, C.A., and Cordaro, J.T. (1983), Determination of the SEM parameters for an aircraft model from the transient surface current, *Electromagnetics 3*, 65-75.

Lytle, R.J., and Lager, D.L. (1976), Using the natural-frequency concept in remote probing of the earth, *Radio Sci. 11*, 199-209.

Miller, E.K. (1981), Natural mode methods in frequency and time domain analysis. In: *Theoretical methods for determining the interaction of electromagnetic waves with structures*, Skwyrzinski, J.K. (Ed.), Sijthoff and Noordhoff, Alphen aan den Rijn, The Netherlands, pp. 173-212.

Miller, E.K., Brittingham, J.N., and Willows, J.L. (1977), Identification of E.M. spectrum by known pole sets, *Electronics Letters 13*, 774-775.

Miller, E.K., and Lager, D.L. (1982), Inversion of one-dimensional scattering data using Prony's method, *Radio Sci. 17*, 211-217.

NAG Fortran Library Manual Mark 11 (1984), Numerical Algorithms Group, Oxford.

Pearson, L.W., and Roberson, D.R. (1980), The extraction of the singularity expansion description of a scatterer from transient surface current response, *IEEE Trans. Antennas Propagat. 28*, 182-190.

Poggio, A.J., Van Blaricum, M.L., Miller, E.K., and Mittra, R. (1978), Evaluation of a processing technique for transient data, *IEEE Trans. Antennas Propagat. 26*, 165-173.

Prony, R. (1795), Essai expérimental et analytique sur les lois de la dilatabilité des fluides élastiques et sur celles de la force expansive de la vapeur de l'eau et de la vapeur de l'alkool, à différentes températures, *Journal de l'École Polytechnique 1*, Paris, Cahier 2, 24-76.

Sarkar, T.K., Dianat, S.A., and Weiner, D.D. (1984), A discussion of various approaches to the linear system identification problem, *IEEE Trans. Acoustics, Speech, and Signal Processing 32*, 654-656.

Sarkar, T.K., Weiner, D.D., and Jain, V.K. (1981), Some mathematical considerations in dealing with the inverse problem, *IEEE Trans. Antennas Propagat. 29*, 373-379.

Sarkar, T.K., Weiner, D.D., Nebat, J., and Jain, V.K. (1982), A discussion of various approaches to the identification/approximation problem, *IEEE Trans. Antennas Propagat. 30*, 89-98.

Van Blaricum, M.L., and Mittra, R. (1975), A technique for extracting the poles and residues of a system directly from its transient response, *IEEE Trans. Antennas Propagat. 23*, 777-781.

Van Blaricum, M.L., and Mittra, R. (1978), Problems and solutions associated with Prony's method for processing transient data, *IEEE Trans. Antennas Propagat. 26*, 174-182.

Weiss, L., and McDonough, R.N. (1963), Prony's method, z-transforms, and Padé approximation, *SIAM Review 5*, 145-149.

INVERSE PROFILING

5. FREQUENCY-DOMAIN TECHNIQUES

5.1 Introduction

In inverse scattering, one attempts to reconstruct the material composition of an object whose interior is otherwise inaccessible to measurements, i.e. its geometrical and/or physical parameters, by probing it from the exterior. The probing is carried out by exciting the object with one or more sources, while the resulting wave field is measured at a number of observation points outside the object. In this and the next chapter, we consider solving the one-dimensional inverse-scattering problem for the inhomogeneous, lossy dielectric slab embedded in vacuum as specified in Subsection 2.2.1 and analyzed further in Sections 2.3 and 3.3. In this inverse-profiling problem, the reflected field caused by one or more plane, electromagnetic waves incident from one or both sides of the slab is assumed to be known. The aim of the inverse-profiling computation is to reconstruct one or both of the constitutive parameters of the slab, i.e. the conductivity $\sigma(z)$ and the dielectric susceptibility $\chi(z)$, from these reflected-field data. In particular, we investigate whether the inverse-scattering problem can be solved by utilizing the same type of integral relations that figured in the analysis of the corresponding direct-scattering problems discussed in Chapters 2 and 3. With a single exception, one and the same method of solution will be applied. According to what type of reflected-field data are available, we distinguish between frequency-domain and time-domain problems. In frequency-domain problems, the reflection coefficient for plane-wave incidence is known for a specified range of real frequencies ($s = i\omega$). In time-domain problems, the known data consist of the (sampled) reflected field caused by a specified incident pulse of finite duration. In the present chapter, we discuss the method of solution in general, and its application to a number of frequency-domain problems. In the next chapter, time-domain problems will come up for discussion. Throughout both chapters, we will simplify the notation by using the dimensionless quantities introduced in Subsection 2.2.1 (see p. 37). As before, we omit the bars.

In order to place our approach into some perspective, let us first review the techniques that have been reported in the literature for the solution of the one-dimensional inverse-scattering problem specified above and the related one of a slab terminated by a perfectly conducting boundary. In view of the large number of papers on this problem, we will not, as in previous chapters, list all the applications to related problems from electrodynamics and acoustics. Only when a suitable reference from electromagnetics is not available will we make an exception to this rule. In addition, we will restrict ourselves to the purely one-dimensional case of plane-wave incidence, and disregard all situations where the source is localized.

The problem of a lossless slab was first theoretically treated by Kay (1955). Through a Liouville transformation, he reduced it to an equivalent, uniquely solvable, quantum-mechanical scattering problem. For the latter problem, a number of integral- and differential-equation approaches have been developed, reviews of which have been given by Chadan and Sabatier (1977), Burridge (1980), Newton (1980), and Sabatier (1983). Hence, the equivalence observed by Kay also permits the reconstruction of the unknown susceptibility profile according to one of these approaches. The practical aspects of applying such an approach have, for instance, been discussed by Berryman and Greene (1980), and by Fawcett (1984). Numerical results obtained along these lines were first reported by Schaubert and Mittra (1977), and Coen (1981). More recent results can be found in Ladouceur and Jordan (1985), Jaggard and Kim (1985), and Jaggard and Olson (1985). The main difficulty in using the correspondence with quantum mechanics appears to be the Liouville transformation, which involves a double space differentiation of the unknown refractive index. Since, for practical applications, we must allow the susceptibility - and hence the index of refraction - to be discontinuous across the slab interfaces, numerical difficulties can be expected in the reconstruction of the equivalent quantum-mechanical scattering potential near those interfaces. A second disadvantage of the Liouville transformation is its typically one-dimensional character, which, from the outset, prevents a possible generalization to more complicated inverse-scattering problems.

The generalization of the quantum-mechanical approach to the case of a lossy slab was first dealt with by Weston (1972,1974), and Weston and Krueger (1973), who applied a Riemann function approach to develop a system of Gelfand-Levitan type equations whose solution yielded the desired profiles. The data for this problem consisted of the time-domain reflection and transmission operators. These results were generalized by Krueger (1976,1978) to include more realistic material profiles. This same author also examined the numerical practicability of the approach (Krueger (1981)). A more accessible procedure for arriving at the scattering kernels was proposed by Corones and Krueger (1983), who obtained these kernels by using invariant embedding. The invariant-embedding approach was coupled with a splitting analysis by Corones et al. (1985). In addition, this reference also describes a numerical solution method and some numerical results. Recently, a series of papers was started which aims at giving a systematic survey of both the analytical and the numerical aspects of the approach (Kristensson and Krueger (1986a,b)). Like Weston's work, all references cited above pertain to the time-domain inverse-scattering problem. The equivalent frequency-domain problem was analyzed from a theoretical point of view by Jaulent (1976, 1982). For this problem, numerical results have not yet been reported. As in the lossless case, the main disadvantage of the approach lies in the necessary Liouville transformation. As far as the time-domain case is concerned, problems may also arise in the determination of the scattering kernels, i.e. the delta-function responses of the slab under consideration. The determination of these responses from the reflected and transmitted fields caused by a practical incident pulse is a deconvolution problem. As is well known, such a deconvolution problem is ill-posed (see e.g. Sarkar et al. (1981)).

In the geophysical literature, one usually follows an entirely different approach and models the medium by a number of homogeneous layers, all of which have a common one-way travel-time (Goupillaud (1961)). Then the system is described completely by the set of reflection coefficients at the interfaces between these layers. A survey of methods for solving these reflection coefficients for lossless layered media systems, which are the geophysical equivalent of lossless layered dielectric systems, can be found in Mendel and Habibi-Ashrafi (1980). The only electromagnetic application of this approach known to the author was carried out by Lee (1982), who considered the more general case of an inhomogeneous, lossy medium. The principal disadvantage of the approach is that, in the piecewise-homogeneous model, a large number of layers should be taken to arrive at an acceptable description of the reflected fields. With each layer, one should associate one or two medium parameters, according to whether the configuration is lossless or lossy. For the inverse-scattering scheme obtained by using the layer model, this has the consequence that a large number of parameters must be reconstructed. As a result, stability and/or ambiguity problems will be encountered when the scheme is applied to process practical reflected fields caused by a smooth incident pulse of finite duration, as presented in Subsections 2.3.5, 2.4.3 and 3.3.3.

In the electromagnetic literature we can, in addition to the quantum-mechanical approach, distinguish two trends in developing solution methods for the one-dimensional inverse-scattering problem. In the first one, the integral relations pertaining to the corresponding direct-scattering problem are being used to draw up additional relations between the known reflected-field data and the unknown fields and constitutive parameters inside the slab. In conjunction with the corresponding direct-scattering integral or differential equations, these relations are then used to resolve the inverse-scattering problem. One way to achieve this objective is to accurately trace the wavefront in the time-domain case. In this manner, a single unknown constitutive parameter can be reconstructed from a single time-domain response. Calculating step by step, in the slab, the field at the front of the pulse, one can make an estimate of the local value of the susceptibility or the conductivity (Bolomey et al. (1979), Lesselier (1978), Bojarski (1980)). This method has the disadvantage that the computed field at the wavefront is afflicted with a large relative error, especially for a nonvanishing susceptibility contrast. As a consequence, the method is restricted to incident pulses with rather steep ramp-like fronts. An alternative is to make some educated guess about the unknown total field inside the slab, and to substitute this guess in the additional relations involving the known reflection data. These relations can then be regarded as one or two integral equations of the first kind, whose solution yields an approximation to the unknown susceptibility and/or conductivity profiles. The first application along these lines is due to Tabbara (1976,1979), who considered the reconstruction of the susceptibility profile of an inhomogeneous, lossless dielectric slab from multi-frequency measurements of the reflection coefficient. It was shown that, in the Born approximation, the reflection coefficient of a normally incident plane wave can be identified with a spectral component of the unknown susceptibility profile. For higher contrasts, appropriate correction formulas were derived from closed-form results for a homogeneous slab. Further on, it will be shown that these correction formulas are basically high-frequency results. As such, they link up with inversion methods based on high-frequency asymptotic approximations (see e.g. Kaiser and Kaiser (1983)). A logical extension of Tabbara's approach is an iterative procedure as applied by Roger et al. (1978), Tijhuis (1981), Tijhuis and Van der Worm (1984), and Uno and Adachi (1985). In such a procedure, successive approximations for the unknown reflected field inside the slab and the unknown constitutive parameters are obtained by alternately solving an approximate direct-scattering problem and an approximate inverse-scattering problem. A more detailed discussion of this procedure will be given below.

The second trend in the solution of one-dimensional, electromagnetic inverse-scattering problems is the use of optimization techniques such as the quasi-Newton method and the conjugate-gradient method (a brief explanation of these methods will be given in Subsection 6.3.2). Such methods are applied to minimize some cost function involving the deviation between the known reflection data and their counterparts in some trial medium characterized by a finite set of parameters. If necessary, *a priori* information can be included in the analysis, either by augmenting the cost function with a suitably chosen regularization term, or by posing additional constraints to the unknown profiles. Successful applications of optimization techniques were described by Mostafavi and Mittra (1972), Coen et al. (1981), Roger (1981), and Lesselier (1982a,b).

The essential difference between these optimization techniques and the Born-type iterative procedure cited above does not lie in the introduction of a cost function. In fact, the solution of the approximate inverse-scattering problem in each step of that iterative procedure also involves the minimization of a cost function. Rather, the difference is in the method of minimization. The standard optimization techniques mentioned above have the advantage of a guaranteed convergence to some - local or global - minimum. However, the rate of convergence may be rather slow. As a consequence, the application of such techniques may require a rather large number of evaluations of the cost function. Since each evaluation requires the solution of at least one approximate direct-scattering problem, this may lead to unacceptably long computation times. The Born-type iterative technique, on the other hand, converges at a much faster rate. Consequently, only a few approximate direct-scattering problems need to be solved in its application. However, this improvement in efficiency is obtained at the cost of losing the guaranteed convergence. This indicates that both approaches may be complementary: when the Born-type iterative procedure converges, it should be applied because of its faster convergence; when it does not, an optimization technique should be applied.

From the literature review given above, it can be observed that the solution of an inverse-profiling problem should, ideally, be carried out in three steps:

i) First, the uniqueness of the solution should be established. As mentioned above, for the lossless slab this was performed by Kay (1955). For the general case, no uniqueness proof is known to the author. When the solution is not unique, the problem should be reformulated by augmenting the known field data with a sufficient amount of a priori information.

ii) Next, the "numerical condition" of the problem should be analyzed. An example of such an analysis was given by Roger et al. (1978) for the configuration of a lossless slab terminated by a perfect conductor. As known data these authors used the angular dependence of the reflection coefficient at a single frequency. For the present configuration, a similar analysis remains to be carried out.

iii) Finally, an algorithm should be devised with which to obtain the solution as well as possible within the constraints found from steps i) and ii).

In practice, we have to accept the fact that it is at this time impossible to effectuate the first two steps. This leaves us no option but to start directly with step iii), i.e. with devising and analyzing a method of solution. In this and the next chapter, such an analysis will be carried out for the Born-type iterative procedure proposed in Tijhuis (1981) and Tijhuis and Van der Worm (1984). By performing numerical experiments and by applying approximate analytical techniques we can, for this particular method, arrive at a good understanding of its reconstructive potentialities. Moreover, the conclusions obtained may give us an indication how to carry out steps i) and ii). Any future effectuation of these steps should at least reproduce these conclusions.

Let us now consider the Born-type iterative procedure in some more detail. A flow diagram of the frequency-domain version of this procedure is given in Figure 5.1.1. As shown in that figure, the starting point of the procedure is an initial estimate for the unknown conductivity and/or susceptibility inside the slab. This estimate should either be available from *a priori* information or from some characteristic features of the known reflected fields. Starting from this estimate we carry out a number of iteration steps. In each step, we first approximate the unknown electric field inside the slab by the field that results from the incident field if the unknown configuration equals the approximation found in the previous step. Next, the approximate field thus obtained is substituted in an integral relation of the type derived in Subsection 2.4.4. The new approximation of the unknown slab configuration is then obtained by minimizing some squared error in the equality sign of the resulting approximate integral relation. In general, it will be necessary to account for additional profile information in this minimization step. The most obvious example is the case where one of the relevant

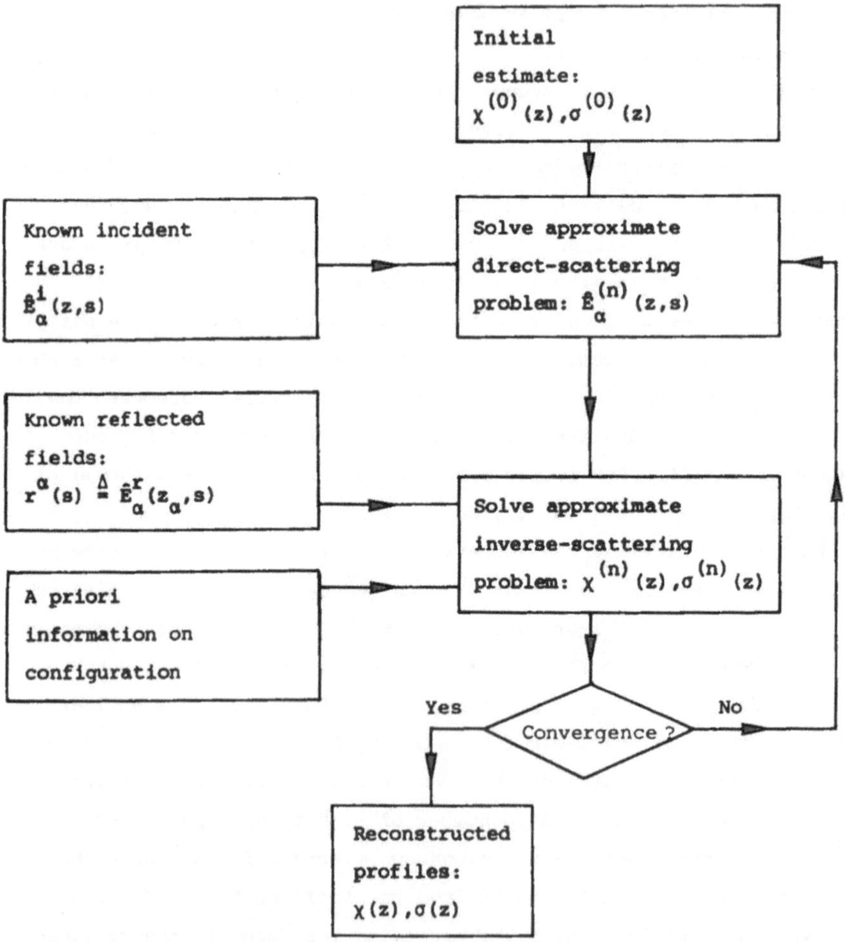

Figure 5.1.1 Flow diagram of the reconstruction of the susceptibility $\chi(z)$ and/or the conductivity $\sigma(z)$ of a dielectric slab from frequency-domain reflected-field data via the Born-type iterative procedure. In the diagram, α is a summation index, z a position coordinate, s a complex frequency, and z_α an observation point. $\hat{E}^{i}_{\alpha}(z,s)$ denotes a unit-amplitude, monochromatic, plane-wave incident field, and $\hat{E}^{(n)}_{\alpha}(z,s)$ and $\hat{E}^{r}_{\alpha}(z,s)$ responses to that field.

constitutive parameters is known. In that case, we need only reconstruct the remaining one. A second example occurs when the short-range behavior of the unknown profiles is supplemented. This can be achieved by representing these profiles as a linear combination of a suitably chosen, finite set of expansion functions, and by augmenting the squared error with suitable regularization terms.

In the above formulation of the iterative scheme, we run up against two types of problems. In the first place, we have to justify two implicit assumptions made in this formulation. Both assumptions pertain to the equations obtained by substituting the field resulting from the incident field in a given, approximate configuration in the integral relations involving the known reflection data. The first one is that these equations suffice for the unique determination of a new approximate configuration, at least within some limitations. The second assumption is that the error in substituting the approximate field is small enough for the new approximate configuration to be a better estimate than the previous one. Only by justifying both assumptions can we show that the method converges. In the second place, we run up against the problem of selecting the most suitable background medium in the integral relations containing the known reflected-field information. At first glance, there seems to be some advantage in selecting a vacuum. Free-space Green's functions are known in closed form in both the time domain and the frequency domain in both one- and multi-dimensional problems. Hence, selecting a vacuum keeps open the possibility of generalizing the method to more complicated inverse-scattering problems. In addition, we have the practical advantage that the discretization of the relevant integral relation may be available from the solution of the corresponding direct-scattering problem, for which the selection of a vacuum background medium seems mandatory (see also Subsection 3.3.4). On the other hand, it is by no means clear whether the selection of a vacuum background medium produces, in each iteration step, the most accurate approximation of the unknown slab configuration.

In the present chapter, both questions broached in the previous paragraph are answered with the aid of the WKB expressions for the field inside the slab derived in Subsection 2.3.3. It will be argued that choosing a vacuum background medium in the integral relations involving the known reflected fields leads to an inherent band limitation in the reconstruction. Due to this band limitation, the agreement between the reconstructed and the actual susceptibility and/or conductivity profiles is confined to a limited wave number k. This means that only the long-range behavior of the unknown profile or profiles can be retrieved with an acceptable accuracy. The short-range behavior turns out to be ambiguous and must be supplied from a priori information.

In addition, a modified procedure that avoids the band limitation will be proposed. The modification consists of taking, in each iteration step, the background medium equal to the previously obtained approximate configuration, and of restricting the analysis to the time-domain field in a specific time interval. Finally, it will be argued that, in view of the more involved implementation of the modified scheme, it would still seem most feasible to start the numerical experimentation by applying the scheme with a free-space background medium. From the results, we can then decide whether the application of an alternative scheme is warranted.

The general aspects of the one-dimensional inverse-scattering problem indicated above are analyzed in detail in Section 5.2. Towards the end of that section, we also pay attention to the consequences of the analysis for the actual numerical implementation of the scheme. The remainder of the chapter is then devoted to this implementation for two special frequency-domain problems. In Section 5.3, we consider the reconstruction of a small susceptibility profile $\chi(z)$ and a small conductivity profile $\sigma(z)$ from the known real-frequency behavior of the reflection coefficient for two different directions of incidence in the *Born* approximation. This problem can be envisaged as the most simple example of an iteration step of the type occurring in our iterative procedure. As such it provides an ideal starting point for devising a numerical implementation of that procedure which is consistent with the conclusions obtained in Section 5.2. For this special configuration, we will also consider the possibility of using reflected fields at oblique incidence. In Section 5.3, we will then reconsider the problem of reconstructing the susceptibility profile of a lossless slab from known values of the frequency-domain reflection coefficient at normal incidence as analized by Tijhuis and Van der Worm (1984). As remarked above, the special attraction of this particular inverse-scattering problem is that its uniqueness has been established. Hence, any ambiguities in the reconstruction must be due to either the ill-posedness of the inverse-scattering problem or the method of solution. The first part of Section 5.4 is devoted to approximate estimation procedures. In particular, Tabbara's estimation procedure will be linked up with the WKB analysis performed in Section 5.2. In the second part, the results of Tijhuis and Van der Worm are interpreted in the light of the results obtained from that analysis. Finally, in Section 5.5, some preliminary conclusions are drawn about the specific frequency-domain application of our Born-type iterative scheme. The final assessment of our method of solution will be postponed to Chapter 6, where the time-domain case and the application of alternative schemes will be taken into consideration.

5.2 General aspects of the one-dimensional inverse-scattering problem

5.2.1 Formulation of the problem

In this chapter, we reconsider the frequency-domain problem of plane-wave scattering by an inhomogeneous, lossy dielectric slab as specified in Subsection 2.2.1 and further analyzed in Sections 2.3 and 2.4. We restrict ourselves to the case where the slab is embedded in a vacuum, i.e. $N_1 = N_3 = 1$ in the expressions given in Section 2.4. For the incident field, we take one or both of the unit-amplitude plane waves

$$\hat{E}^i_+(z,s) = \exp(-sz), \tag{5.2.1a}$$

$$\hat{E}^i_-(z,s) = \exp(s(z - 1)). \tag{5.2.1b}$$

As shown in Figure 5.2.1, the superscripts + and - refer to the direction of the propagation of these waves. In particular, we consider the situation where one or both of the reflection coefficients corresponding to the incident fields specified in (5.2.1) are known and one or both of the constitutive parameters of the slab, i.e. the dielectric susceptibility $\chi(z) = \epsilon_r(z) - 1$ and the conductivity $\sigma(z)$, are to be determined. In this first section, our method of solution is described and an analysis of its potentialities is presented.

As mentioned in the introduction, we aim to solve the inverse-scattering problem with the sole aid of the differential equations for the forward problem and the integral relations written up in Section 2.4. As described in Subsection 2.4.2, for known $\chi(z)$ and $\sigma(z)$, the electromagnetic field inside the slab can be obtained directly by integrating the system of first-order differential equations (2.4.2), or, equivalently, (2.2.6). Substituting (2.4.6) and a related form pertaining to the magnetic field in these equations, we directly obtain

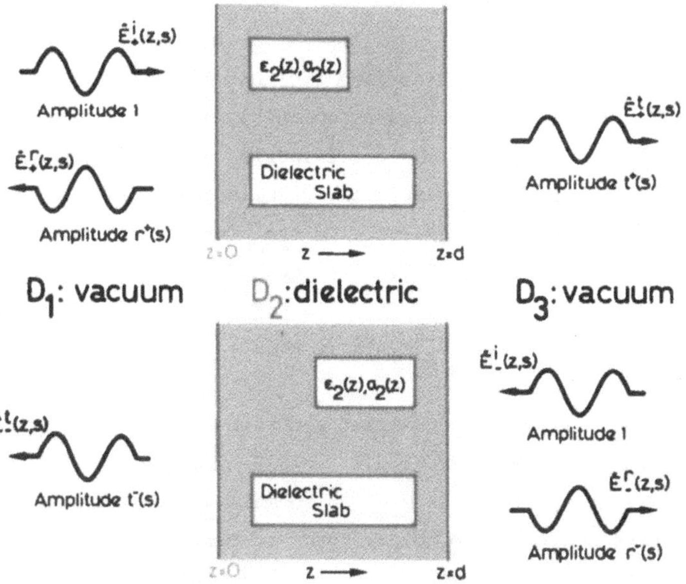

Figure 5.2.1 Monochromatic plane waves normally incident from the left and from the right on an inhomogeneous, lossy dielectric slab embedded in vacuum.

$$\partial_z \hat{H}(z,s) = s(\varepsilon_r(z) + \sigma(z)/s)\hat{E}(z,s),$$
$$\partial_z \hat{E}(z,s) = s\hat{H}(z,s), \tag{5.2.2}$$

with $\hat{H}(z,s)$ being the customary frequency-domain magnetic-field strength. The relevant boundary conditions follow by setting $N_1 = N_3 = 1$ and $F(s) = 1$ in (2.4.4). The known reflected field is accounted for via one or two additional integral relations of the type (2.4.24). In their most general form, these relations can be written as

$$r^{\alpha}(s) - \bar{r}^{\alpha}(s) = -\frac{s}{2}\int_{-\infty}^{\infty} C(z,s)\bar{E}_{\alpha}(z,s)\hat{E}_{\alpha}(z,s)\,dz, \tag{5.2.3}$$

where the contrast function $C(z,s)$ is defined as in (2.4.18) and where the label α stands for either of the labels + or - that occur as subscripts and superscripts both above and in Equations (2.4.21) - (2.4.24). In (5.2.3), $\hat{E}_{\alpha}(z,s)$ and $\bar{E}_{\alpha}(z,s)$ denote harmonic responses excited by the plane-wave incident field $\hat{E}^{i}_{\alpha}(z,s)$ given in (5.2.1) in the actual configuration and in some reference medium with constitutive parameters $\{\bar{\varepsilon}_r(z),\bar{\sigma}(z)\}$, respectively.

As reasoned in the introduction, we will preferably choose the reference medium to be a vacuum. In that case, $\bar{r}^{\alpha}(s)$ vanishes and the additional relation (5.2.3) reduces to

$$r^{\alpha}(s) = -\frac{s}{2}\int_{-\infty}^{\infty} C_0(z,s)\hat{E}^{i}_{\alpha}(z,s)\hat{E}_{\alpha}(z,s)dz, \tag{5.2.4}$$

where $C_0(z,s)$ is now the contrast function with respect to vacuum as defined in (2.2.10). With (5.2.2) - (5.2.4) we now have available the equations that will be used in formulating our method of solving the inverse-scattering problem. However, their present form and the terminology associated with it could give rise to some confusion further on, where we analyze our method in more detail. Namely, in the course of the solution procedure, the exact field $\hat{E}_{\alpha}(z,s)$ in the additional equation (5.2.4) will be replaced by the harmonic response $\bar{E}_{\alpha}(z,s)$ in some reference medium other than the one chosen in obtaining (5.2.4) from the general equation (5.2.3). In order to avoid that confusion, we will, from now on, define the *background medium* as the reference medium chosen to derive the special version of the relation (5.2.3) that is used as "additional relation" in the formulation of the inverse-scattering problem. Similarly, we will denote the contrast function with respect to that background medium as $C_0(z,s)$, where the subscript 0 has been appended since this background medium will usually be a vacuum. The term *reference medium* will be reserved for the current approximate configuration obtained in the solution procedure, with $\bar{E}_{\alpha}(z,s)$ and $C(z,s)$ being the harmonic response and the contrast function associated with that medium, respectively.

5.2.2 Method of solution

The method by which we solve the inverse-scattering problem formulated in Subsection 5.2.1 is based upon the following observations. As outlined in Subsection 2.4.2, for known $\chi(z)$ and $\sigma(z)$, the fields $\hat{E}_\alpha(z,s)$ inside the slab can be determined from Equation (5.2.2). On the other hand, when one or both of the fields $\hat{E}_\alpha(z,s)$ were known inside the slab either of the equations (5.2.3) and (5.2.4) could be regarded as a pair of integral equations of the first kind for the unknown profiles $\chi(z)$ and/or $\sigma(z)$. However, we only know $r^+(i\omega)$ and/or $r^-(i\omega)$ for a given set or range of real-valued frequencies ω, with $s = i\omega$. Therefore, our only chance is to solve the inverse-scattering problem by an iterative procedure that alternately uses both equations to obtain successive approximations to the unknown field inside the slab as well as the unknown configuration.

The starting point for our iteration scheme is an initial estimate for the contrast function in the slab, i.e. $C_0^{(0)}(z) = \chi^{(0)}(z) + \sigma^{(0)}(z)/s$. This estimate should either be available from *a priori* information or be procured from some characteristic features of the known reflected field. The manner in which the latter can be accomplished depends on the specifics of the problem under consideration, and, therefore, will be indicated when the individual problems come up for discussion. Starting from the initial estimate, we carry out a number of iteration steps. In each step, we first approximate the electric field inside the slab by the field that results from the incident field if the unknown configuration equals the approximation found in the previous step. In the n'th step, this field is obtained by integrating the pair of equations

$$\begin{aligned} \partial_z\hat{H}^{(n)}(z,s) &= s[1 + \chi^{(n-1)}(z) + \sigma^{(n-1)}(z)/s]\hat{E}^{(n)}(z,s), \\ \partial_z\hat{E}^{(n)}(z,s) &= s\hat{H}^{(n)}(z,s), \end{aligned} \tag{5.2.5}$$

which follow directly from (5.2.2), subject to the boundary conditions imposed by (2.4.4) and a related form for $\hat{E}_-(z,s)$ and $\hat{H}_-(z,s)$. Next, we define $C_0^{(n)}(z,s)$ as that contrast function of the form (2.2.10) that minimizes some squared error in the equality signs of the available integral relations of the form (5.2.4) if $\hat{E}_\alpha(z,s)$ is replaced by $\hat{E}_\alpha^{(n)}(z,s)$. Specifically, we minimize the squared error

$$\sum_\alpha \int_{-\infty}^{\infty} d\omega\ w^2(\omega)\ \Big|\ r^\alpha(i\omega) + \frac{i\omega}{2}\int_0^1 C_0^{(n)}(z,i\omega)\hat{E}_\alpha^{i}(z,i\omega)\hat{E}_\alpha^{(n)}(z,i\omega)dz\ \Big|^2. \tag{5.2.6}$$

In (5.2.6), the summation index α runs over one or both of the labels $\pm$, and $w(\omega)$ is a real-valued weighting function that can, for example, be identified as the magnitude of the Laplace-transformed incident pulse $F(i\omega)$. In addition, the function $w(\omega)$ can be used to account for the band limitation of the measured reflected-field data. As we shall see further on, defining $C_0^{(n)}(z)$ as above may not suffice to establish this approximation in full detail. In particular, the local behavior of $C_0^{(n)}(z)$ may not be retrievable by minimizing (5.2.6). In that sense, the problem of determining this contrast function is ill-conditioned. This deficiency can be remedied by augmenting (5.2.6) by regularization terms of the type

$$\delta\int_0^1 |d_z^2\chi^{(n)}(z)|dz, \tag{5.2.7}$$

and similar terms pertaining to $\sigma^{(n)}(z)$. In (5.2.7), δ is a small, nonnegative regularization or tuning parameter. Adding (5.2.7) or similar terms to (5.2.6) results in a built-in preference in the reconstruction for the "most linear" profile, i.e. the one with the smallest global curvature, in cases where minimizing (5.2.6) leads to ambiguities. Obviously, the regularization parameter δ should be chosen so small that the determination of those characteristics that do show up in (5.2.6) is hardly disturbed by the regularization. The iteration scheme is terminated when the error thus composed no longer changes significantly with n or when it reaches a previously defined value.

Although the scheme outlined above and most of its properties were first found from numerical experimentation, a closed-form analysis that explains these properties in a more consistent manner can now be given. In particular, this analysis serves to justify two assumptions that were made implicitly in formulating the iterative scheme. In the first place, we have assumed that one or both of the equations that are obtained by substituting either of the labels + and - for α in (5.2.4) suffices for the unique determination of $\chi^{(n)}(z)$ and/or $\sigma^{(n)}(z)$, at least within some approximation. In the second place, a scheme of the type outlined above can only converge if, for $\chi^{(n-1)}(z)$ and $\sigma^{(n-1)}(z)$ sufficiently close to the actual profiles, the error made by replacing $\hat{E}_\alpha(z,s)$ by $\hat{E}_\alpha^{(n)}(z,s)$ in (5.2.4) is small enough for $\chi^{(n)}(z)$ and $\sigma^{(n)}(z)$ to be better approximations. In the next two subsections, we will successively investigate both assumptions and their consequences for the applicability of the iterative procedure.

5.2.3 Information retrievable from the reflected field

In order to investigate both assumptions outlined above, we consider a model iteration step. We restrict ourselves to incidence from the left ($\alpha = +$), since conclusions about incidence from the right ($\alpha = -$) and about the combination of both cases can be drawn from inference. Let the reference medium $\{\bar{\chi}(z),\bar{\sigma}(z)\}$ be a good approximation to the actual constitutive parameters $\{\chi(z),\sigma(z)\}$ and let $\bar{E}_+(z,s)$ and $G(z,z';s)$ be the corresponding solutions defined by (2.4.4) (with $F(s) = 1$ and $N_1 = N_3 = 1$), and by (2.4.19). As suggested at the end of Subsection 5.2.1, the profiles $\{\bar{\chi}(z),\bar{\sigma}(z)\}$ may be thought of as the current approximations $\{\chi^{(n-1)}(z),\sigma^{(n-1)}(z)\}$ as discussed in Subsection 5.2.2 and $\bar{E}_+(z,s)$ as one of the corresponding fields $\hat{E}_\alpha^{(n)}(z,s)$. From (2.4.20), we have

$$\hat{E}_+(z,s) = \bar{E}_+(z,s) - s^2\int_{-\infty}^{\infty} C(z',s)G(z,z';s)\hat{E}_+(z',s)dz' \qquad (5.2.8)$$

for all $z\in\mathbb{R}$, with $C(z,s)$ being the contrast function as defined in (2.4.18), and with $\bar{E}_+(z,s)$ representing the field that $\bar{E}_+^i(z,s)$ generates in the reference medium. Substitution of (5.2.1a) and (5.2.8) in (5.2.4) with $\alpha = +$ results in

$$r^+(s) = r_B^+(s) + \Delta r_B^+(s), \qquad (5.2.9)$$

with

$$r_B^+(s) \triangleq -\frac{s}{2}\int_{-\infty}^{\infty} C_0(z,s)\exp(-sz)\bar{E}_+(z,s)dz, \tag{5.2.10}$$

and

$$\Delta r_B^+(s) \triangleq \frac{s^3}{2}\int_{-\infty}^{\infty}\Big\{C_0(z,s)\exp(-sz) \times \int_{-\infty}^{\infty}[C(z',s)G(z,z';s)\hat{E}_+(z',s)]dz'\Big\}dz. \tag{5.2.11}$$

In (5.2.9) - (5.2.11), the subscript B expresses the fact that replacing $\hat{E}_+(z,s)$ with $\bar{E}_+(z,s)$ in (5.2.4) is actually a Born-type approximation. In the literature, this substitution is therefore also known as the "distorted Born approximation" (Devaney and Oristaglio (1983)). Now, $r_B^+(s)$ is exactly the term which is equated to $r^+(s)$ in the iterative scheme (see (5.2.6)). Hence, $\Delta r_B^+(s)$ can be identified as the term that generates the error in the reconstructed profile due to enforcing the Born-type approximation. This interpretation suggests the following procedure for the verification of the two assumptions made above. First, we will investigate what information about the configuration is available in $r_B^+(s)$. This will be carried out in the remainder of the present subsection. In the next subsection, we will consider whether this information is really retrievable, by estimating the relative magnitude of the error term $\Delta r_B^+(s)$. Only when the results of both steps are known will we be in a position to assess the reconstructive potentialities of the iterative scheme proposed in Subsection 5.2.2.

From (5.2.10), it is observed that a full analysis of $r_B^+(s)$ would require $\bar{E}_+(z,s)$ to be available in closed form. Since, for a general, inhomogeneous reference configuration, such a form is not available, we use the first-order WKB approximation discussed in Subsections 2.3.3 and 2.4.3. Since this high-frequency approximation remains accurate down to surprisingly low frequencies (see e.g. Figure 2.4.4), its substitution will at least provide us with a good qualitative description. Moreover, comparing (2.3.13) and (2.3.19) - (2.3.21) with (2.2.15) - (2.2.17) reveals that, for a homogeneous, lossless slab, the WKB expression is identical to the actual solution. Hence, there exists at least one class of reference configurations for which any conclusion following from the approximate theory holds rigorously. In particular, this applies to possible ambiguities in the reconstruction of $\chi(z)$ and $\sigma(z)$ from $r_B^+(s)$. We should keep in mind, however, that the WKB expressions derived in Subsection 2.3.3 apply only when, inside the slab, $\bar{\chi}(z)$ and $\bar{\sigma}(z)$ are continuous functions of z, whose derivatives are discontinuous at no more than a finite number of points. This restriction may prevent us from extending our conclusions to the reconstruction of arbitrary profiles.

Replacing $\bar{E}_+(z,s)$ in (5.2.10) by its asymptotic expression $\bar{E}_+^a(z,s)$ as specified in (2.3.19), and dividing the result by s results in

$$s^{-1}r_B^+(s) = -\bar{T}_{0a}^+[R_1(s) + R_2(s)]/2\bar{G}_a(s), \tag{5.2.12}$$

where $R_1(s)$ and $R_2(s)$ represent the integrals

$$R_1(s) \triangleq \int_{-\infty}^{\infty} C_0(z,s)\exp(-sz)\bar{u}_a^+(z;0)dz, \tag{5.2.13a}$$

$$R_2(s) \triangleq \bar{D}_a^+\bar{R}_{1a}\exp(-s\bar{\tau})\int_{-\infty}^{\infty} C_0(z,s)\exp(-sz)\bar{u}_a^-(z;1)dz. \tag{5.2.13b}$$

In (5.2.12) and (5.2.13), $\bar{u}_a^{\pm}(z;z_0)$ denote the asymptotic solutions defined in (2.3.13). The definitions of the remaining new quantities can be found in (2.3.20) and (2.3.21). In order to estimate what profile information is retrievable from $s^{-1}r_B^+(s)$, we now determine the corresponding time-domain signal.

To this end, we first analyze the time signals corresponding to $R_1(s)$ and $R_2(s)$. With (2.3.13), (2.3.37), (2.3.38), and (2.2.10), the equality (5.2.13a) can be rewritten as

$$R_1(s) = \int_0^{\infty}\exp[-s\bar{\tau}^+(z)-sz]\bar{D}_a^+(z)[\chi(z) + \sigma(z)/s]dz, \tag{5.2.14}$$

where the location of the front of the slab at z = 0 has been taken into account to truncate the lower limit of the z-integration. The right-hand side of (5.2.14) can almost be identified as a Laplace transform. The only difficulty is the occurrence of the factor of s^{-1} in the conductivity term. Hence, we remove that factor by partial integration. Let $S^+(z)$ and $\zeta^+(z)$ be defined as

$$S^+(z) \triangleq \int_0^z \bar{D}_a^+(z')\sigma(z')dz',$$
$$\zeta^+(z) \triangleq \bar{\tau}^+(z) + z = \int_0^z [\bar{N}(z') + 1]dz'. \tag{5.2.15}$$

Then, for $\mathrm{Re}(s) \geq 0$, (5.2.14) turns into

$$R_1(s) = \int_0^\infty \exp[-s\zeta^+(z)]\left\{\frac{\bar{D}_a^+(z)\chi(z)}{\bar{N}(z) + 1} + S^+(z)\right\}d\zeta^+(z), \tag{5.2.16}$$

which can be recognized as the Laplace transform of a causal signal. By virtue of Lerch's theorem (Widder (1946)), the time-domain signal corresponding to $R_1(s)$ is uniquely given by

$$\mathcal{R}_1(t) = \{\bar{D}_a^+(z)\chi(z)/[\bar{N}(z) + 1] + S^+(z)\}|_{z = z^+(t)}, \tag{5.2.17}$$

where $z^+(t)$ is the solution of the equation

$$\zeta^+(z^+(t)) = t. \tag{5.2.18}$$

The time-domain signal corresponding to $R_2(s)$ as defined in (5.2.13b) can be handled in a similar manner. Introducing, in analogy with (5.2.15), the quantities

$$S^-(z) \triangleq \int_z^1 \bar{D}_a^-(z')\sigma(z')dz',$$
$$\zeta^-(z) \triangleq \bar{\tau}^-(z) - (1 - z) = \int_z^1 [\bar{N}(z') - 1]dz', \tag{5.2.19}$$

we arrive at the equivalent time signal

$$\mathcal{R}_2(t) = \bar{D}_a^+\bar{R}_{1a}\{\bar{D}_a^-(z)\chi(z)/[\bar{N}(z) - 1] + S^-(z)\}|_{z = z^-(t)}$$
$$+ \bar{D}_a^+\bar{R}_{1a}S^-(0)\,\mathcal{U}(t - 2\bar{\tau}), \tag{5.2.20}$$

where $\mathcal{U}(t)$ denotes the unit time-step function. In (5.2.20), the first term on the right-hand side is understood to vanish outside the time interval $\bar{\tau} + 1 < t < 2\bar{\tau}$. For t lying in that interval, $z^-(t)$ is the solution of the equation

$$\zeta^-(z^-(t)) = t - \bar{\tau} - 1. \tag{5.2.21}$$

The second term extends the end value of the conductivity integral $S^-(z^-(t))$ at $t = 2\bar{\tau}$ into the interval $2\bar{\tau} < t < \infty$. In that interval, its effect is similar to that of the integral $S^+(z^+(t))$ in (5.2.17). Finally, it should be observed that the apparent discontinuities in $\mathcal{R}_1(t)$ and $\mathcal{R}_2(t)$ at $t = \bar{\tau} + 1$ cancel out, so that the sum of these two time signals is only discontinuous at $t = 0$ and $t = 2\bar{\tau}$.

Now that the time-domain signals corresponding to $R_1(s)$ and $R_2(s)$ in (5.2.12) are known, we can directly determine the total time-domain signal corresponding to $s^{-1}r_B^+(s)$. This signal can be obtained by expanding the factor of $1/\bar{G}_a(s)$ on the right-hand side of (5.2.12) in a similar manner as in the derivation of the asymptotic solution found in (2.3.35). We end up with

$$\begin{aligned} s^{-1}r_B^+(s) \;\leftrightarrow\; -\tfrac{1}{2}\, \bar{T}_{0a}^{+} \textstyle\sum_{k=0}^{\infty} (\bar{R}_{0a}\bar{D}_a^{+}\bar{R}_{1a}\bar{D}_a^{-})^k \\ \times [\mathcal{R}_1(t - 2k\bar{\tau}) + \mathcal{R}_2(t - 2k\bar{\tau})]. \end{aligned} \tag{5.2.22}$$

Because of the factor of s^{-1} on the left-hand side, the time-domain signal in (5.2.22) can be envisaged as an approximate step response of the unknown slab configuration. The corresponding delta-function response can be obtained from (5.2.22) by differentiating its right-hand side with respect to t. Obviously, the resulting derivatives should be understood in the sense of generalized functions. The responses to sine-squared incident pulses as displayed in Subsections 2.3.5, 2.4.3 and 3.3.3 may then be regarded as smoothed versions of this delta-function response resulting from its convolution with the sine-squared incident pulses.

From (5.2.17), (5.2.20) and (5.2.22), it would appear that the reflected field for a single direction of incidence contains sufficient information for the approximate determination of both $\chi(z)$ and $\sigma(z)$. Namely, the approximate impulse response consists of two independent linear combinations of distorted versions of $d_z\chi(z)$ and $\sigma(z)$, which are separated in time and repeated with a period of $2\bar{\tau}$ with an exponentially decaying amplitude. However, the information contained in $R_2(t)$ only shows up in terms that contain at least one factor $\bar{R}_{1a} = [\bar{N}(1) - 1]/[\bar{N}(1) + 1]$. Since we want to include the case $\bar{N}(1) \approx 1$ in our analysis, we cannot base our solution procedure on drawing information from these terms. This leaves us with $R_1(t)$ only, from which we can at best reconstruct one of the unknown profiles $\chi(z)$ and $\sigma(z)$. In such a reconstruction, the other profile should be known.

An independent linear functional of $\chi(z)$ and $\sigma(z)$ is always available from the reflection coefficient for incidence from the opposite direction, i.e. $r^-(s)$. This is observed when the above analysis is repeated for (5.2.4) with $\alpha = -$. The results can be obtained directly from (5.2.17), (5.2.20) and (5.2.22) by replacing z with 1 - z in the arguments of all functions and interchanging the subscripts and superscripts + and 0 with - and 1, respectively. This suggests that a reconstruction of $\chi(z)$ and $\sigma(z)$ from the known values of both reflection coefficients $r^+(s)$ and $r^-(s)$ may be possible.

The special form of (5.2.16) also provides us with an indication of how the profile information it contains is distributed over the real-valued frequencies ω, with $s = i\omega$. For these frequencies, (5.2.16) reduces to a Fourier transform of the real-valued function $R_1(t)$ given in (5.2.17). As $|\omega| \to \infty$, this Fourier transform is, up to $O(\omega^{-1})$, determined by the discontinuities in $R_1(t)$ and by the time difference in between them. Now, we should keep in mind the conditions for the validity of the asymptotic expression $\bar{E}^a_+(z,s)$, which was used in deriving (5.2.17). Under these conditions, the discontinuities of $R_1(t)$ are completely given by the profile parameters $\chi(0)$ and $\chi(1)$, and the approximate profile parameters $\bar{\chi}(0)$, $\bar{\chi}(1)$, $\bar{\tau}$ and

$$\int_0^1 \bar{\sigma}(z)/2\bar{N}(z)dz = \int_0^1 \bar{\sigma}(z)/2[1 + \bar{\chi}(z)]^{\frac{1}{2}}dz. \tag{5.2.23a}$$

When the iterative scheme converges, the barred quantities listed above will approach the corresponding actual ones. Upon convergence, the high-frequency behavior of $r^+(i\omega)$ will therefore primarily supply information on the profile parameters $\chi(0)$, $\chi(1)$, τ and

$$\int_0^1 \sigma(z)/2N(z)dz. \tag{5.2.23b}$$

This conclusion is also reached by inspecting the asymptotic reflection coefficient $r_a^+(s)$ specified in (2.3.20) and (2.3.21). It is observed that this first-order approximation to the actual reflection coefficient depends on exactly the same configuration parameters. Hence, all other information about the constitutive coefficients $\chi(z)$ and $\sigma(z)$ must be reconstructed primarily from the low-frequency behavior of $r^+(i\omega)$.

5.2.4 The error in the Born-type approximation

With the analysis of the previous subsection, we have obtained some idea as to what profile information is contained in the approximate reflection coefficient $r_B^+(s)$. Whether that information can really be retrieved by equating $r_B^+(s)$ with the known, exact reflection coefficient $r^+(s)$, as happens in the iterative scheme proposed in Subsection 5.2.2, depends on the error made by neglecting the second term on the right-hand side, i.e. $\Delta r_B^+(s)$ as specified in (5.2.11). The estimation of that term runs up against the same difficulty as the analysis of $r_B^+(s)$ in the previous subsection, namely that closed-form solutions for the Green's function $G(z,z';s)$ for the reference medium and for the exact field $\hat{E}_+(z,s)$ are not available. As in the previous subsection, asymptotic considerations of the WKB type can serve to obtain at least a good qualitative estimate. The first step is to replace $\hat{E}_+(z,s)$ and $G(z,z';s)$ by their first-order WKB approximations. For the former, we again use the form given in (2.3.19). The latter is obtained by a derivation similar to the one resulting in (2.3.19). For z in $\mathcal{D}_2$, i.e. $0 < z < 1$, we end up with:

$$G^a(z,z';s) = [\bar{u}_a^+(z;z')U(z-z') + \bar{u}_a^-(z;z')U(z'-z)]/2s\bar{N}(z')$$
$$+ \bar{R}_{0a}\bar{n}_1^a(z';s)\bar{u}_a^+(z;0)/\bar{d}_a(z';s) \qquad (5.2.24)$$
$$+ \bar{R}_{1a}\bar{n}_2^a(z';s)\bar{u}_a^-(z;1)/\bar{d}_a(z';s),$$

where U(z) denotes the unit space-step function and where the asymptotic numerator functions $\bar{n}_{1,2}^a(z';s)$ and the asymptotic denominator $\bar{d}_a(s)$ are given by

$$\bar{n}_1^a(z';s) = \bar{u}_a^-(0;z') + \bar{R}_{1a}\bar{u}_a^+(1;z')\bar{u}_a^-(0;1),$$
$$\bar{n}_2^a(z';s) = \bar{u}_a^+(1;z') + \bar{R}_{0a}\bar{u}_a^-(0;z')\bar{u}_a^+(1;0), \qquad (5.2.25)$$
$$\bar{d}_a(z';s) = 2s\bar{N}(z')\bar{G}_a(s),$$

where $\bar{G}_a(s)$ is the asymptotic slab denominator for the approximate medium, as defined on the last line of (2.3.20). As in the case of the excited field $\hat{E}_+(z,s)$, the form of the asymptotic Green's function $G^a(z,z';s)$ closely resembles the corresponding homogeneous-slab solution as specified in (2.4.27) - (2.4.29). Hence the physical interpretation of the expressions listed above follows from analogy.

Substitution of (2.3.19) and (5.2.24) in (5.2.11) results in a series of double integrals over z and z', all of which can be estimated in a similar manner. For example, the contribution of the first term on the right-hand side of (5.2.24) can, with the aid of the relation

$$u_a^+(z;z')/N(z') = u_a^+(z;0)u_a^-(z';0)/N(0), \qquad (5.2.26)$$

which follows directly from the definition (2.3.13), be rewritten as

$$\Delta r_1(s) \triangleq \frac{s^2}{4\bar{N}(0)\bar{G}_a(s)} \int_0^1 C_0(z,s)\exp(-sz)\bar{u}_a^+(z;0)Q_1(z,s)dz. \qquad (5.2.27)$$

In (5.2.27), we have introduced the abbreviation

$$Q_1(z,s) \triangleq a_a^+(s)\int_0^z C(z';s)\bar{u}_a^-(z';0)u_a^+(z';0)dz' + b_a^+(s)\int_0^z C(z';s)\bar{u}_a^-(z';0)u_a^-(z';1)dz', \tag{5.2.28}$$

with $a_a^+(s)$ and $b_a^+(s)$ as defined in (2.3.20). The behavior of $\Delta r_1(s)$ as $|s| \to \infty$ is estimated as follows. Since the reference configuration was assumed to be a good approximation to the actual one, we may approximate the asymptotic solutions $u_a^\pm$ in (5.2.28) by their counterparts for the reference configuration. Subsequently, we can use the identities

$$u_a^-(z';0)u_a^+(z';0) = N(0)/N(z'), \quad \text{and} \tag{5.2.29a}$$

$$u_a^-(z';0)u_a^-(z';1) = u_a^-(1;0)u_a^-(z';1)^2, \tag{5.2.29b}$$

which also follow from (2.3.13). Substitution of (5.2.29b) in the second line of (5.2.28) results in a Fourier-type integral of a form similar to that we encountered in the previous subsection. For a general, discontinuous $C(z,s)$, such an integral is of $O(s^{-1})$ as $|s| \to \infty$. This leaves us with

$$Q_1(z,s) = \bar{N}(0)a_a^+(s)\int_0^z [\chi(z') - \bar{\chi}(z')]/\bar{N}(z')dz', \tag{5.2.30}$$

which holds up to $O(s^{-1})$ as $|s| \to \infty$. Now, the right-hand side of (5.2.30) is a continuous function of z of $O(\chi - \bar{\chi})$. Substituting this result in (5.2.27) allows us to recognize that integral, in turn, as a Fourier-type integral. Thus, we arrive at

$$\Delta r_1(s) = sO(\chi)O(\chi - \bar{\chi}), \tag{5.2.31}$$

which again holds as $|s| \to \infty$.

With this estimate and similar estimates for the integrals involving the remaining three terms in (5.2.24), it follows that

$$\Delta r_B^+(s) = sO(\chi)O(\chi - \bar{\chi}) \quad \text{as } |s| \to \infty. \tag{5.2.32}$$

The conductivity terms do not show up in this high-frequency estimate because of the factor of s^{-1} in $C(z,s)$ (see (2.4.18)). Note that the error made by replacing $\hat{E}_+(z,s)$ with $\bar{E}_+(z,s)$ in (5.2.4) is of the same order of magnitude as the improvements $\chi(z) - \bar{\chi}(z)$ and $\sigma(z) - \bar{\sigma}(z)$ that one would like to obtain. Moreover, as $|s| \to \infty$, this error dominates over the exact reflection coefficient $r^+(s)$, which is of $\mathcal{O}(1)$. Hence, a computation neglecting this error can at best yield a band-limited approximation of these improvements. This provides an additional argument for regarding only the dominant term corresponding to $R_1(t)$. Furthermore, the operator producing $\Delta r_B^+(s)$ from $\chi(z) - \bar{\chi}(z)$ and $\sigma(z) - \bar{\sigma}(z)$ becomes more and more independent of $\bar{\chi}(z)$ and $\bar{\sigma}(z)$ as these approach the actual profiles. Consequently, if the iterative scheme outlined in Subsection 5.2.2 converges, the band limitation of the approximate solutions $\chi^{(n)}(z)$ and $\sigma^{(n)}(z)$ as well as the rate of convergence will become constant in the final stage of the procedure.

Finally, we will summarize the consequences of the analysis in this and the previous subsection for the implementation of the iterative scheme.

- It seems possible to reconstruct one unknown constitutive parameter from one of the reflection coefficients $r^+(s)$ and $r^-(s)$. For the determination of $\chi(z)$ as well as $\sigma(z)$, both of them seem to be needed.
- In order to obtain convergence, we must represent $\chi^{(n)}(z)$ and $\sigma^{(n)}(z)$ by a linear combination of expansion functions which prescribes the short-range behavior of these profiles. Thus, we supplement the band-limited information available in the iterative scheme with the necessary *a priori* information.
- Because of the band limitation, rapidly varying profiles (e.g. those with discontinuities) will be hard to recover.
- The optimum choice for the weighting function $w(\omega)$ in (5.2.6) is governed by the frequency behavior of the error $\Delta r_B(i\omega)$ as well as by the distribution of the profile information over the frequency range. The error $w(i\omega)|\Delta r_B(i\omega)|$ must be of the same order of magnitude over the entire frequency range. Moreover, for high frequencies, $r^+(i\omega)$ mainly depends on a few select properties of the

slab. The remaining properties of the slab, which are of interest in the inverse-scattering problem, only influence the correction term of $O(\omega^{-1})$. Both these observations indicate that $w(\omega)$ must be of $O(\omega^{-1})$ as $|\omega| \to \infty$.

- Because of the band limitation that is inherent in our scheme, we need not concern ourselves with the high-frequency or, equivalently, short-time behavior of the reflected fields.
- In the time-domain case, we can terminate, in each iteration step, the computation at $t = \tau^{(n-1)} + 1$, with $\tau^{(n)}$ being the travel time $\bar{\tau}$ corresponding to $\chi^{(n)}(z)$. This can be explained from the fact that we treat our inverse-scattering problem as the inverse-source problem of determining a source distribution $J(z,t)$ of the prescribed form

$$J(z,t) = \sigma(z)E(z,t) + \chi(z)\partial_t E(z,t), \tag{5.2.33}$$

located in vacuum from the field it excites. The time $\tau^{(n-1)} + 1$ is then the starting time for $J(1,t)$, augmented by the vacuum travel time from $z = 1$ to $z = 0$.

5.2.5 An improved method of solution

The most striking flaw in the scheme as it has been described until now appears to be the inherent band limitation of the resulting approximation. The error estimate (5.2.32) suggests that this problem can possibly be resolved by considering, in each iteration step, the approximate configuration found in the previous step as the background medium. The relation (5.2.4) is then replaced by (5.2.3). Substituting (5.2.8) in the latter results in

$$r^+(s) - \bar{r}^+(s) = -\frac{s}{2}\int_{-\infty}^{\infty} C(z,s)\bar{E}_+(z,s)^2 dz + \Delta r_B^+(s), \tag{5.2.34}$$

with

$$\Delta r_B^+(s) = \frac{s^3}{2}\int_{-\infty}^{\infty}\Big\{C(z,s)\bar{E}_+(z,s) \times \int_{-\infty}^{\infty}[C(z',s)G(z,z';s)\hat{E}_+(z',s)]dz'\Big\}dz. \tag{5.2.35}$$

Note that these equations are of the same type as Equations (5.2.10) and (5.2.11) of Subsection 5.2.3. The main difference is that the quantities $C_0(z,s)$ and $\exp(-sz)$, which originate from taking a vacuum background medium in (5.2.4), have been replaced by their counterparts pertaining to the reference medium $\{\bar{\chi}(z),\bar{\sigma}(z)\}$. In particular, this refers to the contrast function $C_0(z,s)$ occurring in the error $\Delta r_B^+(s)$ defined in (5.2.11), which led to the prediction that the iterative procedure will at best yield a band-limited approximation to the unknown profiles $\chi(z)$ and/or $\sigma(z)$.

Now, we can repeat the analysis of Subsections 5.2.3 and 5.2.4 for Equations (5.2.34) and (5.2.35). Replacing $\bar{E}_+(z,s)$ with its first-order WKB approximation

$$\bar{E}_+^a(z,s) = \bar{a}_a^+(s)\bar{u}_a^+(z;0) + \bar{b}_a^+(s)\bar{u}_a^-(z;1) \tag{5.2.36}$$

results in four integrals of the same type as occurring in the definitions of $R_1(s)$ and $R_2(s)$ in (5.2.13), each of them divided by the square of the approximate slab denominator $\bar{G}_a(s)$. As in Subsection 5.2.3, we can distinguish a dominant term. In the present case, that term corresponds to the time-domain signal

$$- \frac{\bar{T}_{0a}^{+\,2}}{2}\left\{ \bar{D}_a^+(z)^2 \frac{\chi(z) - \bar{\chi}(z)}{2\bar{N}(z)} + \int_0^z \bar{D}_a^+(z')^2[\sigma(z') - \bar{\sigma}(z')]dz' \right\}\Bigg|_{z = z^+(t)}, \tag{5.2.37}$$

where $z^+(t)$ satisfies the equation

$$2\bar{\tau}^+(z^+(t)) = t. \tag{5.2.38}$$

An analysis of the exponential factors contained in the remaining terms reveals that, for $0 < t < 2\bar{\tau}$, the relevant Laplace inversion integrals can be closed to the right and, hence, the corresponding time-domain signals vanish. With (5.2.37), this observation leads to the conclusion that the time-domain signal corresponding to $r^+(s) - \bar{r}^+(s)$, i.e. the difference between the impulse responses for the actual and the reference medium, is, for $0 < t < 2\bar{\tau}$, just a linear combination of a distorted version of $d_z[\chi(z) - \bar{\chi}(z)]$ and a distorted version of $\sigma(z) - \bar{\sigma}(z)$. Similarly, the difference between the reflected fields caused by a sine-squared incident pulse can be regarded as the result of convolving the relevant linear combination with that incident pulse. This is in agreement with the observation made in connection with Figures 3.3.3 - 3.3.5. By regarding the lossless medium specified in Figure 3.3.3, for which we have $\bar{D}_a^+(z) = 1$ and $\bar{N}(z) = 1.5$, as the reference medium, we can indeed interpret the negative of the reflected field in between the primary and the secondary reflection as a smoothed linear combination of $d_z\chi(z)$ and $\sigma(z)$. As in the case discussed in Subsection 5.2.3, the remaining terms in the time-domain signal corresponding to $s^{-1}[r_a^+(s) - \bar{r}_a^-(s)]$ contain at least one factor of $\bar{R}_{1a}$, which makes them less suitable as possible sources of information on the configuration. This observation, too, is confirmed by the numerical results displayed in Subsection 3.3.3, and further by those shown in Subsections 2.2.4, 2.3.5 and 2.4.3. In most of the transient reflected fields presented in these subsections, the field becomes negligible after the arrival of the secondary reflection. In a numerical reconstruction procedure, we should, therefore, not aim at reconstructing more than one constitutive parameter from one reflection coefficient.

The estimation of the error $\Delta r_B^+(s)$ appearing in (5.2.34) requires a more intricate analysis than given in Subsection 5.2.4. The complication is due to the fact that the incident field $\exp(-sz)$, which travels to the right, is now replaced by $\bar{E}_+(z,s)$, which consists of waves traveling to the left as well as to the right. Substitution of the relevant asymptotic expressions in (5.2.35) results, as in Subsection 5.2.4, in a series of double integrals over z and z' that can all be estimated in a similar manner. Limiting ourselves again to the contribution from the first term on the right-hand side of (5.2.24), we arrive at

$$\Delta r_1(s) = \Delta r_{11}(s) + \Delta r_{12}(s), \tag{5.2.39}$$

with

$$\Delta r_{11}(s) \triangleq \frac{s^2\bar{a}_a^+(s)}{4\bar{N}(0)\bar{G}_a(s)} \int_0^1 C(z,s)\bar{u}_a^+(z;0)^2 Q_1(z,s)\,dz,$$

$$\Delta r_{12}(s) \triangleq \frac{s^2\bar{b}_a^+(s)}{4\bar{N}(0)\bar{G}_a(s)} \int_0^1 C(z,s)\bar{u}_a^-(z;1)\bar{u}_a^+(z;0)Q_1(z,s)\,dz, \tag{5.2.40}$$

with $Q_1(z,s)$ as defined in (5.2.28). The term $\Delta r_{11}(s)$ can be handled by the approach outlined in Subsection 5.2.4, which directly leads to the estimate

$$\Delta r_{11}(s) = sO[(\chi - \bar{\chi})^2]. \tag{5.2.41}$$

In the analysis of $\Delta r_{12}(s)$, however, the identity

$$\bar{u}_a^-(z;1)\bar{u}_a^+(z;0) = \bar{D}_a^-(z)\bar{D}_a^+(z)\exp(-s\bar{\tau}) \tag{5.2.42}$$

does not allow an interpretation as a Fourier-type integral. Hence, a more careful estimate of the first integral on the right-hand side of (5.2.28) is required. Keeping the discussion of the second term as it is, we can refine (5.2.30) to

$$Q_1(z,s) = \bar{N}(0)a_a^+(s)\int_0^z \left\{ \frac{D_a^+(z;0)[\chi(z') - \bar{\chi}(z')]}{\bar{D}_a^+(z;0)\bar{N}(z')} \times \exp[-s(\tau^+(z') - \bar{\tau}^+(z'))] \right\} dz', \tag{5.2.43}$$

which also holds up to $O(s^{-1})$. Now, $Q_1(z,s)$ can be written in the form of a Fourier integral by taking a factor of $[N(z') - \bar{N}(z')]$ out of the integrand. This leads to the estimate

$$\Delta r_{12}(s) = sO(\chi - \bar{\chi})O(\chi), \tag{5.2.44}$$

which annuls the improvement achieved by replacing $C_0(z,s)$ in (5.2.11) by $C(z,s)$ in (5.2.35). Estimates similar to the ones given in (5.2.41) and (5.2.44) can also be given for the contributions from the remaining terms on the right-hand side of (5.2.24). In the frequency domain, therefore, we have

$$\Delta r_B^+(s) = sO(\chi - \bar{\chi})O(\chi), \qquad (5.2.45)$$

and the inherent band limitation is not removed. However, all of the terms leading to estimates of the type (5.2.44) contain at least one factor $\bar{R}_{0a}$ or one factor $\bar{R}_{1a}$. In fact, the $O(\chi)$ term in (5.2.44) and (5.2.45), which takes effect as $\chi \to 0$ only, stems from these reflection coefficients. For contrasts which are not too large, therefore, the band limitation may be less severe than the one observed in the results of the original scheme. A removal of the band limitation can be attained by transforming the approximate equations (5.2.34) and (5.2.35) back to the time domain. From (2.3.20) and (5.2.42), it is observed that the error $\Delta r_{12}(s)$ contains a time-delay factor $\exp(-2s\bar{\tau})$. Similar time-delay factors are also present in the total contributions from the last three terms on the right-hand side of (5.2.24). In the time domain, these contributions will therefore only show up for $2\bar{\tau} < t < \infty$. Consequently, they will not affect the reconstruction of profile parameters from the time-domain reflected field in the interval $0 < t < 2\bar{\tau}$, where only $\Delta r_{11}(s)$ contributes to the error. In that interval, the factor of $O[(\chi - \bar{\chi})^2]$ in (5.2.41) will remove the band limitation in the reconstructed profiles as the iterative procedure progresses.

The estimate given above for the error made during the time interval $0 < t < 2\bar{\tau}$ can be used to interpret (5.2.37) in an alternative manner. With the aid of this estimate, the terms proportional to $\chi(z) - \bar{\chi}(z)$ and $\sigma(z) - \bar{\sigma}(z)$ can be envisaged as the derivatives of the approximate step response with respect to the relevant constitutive parameters. For the lossless case treated in Section 5.4, this interpretation provides one way to arrive at an initial estimate of the unknown susceptibility profile. Integrating the profile derivative in closed form results, in that case, in an approximate expression for the time-domain response corresponding to $s^{-1}r(s)$. Equating that expression with the actual step response yields the desired estimate of $\chi(z)$.

Note that in the scheme proposed above, the unknown slab configuration would be determined in full detail from the reflected fields for $0 < t < 2\tau$. Comparing this result with the one derived in Subsection 5.2.3, we observe that the possible improvement may be explained from the fact that the time it takes for the field reflected at one slab interface to travel to the opposite interface is now taken into account properly. In this sense, the inverse-scattering problem is now really treated as an inverse-profiling problem. A second difference between both schemes is that (5.2.37), which describes the time-domain reflected field for $0 < t < 2\bar{\tau}$, does not contain a term resembling (5.2.20). Along with the error estimates (5.2.32) and (5.2.41), this observation suggests that the term $\mathcal{R}_2(t)$ in (5.2.22) may originate from the Born error $\Delta r_B^+(s)$ rather than from the actual reflection coefficient $r^+(s)$. A limiting factor in the application of the scheme may be the additional attenuation factor $\bar{D}_a^+(z)$ in (5.2.37), which represents the additional damping accumulated during the return journey of the reflected field. Especially for large $\sigma(z)$, this factor may prevent a successful reconstruction.

Contemplating the results obtained in this section, we conclude that we have obtained two variants of an iterative procedure for the reconstruction of one or both of the constitutive parameters $\chi(z)$ and $\sigma(z)$ from the reflected fields for one or two directions of incidence. Each variant has its specific advantages and disadvantages. The procedure outlined in Subsection 5.2.2 has the disadvantage of an inherent band limitation in the reconstruction. On the other hand, the relevant equations are available in closed form in the time domain as well as in the frequency domain, since they follow directly from the corresponding direct-scattering integral equation. Moreover, the procedure would seem to be capable of being generalized to more complicated inverse-scattering problems. When carried out in the frequency domain, the procedure outlined in the present subsection will also suffer from a band limitation. This disadvantage is removed when this modified procedure is carried out in the time domain. However, a time-domain equivalent of (5.2.34) can only be obtained by numerically evaluating either a Fourier inversion integral or a time-domain convolution integral. In view of the fact that, in a practical situation, the known reflection data will always be band-limited, it would therefore seem most feasible to start the numerical experimentation with implementing the simple scheme proposed in Subsection 5.2.2. From the results, we can then decide whether the severity of the band limitation inherent in that scheme warrants the application of an alternative scheme.

5.3 The one-dimensional inverse-scattering problem for low contrast and oblique incidence

5.3.1 Formulation of the problem

The most simple example of an iteration step as discussed in Section 5.2 is the reconstruction of a small $\chi(z)$ and $\sigma(z)$ from the known real-frequency behavior of the reflection coefficient for two different directions of incidence in the *Born approximation*. This approximation amounts to choosing both the background medium and the reference medium to be a vacuum in a single iteration step of either type discussed in Section 5.2. Hence, the small-contrast problem constitutes an ideal starting point for gaining insight into the type of difficulties that we can expect in implementing the full iterative scheme. In contrast with the slab problems discussed up to now, we allow, in this particular problem, the incident field to be an E-polarized monochromatic plane wave of oblique incidence. This allows us to state the reasons for restricting ourselves to the case of normal incidence in the slab problems discussed elsewhere in this and the next chapter. In the case of oblique incidence, the electric-field intensity and the magnetic-field intensity of the incident field can be rewritten as

$$\begin{aligned} \underline{E}^{i} &= F(t - x\sin\theta/c_0 - z\cos\theta/c_0)\underline{i}_y, \\ \underline{H}^{i} &= -Y_0\cos\theta F(t - x\sin\theta/c_0 - z\cos\theta/c_0)\underline{i}_x \\ &\quad + Y_0\sin\theta F(t - x\sin\theta/c_0 - z\cos\theta/c_0)\underline{i}_z, \end{aligned} \qquad (5.3.1)$$

where $Y_0 = (\varepsilon_0/\mu_0)^{\frac{1}{2}}$ and $c_0 = (\varepsilon_0\mu_0)^{-\frac{1}{2}}$. Note that, for $\theta = 0$, the incident field specified in (5.3.1) reduces to the normally incident field specified in (2.2.1). For $|\theta| < \pi/2$, it originates from region $\mathcal{D}_1$ and for $\pi/2 < |\theta| < \pi$ from region $\mathcal{D}_3$. In what follows these regions will be referred to as the *region of incidence* $\mathcal{D}_i$. The half-space on the opposite side of the slab will be called the *region of transmission* $\mathcal{D}_t$. As in (2.5.2), we restrict the unknowns in the total electromagnetic field to those quantities that are continuous at any discontinuity in $\chi(z)$ and $\sigma(z)$. This yields the representation

$$\underline{E} = E(z,t - x\sin\theta/c_0;\theta)\underline{i}_y,$$
$$\underline{H} = H(z,t - x\sin\theta/c_0;\theta)\underline{i}_x + Y_0\sin\theta E(z,t - x\sin\theta/c_0;\theta)\underline{i}_z. \tag{5.3.2}$$

Next, we go through the same Laplace-transform and normalization procedures as in Subsections 2.2.1 and 2.4.1. Thus, we arrive at the Laplace-transformed, normalized, source-free electromagnetic-field equations

$$\partial_z H(z,s;\theta) = s[\cos^2\theta + \chi(z) + \sigma(z)/s]E(z,s;\theta), \tag{5.3.3a}$$

$$\partial_z E(z,s;\theta) = s\, H(z,s;\theta), \tag{5.3.3b}$$

which are the generalization of (2.2.1). The corresponding second-order differential equation is given by

$$[\partial_z^2 - s^2(\cos^2\theta + \chi(z) + \sigma(z)/s)]E(z,s;\theta) = 0. \tag{5.3.4}$$

As in the case of normal incidence, an integral relation can be derived which is equivalent to (5.3.4). With the aid of the one-dimensional Green's function

$$G(z,z';s,\theta) \triangleq (2s|\cos\theta|)^{-1}\exp[-s|(z - z')\cos\theta|], \tag{5.3.5}$$

where the absolute values have been taken to enforce causality, we obtain

$$E(z,s;\theta) = E^i(z,s;\theta) - s^2\int_{-\infty}^{\infty} C_0(z',s)G(z,z';s,\theta)E(z',s;\theta)dz', \tag{5.3.6}$$

with $C_0(z,s) = \chi(z) + \sigma(z)/s$ being, as before, the contrast function with respect to vacuum. The form of the electromagnetic field outside the slab follows directly from (5.3.6). For the electric field, we have

$$E(z,s;\theta) = \begin{cases} F(s)[\exp(-sz\cos\theta) + r(s;\theta)\exp(sz\cos\theta)] & \text{in } \mathcal{D}_i, \\ F(s)[t(s;\theta)\exp(-sz\cos\theta)] & \text{in } \mathcal{D}_t. \end{cases} \tag{5.3.7}$$

The corresponding expression for the magnetic field $H(z,s;\theta)$ is found directly with the aid of (5.3.3b). In (5.3.7), we have implicitly introduced the angular reflection and transmission coefficients $r(s;\theta)$ and $t(s;\theta)$. For normal incidence, these coefficients may differ by a phase factor from the ones introduced in (2.4.21). This is due to the uniform choice of $z = 0$ as the reference plane in the plane-wave solutions that occur in (5.3.7). In fact, we have the connecting formulas

$$\begin{aligned} r^+(s) &= r(s;0), \\ t^+(s) &= \exp(-s)t(s;0), \\ r^-(s) &= \exp(-2s)r(s;\pi), \\ t^-(s) &= \exp(-s)t(s;\pi). \end{aligned} \tag{5.3.8}$$

For a general, inhomogeneous slab problem, the distribution of $\{E(z,s;\theta),H(z,s;\theta)\}$ inside the slab and the reflection and transmission coefficients have to be computed numerically. This can be carried out by numerically integrating the system of equations (5.3.3) according to the procedure explained in Subsection 2.4.2, starting from the interface of the slab with the region of transmission $\mathcal{D}_t$. The necessary initial conditions for the field components $\{E(z,s;\theta), H(z,s;\theta)\}$ at that interface are obtained from (5.3.7) and the corresponding expression for the magnetic-field strength in the same manner as (2.4.8) was found from (2.4.4). Thus, the reflection coefficients can be determined with an accuracy similar to that specified in Tables 2.4.2 - 2.4.4. This is considerably more accurate than some of the approximations that will have to be made further on to arrive at a solution procedure for the inverse-scattering problem. In the context of that problem we may, therefore, consider the solution of the corresponding direct-scattering problem for a given, inhomogeneous slab configuration to be known exactly.

In terms of the angle-dependent reflection coefficient introduced in (5.3.7), we can now formulate the problem at hand in a more exact manner. In the present section, our aim is to reconstruct, as well as possible, in a single iteration step of the type described in Section 5.2, both constitutive profiles $\chi(z)$ and $\sigma(z)$ from the known values of $r(i\omega;\theta_1)$ and $r(i\omega;\theta_2)$ with $\omega\in\mathbb{R}$ and $\cos\theta_1 \neq \cos\theta_2$. In the numerical implementation, we will also concern ourselves with the more practical situation where these reflection coefficients are known over a finite frequency range only.

5.3.2 Born approximation: elaboration and error estimates

As indicated above, we wish to determine $\chi(z)$ and $\sigma(z)$ in a single iteration step of the type discussed in Section 5.2. To this end, we need to generalize the additional relation (5.2.4) to the case of oblique incidence. This can be achieved with the aid of the property

$$(z' - z)\cos\theta > 0, \tag{5.3.9}$$

which holds for any combination of a $z'\in\mathcal{D}_2$ and a $z\in\mathcal{D}_1$. Substituting this property and the first line of (5.3.7) in (5.3.6), we arrive at

$$r(s;\theta) = -\frac{s}{2|\cos\theta|}\int_{-\infty}^{\infty} C_0(z,s)\hat{E}^i(z,s;\theta)\hat{E}(z,s;\theta)dz. \tag{5.3.10}$$

In (5.3.10),

$$\hat{E}^i(z,s;\theta) = \exp(-sz\cos\theta) \tag{5.3.11}$$

constitutes the normalized electric-field strength associated with a monochromatic, E-polarized incident plane wave of unit amplitude and oblique incidence.

Now the *Born approximation* can be phrased as neglecting the difference between the actual reflection coefficient $r(s;\theta)$ and its appropriate counterpart $r_B(s;\theta)$ obtained by replacing, in (5.3.10), the unknown plane-wave response $\hat{E}(z,s;\theta)$ by the known incident field $\hat{E}^i(z,s;\theta)$. With the definition of $C_0(z,s)$, this approximate reflection coefficient can be expressed as

$$r_B(s;\theta) \triangleq -\frac{s}{2|\cos\theta|}\int_{-\infty}^{\infty}[\chi(z) + \sigma(z)/s]\exp(-2sz\cos\theta)dz. \qquad (5.3.12)$$

As in Equation (5.2.14), the right-hand side of (5.3.12) can be recognized as a Laplace transform. In fact, from the special equation (5.3.12), closed-form expressions can be derived for the unknown profiles $\chi(z)$ and $\sigma(z)$. Let, in analogy with (3.4.21), the spatial Fourier transforms of these profiles be given by

$$\begin{aligned}\tilde{\chi}(k) &\triangleq \int_{-\infty}^{\infty}dz\ \exp(-ikz)\chi(z),\\ \tilde{\sigma}(k) &\triangleq \int_{-\infty}^{\infty}dz\ \exp(-ikz)\sigma(z).\end{aligned} \qquad (5.3.13)$$

Then, (5.3.12) can be rewritten as

$$r_B\left(\frac{ik}{2\cos\theta},\theta\right) = \frac{-ik\tilde{\chi}(k)}{4\cos\theta|\cos\theta|} - \frac{\tilde{\sigma}(k)}{2|\cos\theta|}, \qquad (5.3.14)$$

which constitutes one inhomogeneous, linear equation for these two Fourier coefficients. From (5.3.14), it follows that either $\chi(z)$ or $\sigma(z)$ can be reconstructed from the real-frequency behavior of the approximate reflection coefficient $r_B(s;\theta)$ at a single, fixed angle of incidence θ, provided that the remaining constitutive parameter is known. In Section 5.4, we will come back to this observation for the case of the lossless slab. For the determination of both $\chi(z)$ and $\sigma(z)$, we need at least two independent equations of the type (5.3.14). Consequently, we need to know the real-frequency behavior of $r_B(s;\theta)$ for at least two angles of incidence $\theta = \theta_1$ and $\theta = \theta_2$, with $\cos\theta_1 \neq \cos\theta_2$. In that case, we have from (5.3.14)

$$\tilde{\chi}(k) = \frac{4}{ik}\,\frac{\cos\theta_1\cos\theta_2}{\cos\theta_2 - \cos\theta_1}\left\{|\cos\theta|\,r_B\left(\frac{ik}{2\cos\theta},\theta\right)\right\}\Big|_{\theta_1}^{\theta_2}, \qquad (5.3.15)$$

and

$$\tilde{\sigma}(k) = \frac{2}{\cos\theta_1 - \cos\theta_2}\left\{\cos\theta|\cos\theta|\,r_B\left(\frac{ik}{2\cos\theta},\theta\right)\right\}\Big|_{\theta_1}^{\theta_2}. \qquad (5.3.16)$$

Equations (5.3.15) and (5.3.16) hold rigorously for any pair of angles θ_1 and θ_2 with $\cos\theta_1 \neq \cos\theta_2$. Nevertheless, they only lead to an approximate solution of the inverse-scattering problem, where the actual reflection coefficient $r(s;\theta)$ is known instead of its approximate counterpart $r_B(s;\theta)$. Hence, we must replace the reflection coefficient r_B in (5.3.15) and (5.3.16) by the actual reflection coefficient r. This results in the Born-approximated Fourier coefficients

$$\tilde{\chi}_B(k) = \frac{4}{ik}\frac{\cos\theta_1\cos\theta_2}{\cos\theta_2 - \cos\theta_1}\left\{|\cos\theta|\,r\left(\frac{ik}{2\cos\theta};\theta\right)\right\}\Big|_{\theta_1}^{\theta_2}, \tag{5.3.17}$$

and

$$\tilde{\sigma}_B(k) = \frac{2}{\cos\theta_1 - \cos\theta_2}\left\{\cos\theta|\cos\theta|\,r\left(\frac{ik}{2\cos\theta};\theta\right)\right\}\Big|_{\theta_1}^{\theta_2}, \tag{5.3.18}$$

where the subscript B has been appended in line with the notation used for the reflection coefficients. Obviously, the approximate susceptibility and conductivity profiles are then defined via the inverse Fourier transformations

$$\begin{aligned}\chi_B(z) &= (2\pi)^{-1}\int_{-\infty}^{\infty}dk\ \exp(ikz)\tilde{\chi}_B(k),\\ \sigma_B(z) &= (2\pi)^{-1}\int_{-\infty}^{\infty}dk\ \exp(ikz)\tilde{\sigma}_B(k).\end{aligned} \tag{5.3.19}$$

With (5.3.17) - (5.3.19) we have at our disposal the set of estimation formulas that allows us to resolve the inverse-scattering problem posed in Subsection 5.3.1 in the Born approximation.

As remarked in Subsection 5.2.3, the error in the results obtained with the aid of these formulas is generated by the deviation between the exact reflection coefficient $r(s;\theta)$ and its Born approximation $r_B(s;\theta)$. This deviation can be estimated by substituting the integral equation (5.3.6) in the additional relation (5.3.10) and proceeding in a manner similar to that in Subsections 5.2.4 and 5.2.5. The only difference is the fact that the error now depends on the angle of incidence θ. Comparing (5.3.6) and (5.3.10) with the corresponding equations for the case of normal incidence, i.e. (5.2.8) and (5.2.4), reveals that the desired estimate can be obtained from (5.2.32) by changing

$$\begin{aligned} s &\to s|\cos\theta|, \\ \chi(z) &\to \chi(z)/\cos^2\theta. \end{aligned} \tag{5.3.20}$$

Thus, we directly have

$$r(s;\theta) - r_B(s;\theta) = s|\cos\theta|\mathcal{O}(\chi^2/\cos^4\theta). \tag{5.3.21}$$

As in Subsection 5.2.4, the estimate (5.3.21) holds for a general, discontinuous susceptibility profile and for $\chi(z)$ and $\sigma(z)$ being of the same order of magnitude. The estimate (5.3.21) could also have been obtained from the correspondence of Equations (5.3.6) and (5.3.10) with Equations (5.2.8) and (5.2.3). In that case, the change (5.3.20) must be made in either of the estimates (5.2.41) and (5.2.45).

With the aid of (5.3.21), we can now determine the size of the error in the approximate Fourier coefficients $\tilde{\chi}_B(k)$ and $\tilde{\sigma}_B(k)$. For $\tilde{\chi}_B(k)$, this is achieved by subtracting (5.3.17) from (5.3.15), and substituting (5.3.21) and the identity $s = ik/2\cos\theta$, which we encountered in all of the equations (5.3.13) - (5.3.18). For $\tilde{\sigma}_B(k)$, we follow the same procedure, starting from (5.3.16) and (5.3.18). We end up with

$$\tilde{\chi}(k) - \tilde{\chi}_B(k) = \Omega_\chi(\theta_1,\theta_2)\mathcal{O}(\chi^2), \tag{5.3.22a}$$

$$\tilde{\sigma}(k) - \tilde{\sigma}_B(k) = k\Omega_\sigma(\theta_1,\theta_2)\mathcal{O}(\chi^2), \tag{5.3.22b}$$

as $|k| \to \infty$, where the angle-dependent size functions $\Omega_{\chi,\sigma}(\theta_1,\theta_2)$ are defined as

$$\Omega_\chi(\theta_1,\theta_2) \triangleq \frac{\cos\theta_1\cos\theta_2}{\cos\theta_1 - \cos\theta_2}\left\{\frac{1}{|\cos^3\theta_1|} + \frac{1}{|\cos^3\theta_2|}\right\}, \tag{5.3.23a}$$

$$\Omega_\sigma(\theta_1,\theta_2) \triangleq \frac{1}{\cos\theta_1 - \cos\theta_2}\left\{\frac{1}{\cos^2\theta_1} + \frac{1}{\cos^2\theta_2}\right\}. \tag{5.3.23b}$$

From (5.3.23), it is observed that the best reconstruction is obtained for $\theta_1 = 0$ and $\theta_2 = \pi$, i.e. for the case of normal incidence discussed in Section 5.2. In that case, the Born error given in (5.3.21) is minimal while the two linear equations of the form (5.3.14) are maximally independent. Because of this observation, the definitions in (5.3.23) have been chosen such that

$$\Omega_{\chi,\sigma}(0,\pi) = 1. \tag{5.3.24}$$

With this normalization, the size functions $\Omega_{\chi,\sigma}(\theta_1,\theta_2)$ can be envisaged as the magnification of the error in the approximate Fourier coefficients $\tilde{\chi}_B(k)$ and $\tilde{\sigma}_B(k)$ compared with the "ideal" case of normal incidence. The angular dependence of these size functions is illustrated in more detail in Table 5.3.1, which contains values of $\Omega_{\chi,\sigma}(\theta_1,\theta_2)$ for a fixed $\theta_1 = 0$ and for a varying θ_2. The first three columns of this table refer to the case of a normally and an obliquely incident wave, both arriving from the left. In this case, we apparently have a trade-off between the error due to the dependence between the two linear equations of the type of (5.3.14) and the Born error specified in (5.3.21). As a consequence, the errors in $\tilde{\chi}_B(k)$ and $\tilde{\sigma}_B(k)$ remain almost constant in magnitude over the range $35° \le \theta_2 \le 70°$. Compared with the case of normal incidence, however, these errors are magnified by a factor of 10. The last three columns of Table 5.3.1 deal with the case of normal incidence from the left and oblique incidence from the right. As could be expected, the best reconstruction is obtained when the wave arriving from the left is also normally incident. However, over the entire interval $130° \le \theta_2 \le 180°$, the Born errors are at worst magnified by a factor of 2.

Table 5.3.1 Magnification of the errors in $\tilde{\chi}_B(k)$ and $\tilde{\sigma}_B(k)$ for $\theta_1 = 0°$ and varying θ_2 as specified in (5.3.22) compared with the "ideal" case where $\theta_2 = 180°$ (θ_2 in degrees).

θ_2	$\Omega_\chi(0,\theta_2)$	$\Omega_\sigma(0,\theta_2)$	θ_2	$\Omega_\chi(0,\theta_2)$	$\Omega_\sigma(0,\theta_2)$
30	16.4	17.4	120	3.00	3.33
35	12.8	13.8	130	1.86	2.08
40	10.6	11.6	140	1.40	1.53
45	9.2	10.2	150	1.18	1.25
50	8.6	9.6	160	1.07	1.10
55	8.5	9.5	170	1.02	1.02
60	9.0	10.0	180	1.00	1.00
65	10.4	11.4			
70	13.5	14.5			
75	20.5	21.5			

In conclusion, it can be stated that, in the Born approximation, the unknown constitutive parameters $\chi(z)$ and $\sigma(z)$ should preferably be reconstructed from reflection data associated with incident fields originating from both sides of the slab. Once it has been established from which sides the incident waves should hit the slab, the precise value of the angle of incidence turns out to be less critical.

It should be remarked that the angular dependence discussed above remains the same when the general problem discussed in Section 5.2 is extended to the case of oblique incidence. In fact, the changes (5.3.20) form part of a transformation of the relevant direct-scattering problem to a corresponding one at normal incidence (see also Tijhuis (1981)). It was precisely this similarity which was at the back of the decision to confine the analysis of the case of oblique incidence to the present section and to deal solely with the "ideal" case of normal incidence elsewhere.

Finally, some discussion about the dependence of the errors in the approximations $\tilde{\chi}_B(k)$ and $\tilde{\sigma}_B(k)$ on the spectral wave number k would seem to be in order. As (5.3.22a) shows, the error in $\tilde{\chi}_B(k)$ remains approximately constant in magnitude over the entire range of k. The error in $\tilde{\sigma}_B(k)$, on the other hand, increases linearly in magnitude with increasing $|k|$. As mentioned above, the estimates (5.3.22) were derived for the case where the susceptibility $\chi(z)$ is discontinuous for at least one value of z, and where $\chi(z)$ and $\sigma(z)$ are of the same order of magnitude. A different situation arises when $\chi(z)$ is a continuous function of z with a derivative of bounded variation. In that case, we may integrate by parts in the error integrals resulting from the substitution of (5.3.6) in (5.3.10). The partial integration has the same effect as in the estimation of the magnitude of a Fourier integral when the transform variable becomes large (see Titchmarsh (1950)), namely it results in an additional factor of k^{-1} in each of the estimates on the right-hand sides of (5.3.21) and (5.3.22). For the conductivity profile, in particular, this implies that the short-range behavior can be reconstructed more accurately than in the general case of a discontinuous $\chi(z)$.

5.3.3 Born approximation: numerical implementation and results

In order to obtain insight into the applicability of the Born approximation in practice, the theory of the previous subsection was implemented numerically. In the numerical experiments, we considered the situation where sufficient reflection data were available to determine $\tilde{\chi}_B(k)$ and $\tilde{\sigma}_B(k)$ according to (5.3.17) and (5.3.18) over some finite interval $-k_{max} < k < k_{max}$. In that case, the Fourier inversion integrals in (5.3.19) must be truncated and we end up with the approximate expressions

$$\chi_B(k) = (2\pi)^{-1}\int_{-k_{max}}^{k_{max}} dk \, \exp(ikx)\tilde{\chi}_B(k), \tag{5.3.25a}$$

$$\sigma_B(z) = (2\pi)^{-1}\int_{-k_{max}}^{k_{max}} dk \, \exp(ikx)\tilde{\sigma}_B(k). \tag{5.3.25b}$$

In spite of the factor of k^{-1} in (5.3.17), the integral on the right-hand side of (5.3.25a) need not be considered as a principal-value integral, since from (5.3.10) we have

$$r(s;\theta) = \mathcal{O}(s) \quad \text{as } s \to 0. \tag{5.3.26}$$

The spectral integrals in (5.3.25) were evaluated by breaking up the domain of integration into (2K + 1) subintervals of equal length and applying a four-point Gauss-Legendre quadrature rule in each subinterval. In the computation, only the values of $\tilde{\chi}_B(k)$ and $\tilde{\sigma}_B(k)$ for $k \geq 0$ need to be used since the composite quadrature rule is even in k and since it follows from (5.3.17) and (5.3.18) that

$$\tilde{\chi}_B(-k) = \tilde{\chi}_B^*(k), \qquad \tilde{\sigma}_B(-k) = \tilde{\sigma}_B^*(k) \tag{5.3.27}$$

for $k \in \mathbb{R}$. The choice of repeating the four-point rule an odd number of times was made to avoid possible numerical round-off errors in the determination of $\tilde{\chi}_B(k)$ at the abcissae nearest to $k = 0$. Finally, the use of the same quadrature rule for the evaluation of both spectral integrals in (5.3.25) was motivated by our starting point of using exactly the same reflection data in the approximate determination of both profiles.

As a first example, we present, in Figures 5.3.1 and 5.3.2, results of reconstructing the susceptibility and the conductivity of a homogeneous, lossy dielectric slab as discussed in Section 2.2 from the reflected fields at normal incidence. The relevant reflection coefficients $r(s;0)$ and $r(s;\pi)$ are obtained directly from the plane-wave reflection coefficients $r^{\pm}(s)$ introduced in (2.4.21) via the correspondence relation (5.3.8). In turn, the latter reflection coefficients are given by the closed-form expression on the first line of (2.2.17). For this configuration, therefore, the values of $\tilde{\chi}_B(k)$ and $\tilde{\sigma}_B(k)$ are available up to machine precision. Hence, any errors in the estimates $\chi_B(z)$ and $\sigma_B(z)$ obtained can be attributed to the

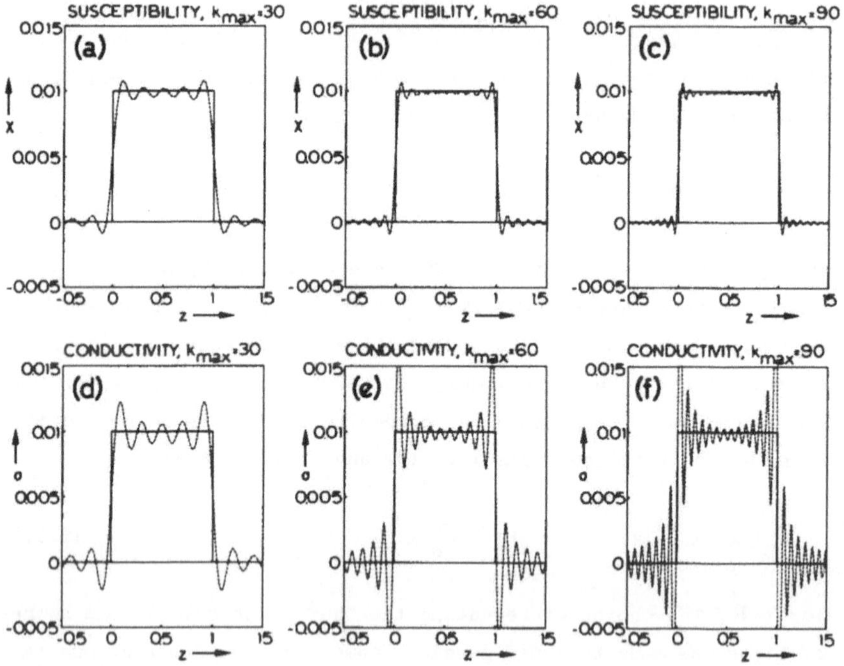

Figure 5.3.1 Results of the simultaneous Born reconstruction of both $\chi(z)$ and $\sigma(z)$ of a homogeneous slab with $\chi(z) = \sigma(z) = 0.01$ from the reflection coefficients at normal incidence. Solid lines: actual profiles. Dashed lines: results from evaluating the integrals in (5.3.25) with $k_{max} = 30,60,90$ and $K = 6,12,18$, respectively.

Born errors given in (5.3.22) and to the truncation of the spectral integrals performed in (5.3.25). From left to right, Figure 5.3.1 shows results for k_{max} = 30,60,90 and K = 6,12,18, respectively. In the upper half of the figure, the reconstructed susceptibility profile is compared with the actual one. Apparently, the reconstruction is accurate apart from a local oscillation near the discontinuities of $\chi(z)$. Moreover, increasing k_{max} raises the rate of the oscillation, while its amplitude remains constant in magnitude. Hence, we can only conclude that this oscillation is simply caused by Gibbs' phenomenon. In the lower half of Figure 5.3.1, the above comparison is repeated for the actual and the reconstructed conductivity profiles. We observe a similar (unwanted) oscillation as in the reconstructed susceptibility profiles. However, the amplitude of the oscillation is larger and increases with increasing k_{max}. Both these observations are in agreement with (5.3.22), which states that the error in $\tilde{\chi}_B(k)$ remains constant in magnitude as $|k| \to \infty$, while that in $\tilde{\sigma}_B(k)$ increases linearly in magnitude. A more definite confirmation of the error estimates in (5.3.22) is obtained from Figure 5.3.2, which compares the sampled values of $\tilde{\chi}_B(k)$ and $\tilde{\sigma}_B(k)$ that were used in the numerical evaluation of the spectral integrals in (5.3.25) for k_{max} = 30 and K = 6 with the corresponding actual Fourier coefficients $\tilde{\chi}(k)$ and $\tilde{\sigma}(k)$ as defined in (5.3.13). Clearly, the results agree with (5.3.22).

The numerical experiments leading to Figures 5.3.1 and 5.3.2 were repeated for several inhomogeneous slab configurations and for a wide range of angles of incidence. The required values of the reflection coefficient $r(s;\theta)$ were computed by numerically integrating the system of equations (5.3.3) as outlined in Subsection 5.3.1. As in the experiments of Figures 5.3.1 and 5.3.2, the errors in the results behaved as predicted in Subsection 5.3.2. Some representative examples will be encountered in the next subsection, where they are compared with the results of a modified Born-type reconstruction procedure.

As predicted toward the end of Subsection 5.3.2, the Born approximation works best when the unknown susceptibility profile is continuous and differentiable as a function of z with a derivative of

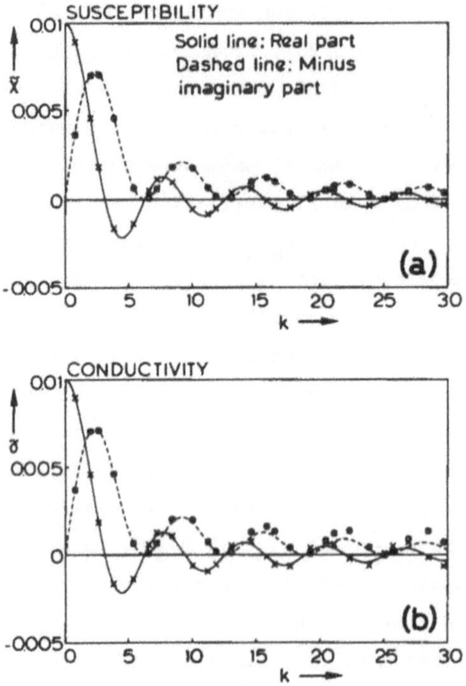

Figure 5.3.2 Sampled Born approximations $\tilde{\chi}_B(k)$ and $\tilde{\sigma}_B(k)$ as used in the computation of the results shown in Figures 5.3.1a,d compared with the actual spatial Fourier transforms $\tilde{\chi}(k)$ and $\tilde{\sigma}(k)$. (a): Solid and dashed lines: $\mathrm{Re}(\tilde{\chi}(k))$ and $-\mathrm{Im}(\tilde{\chi}(k))$; x: $\mathrm{Re}(\tilde{\chi}_B(k))$; o: $-\mathrm{Im}(\tilde{\chi}_B(k))$. (b): Same notation for $\tilde{\sigma}(k)$ and $\tilde{\sigma}_B(k)$.

bounded variation. In that case, no significant oscillations are observed in the reconstructed susceptibility profile $\chi_B(z)$, while the oscillations in the recovered conductivity profile behave in conformity with Gibbs' phenomenon. As a second example, we present, in Figures 5.3.3 and 5.3.4, results for such a configuration. Figure 5.3.3 contains results of reconstructing, for varying contrast, a sine-squared susceptibility profile and a linear conductivity profile from reflection data at normal incidence. The appearance of the extra factor of k^{-1} in the error estimates for $\tilde{\chi}_B(k)$ and $\tilde{\sigma}_B(k)$ is verified in Figure 5.3.4, which repeats the comparison of Figure 5.3.2 for case B as specified in Figure 5.3.3. In particular, the improvement is noticed in Figure 5.3.4b, where the deviation between

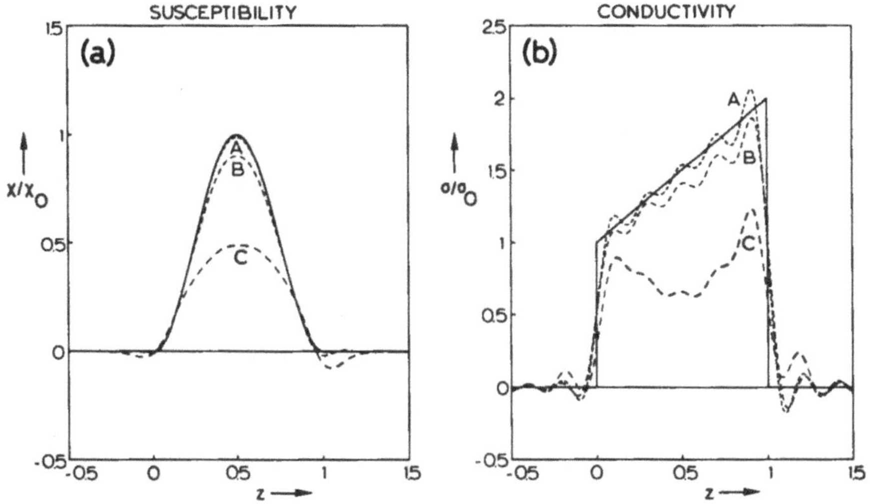

Figure 5.3.3 Results of the simultaneous Born reconstruction of the sine-squared susceptibility profile $\chi(z) = \chi_0 \sin^2(\pi z)$ and the linear conductivity profile $\sigma(z) = \sigma_0(1 + z)$ from the reflection coefficients at normal incidence for $k_{max} = 30$, $K = 6$ and for varying χ_0 and σ_0. Solid lines: actual profiles. Dashed lines: reconstructed profiles. Contrasts: (A): $\chi_0 = \sigma_0 = 0.01$; (B): $\chi_0 = \sigma_0 = 0.1$; (C): $\chi_0 = \sigma_0 = 1$.

$\tilde{\sigma}(k)$ and $\tilde{\sigma}_B(k)$ now remains constant in magnitude over the entire range of observation.

Since the numerical experiment of Figure 5.3.3 constitutes an "ideal" application of the Born approximation, the results displayed may also be used to assess the maximum reliability range of that approximation. Apparently the reconstruction obtained from (5.3.17) - (5.3.19) can no longer be trusted when either $\chi_B(z)$ or $\sigma_B(z)$ becomes much larger than 0.1. However, the results for configuration C indicate that, in that case, the application of the iterative procedure outlined in Subsection 5.2.2 may lead to an acceptable reconstruction.

5.3.4 A modified method of solution

The most annoying disadvantage of the estimation procedure described in the previous two subsections is that the agreement between the

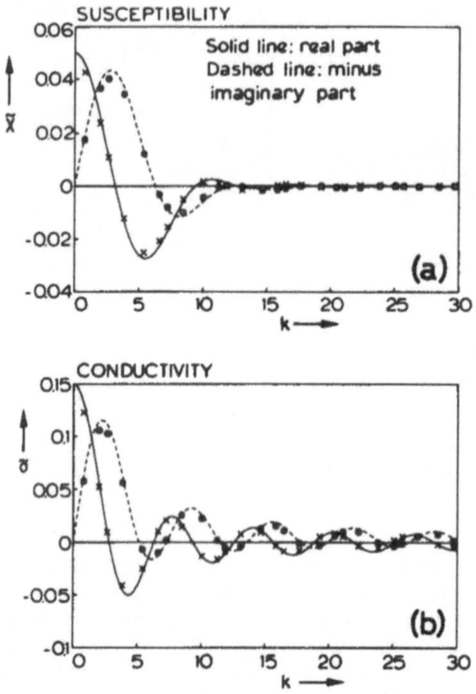

Figure 5.3.4 Sampled Born approximations $\tilde{\chi}_B(k)$ and $\tilde{\sigma}_B(k)$ as used in the computation of the results shown as case (B) in Figure 5.3.3 with the actual spatial Fourier transforms $\tilde{\chi}(k)$ and $\tilde{\sigma}(k)$. Symbols as given in Figure 5.3.2.

reconstructed and the actual profiles is confined to a limited range of the spatial wave number k. For the susceptibility profile, this restriction is due to the truncation of the integration interval in (5.3.25). For the conductivity profile, the error estimated in (5.3.22b) induces an inherent band limitation that remains present even when $k_{max} \to \infty$. As Figures 5.3.1 and 5.3.3 show, the consequence of these restrictions is that the reconstructed profiles do not exhibit the correct short-range behavior.

One way to remedy this situation is to supply additional *a priori information* about this short-range behavior. When the unknown profiles are continuous within the slab, this can be achieved by representing them by piecewise-linear expansions of the type

$$\chi(z) = \sum_{m=0}^{M} \chi_m \phi_m(z), \qquad \sigma(z) = \sum_{m=0}^{M} \sigma_m \phi_m(z), \tag{5.3.28}$$

with the $\{\phi_m(z)\}$ being triangular expansion functions given by

$$\phi_m(z) = \begin{cases} 0 & -\infty < z \le (m-1)/M, \\ 1 - M|z - m/M| & (m-1)/M \le z \le (m+1)/M, \\ 0 & (m+1) \le z < \infty, \end{cases} \tag{5.3.29}$$

for $0 < m < M$ and by versions truncated at the slab's interfaces for $m = 0,M$. In (5.3.28), the expansion coefficients $\{\chi_m\}$ and $\{\sigma_m\}$ can be interpreted as being approximations to values of the corresponding profiles at $z = m/M$. Compared with a global representation such as a truncated power or Fourier series, the representation (5.3.28) has the advantage that the expansion coefficients are all of the same order of magnitude, and hence, will all be determined up to the same degree of accuracy. Moreover, the maximum permissible local variation in $\chi(z)$ and $\sigma(z)$ can be adjusted via the parameter M.

Taking into account the information that the unknown profiles can be represented by (5.3.28) requires a different inversion procedure than the Fourier inversion used in (5.3.19) and (5.3.25). While keeping the identification of $\tilde{\chi}_B(k)$ and $\tilde{\sigma}_B(k)$ as specified in (5.3.17) and (5.3.18), we should now obtain the coefficients $\{\chi_m\}$ and $\{\sigma_m\}$ from the approximate equations

$$\sum_{m=0}^{M} \chi_m \tilde{\phi}_m(k) = \tilde{\chi}_B(k), \tag{5.3.30a}$$

$$\sum_{m=0}^{M} \sigma_m \tilde{\phi}_m(k) = \tilde{\sigma}_B(k), \tag{5.3.30b}$$

where k ranges over the abcissae that are also used in the numerical evaluation of the spectral integrals in (5.3.25). As mentioned above, the Born errors estimated in (5.3.22) prevent the equations in (5.3.30) from holding rigorously even when the representation (5.3.28) is exact. Moreover, the number of abcissae used in the evaluation of the integrals in (5.3.25) will generally be larger than M. Hence, we can only resolve the systems of equations

(5.3.30a) and (5.3.30b) in some least-squares sense by determining those values of $\{\chi_m\}$ and $\{\sigma_m\}$ that minimize some cost function.

The most significant aspect of such a linear-least-squares inversion is the selection of the proper cost function. In the present application, this selection should be based on three grounds. In the first place, the variation of the error estimates (5.3.22) with k should be taken into account. In the second place, we should reckon with the distribution of the profile information contained in $r(s;\theta)$ over the relevant frequency range. In the third place, we should have sufficient flexibility in the piecewise-linear expansions (5.3.28) to handle that profile information.

Let us first draw up the cost function for Equation (5.3.30a), i.e. for the determination of the approximate susceptibility profile. As the estimate (5.3.22a) indicates, the error in $\tilde{\chi}_B(k)$ is of the same order of magnitude over the entire relevant spectral domain. This suggests that (5.3.30a) is already properly normalized and that no additional weighting function need be attributed. The same conclusion is arrived at when we consider the profile information contained in the reflection coefficients constituting $\tilde{\chi}_B(k)$, since the factor of k^{-1} in (5.3.17) already compensates for the relative decrease in that information as $|s| \to \infty$, as signaled in Subsection 5.2.3. This leaves us with selecting the number of subintervals in the piecewise-linear expansion. This choice should be made such that the additional information about the short-range behavior of $\chi(z)$ supplied in that expansion links up with the information about the long-range behavior available from $\tilde{\chi}_B(k)$. From that point of view, it seems sensible to take the length of the subintervals in the piecewise-linear expansion approximately equal to half of the wavelength associated with the maximum value of k for which $\tilde{\chi}_B(k)$ is available, i.e. $M \approx k_{max}/\pi$. For an M in that range, however, the local oscillations in the Born-approximation results as observed in Figure 5.3.1 may pass into an unwanted alternating behavior in the coefficients $\{\chi_m\}$ obtained. This is an ambiguity effect similar to the one observed in Section 3.2 for the numerical solution of the time-domain direct-scattering problem. It can be removed by including in the cost function small regularization terms that constrain the component in the solution vector $\{\chi_m\}$ along the vector $\{\nu_m\}$ with $\nu_m = (-1)^m$, which generates the most rapidly oscillating solution of the form (5.3.28). Obviously, these regularization terms must be so small that they lose their effect when M becomes small enough to render them superfluous.

The combination of the considerations listed above led to the selection of the cost function

$$\begin{aligned} &\int_0^{k_{max}} \left|\sum_{m=0}^{M} \chi_m \tilde{\phi}_m(k) - \tilde{\chi}_B(k)\right|^2 dk \\ &+ \delta \sum_{m=1}^{M-1} (\chi_{m-1} - 2\chi_m + \chi_{m+1})^2 \qquad (5.3.31) \\ &+ \delta(2\chi_0 - 3\chi_1 + \chi_3)^2 + \delta(2\chi_M - 3\chi_{M-1} + \chi_{M-3})^2 . \end{aligned}$$

The spectral integral in (5.3.31) should be understood in a discretized sense. Note that, as $k_{max} \to \infty$, the integral becomes, by Parseval's theorem, proportional to the integrated squared deviation between $\tilde{\chi}_B(z)$ as defined in (5.3.19) and the piecewise-linear approximation given in (5.3.28). In the remaining terms, δ is a small, nonnegative regularization parameter. In these terms, the quantities between brackets are proportional in magnitude to the deviation between the coefficient χ_m and an approximation obtained by linear interpolation or extrapolation of the corresponding values at sample points where an error along $\{\nu_m\}$ would have the opposite sign. Hence, the regularization terms in (5.3.31) can be envisaged as favoring the "most linear" solution in case of ambiguity. Finally, the parallel should be indicated between the second line of (5.3.31) and the smoothing operation (3.2.30), which serves a similar purpose in eliminating an unwanted alternating behavior in the solution obtained by the modified marching-on-in-time method formulated in Subsection 3.2.3.

The cost function for Equation (5.3.30b) is selected along the same lines as the one for (5.3.30a). From (5.3.22b), it is observed that the error in $\tilde{\sigma}_B(k)$ increases linearly in magnitude with increasing $|k|$. Hence (5.3.30b) should be normalized by a weighting function $w(k)$ which is of $O(k^{-1})$ as $|k| \to \infty$. Such a weighting function also does justice to the amount of profile information contained in $r(s;\theta)$, since, in (5.3.18), that reflection coefficient is weighted by a factor that does not depend on k or, equivalently, on s. Special care is needed in the selection of the number of subintervals M. When the amplitude of the oscillation observed in the Born reconstruction of the conductivity profile exceeds the corresponding amplitude in the reconstructed susceptibility profile, it may become necessary to reduce the flexibility in the piecewise-linear expansion. This can be achieved by reducing M and/or increasing the regularization parameter δ. In view of these considerations, we selected the cost function

$$\int_0^{k_{max}} \left\{ \frac{k_0^2}{k^2 + k_0^2} \left| \sum_{m=0}^{M} \sigma_m \tilde{\phi}_m(k) - \tilde{\sigma}_B(k) \right|^2 \right\} dk$$

$$+ \delta \sum_{m=1}^{M-1} (\sigma_{m-1} - 2\sigma_m + \sigma_{m+1})^2 \tag{5.3.32}$$

$$+ \delta(2\sigma_0 - 3\sigma_1 + \sigma_3)^2 + \delta(2\sigma_M - 3\sigma_{M-1} + \sigma_{M-3})^2 .$$

In (5.3.32), k_0 is a parameter that allows adjusting the change-over between the low-frequency range, where $w(k) = O(1)$, and the high-frequency range, where $w(k) = O(k^{-1})$.

In the numerical implementation of the linear-least-squares inversion, the normalized versions of the equations in (5.3.30) were, for each relevant value of k, broken up into their real and imaginary parts. Next, each part was treated as a single equation with real-valued coefficients for the expansion coefficients $\{\chi_m\}$ or $\{\sigma_m\}$. Similarly, the regularization terms in (5.3.31) and (5.3.32) were handled by equating the expressions in these terms between brackets with zero and multiplying the extra equations thus obtained by a factor of $\delta^{1/2}$. Thus, we arrived at an overdetermined system of real-valued equations which was solved with the aid of a linear-least-squares inversion procedure based on Housholder transformations, which is available from the NAG library (NAG (1984), subroutine F04AMF). This is the same procedure that was used in Chapter 4 to solve the systems of equations (4.2.6) and (4.2.2).

Some illustrative examples are given in Figures 5.3.5 and 5.3.6. Figure 5.3.5 shows results of reconstructing a quadratic susceptibility profile and a quadratic conductivity profile both in the Born approximation and by the modified method outlined in the present subsection. As in Figures 5.3.1 - 5.3.4, the reflection coefficient $r(s;\theta)$ was considered to be known at normal incidence. Figures 5.3.5a,b pertain to the case where the regularization terms have

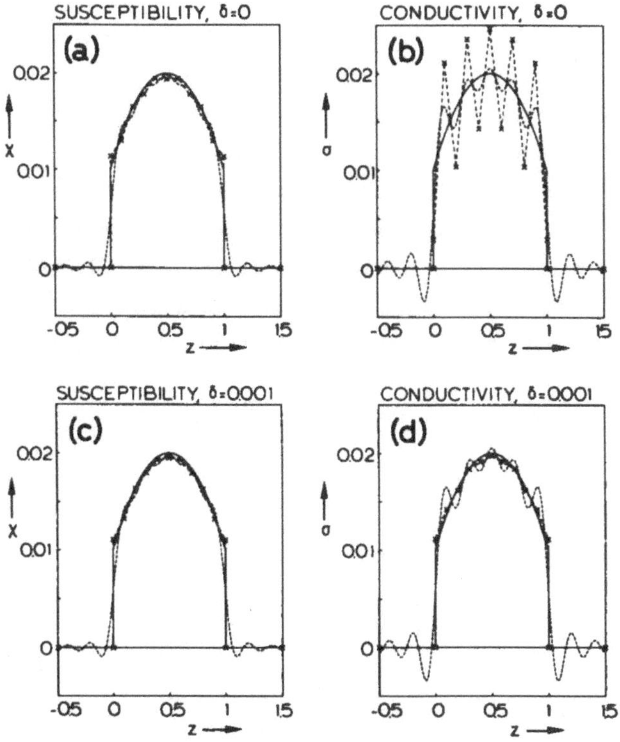

Figure 5.3.5 Results of the simultaneous Born-type reconstruction of both $\chi(z)$ and $\sigma(z) = 0.02 - 0.04(z - 0.5)^2$ from the reflection coefficients at normal incidence by the methods outlined in Subsections 5.3.2 and 5.3.4. Solid lines: actual profiles. Unmarked dashed lines: results from evaluating (5.3.25) with $k_{max} = 30$ and $K = 6$. Marked dashed lines: results from minimizing (5.3.31) and (5.3.32) with $M = 10$, $k_0 = 5$ and δ as indicated in the figure captions. The x's mark the values of the coefficients $\{\chi_m\}$ and $\{\sigma_m\}$ as introduced in (5.3.28).

been dropped from the cost functions, i.e. $\delta = 0$ in (5.3.31) and (5.3.32). Clearly, the reconstructions obtained by minimizing these cost functions are no better than the Born results. Figures 5.3.5c,d present the results of repeating the computation with $\delta = 0.001$. Now, the results obtained by minimizing (5.3.31) and (5.3.32) are considerably better. This numerical experiment confirms the effectiveness of the modified inversion method as well as the ambiguity in the representation (5.3.28) for $M \approx k_{max}/\pi$. Figure 5.3.6 shows the results of applying the modified method to reconstruct two parabolic profiles of half the contrast from reflection data at the angles of incidence $\theta_1 = 0°$ and $\theta_2 = 45°$. In this case, the regularization parameter had to be increased to $\delta = 0.01$, probably due to the increased dependence between the linear equations of the type (5.3.14) that lead to the approximate expressions (5.3.17) and (5.3.18). As could be expected from the error estimates in (5.3.22) and Table 5.3.1, the relative deviation between the actual and the reconstructed profiles is larger than in Figures 5.3.5c,d. In particular, the reconstruction deteriorates as z increases. This corresponds with the fact that, in the situation considered in Figure

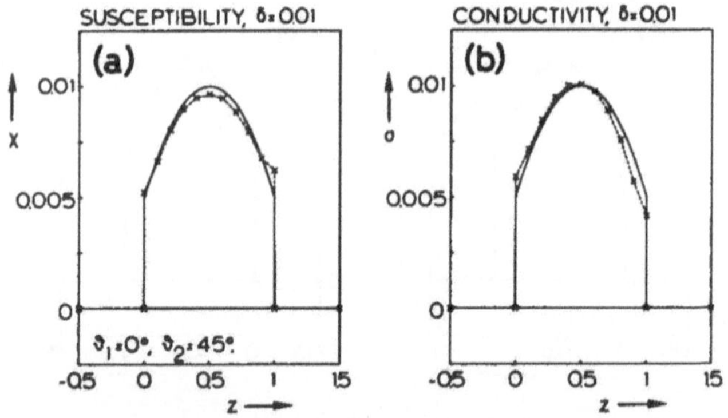

Figure 5.3.6 Results of the simultaneous reconstruction of both $\chi(z) = \sigma(z) = 0.01 - 0.02(z - 0.5)^2$ from the reflection coefficients at $\theta_1 = 0°$ and $\theta_2 = 45°$ by the method outlined in Subsection 5.3.4. The spectral integrals in (5.3.31) and (5.3.32) were evaluated with $k_{max} = 30$ and $K = 6$ and the linear-least-squares inversion was carried out for $M = 10$, $k_0 = 5$ and $\delta = 0.01$. Symbols as given in Figure 5.3.5.

5.3.6, the deviation in amplitude and phase between the actual fields $\hat{E}(z,s;\theta)$ and the incident fields $\hat{E}^i(z,s;\theta)$ increases with increasing z. Consequently, the error in the result of solving an approximate equation obtained by replacing the former field with the latter will behave as observed above.

The representative results displayed in Figures 5.3.5 and 5.3.6 may also serve as an additional justification for our decision to restrict ourselves, in the analysis and the application of the iterative schemes formulated in Section 5.2, to the "ideal" case of normal incidence. Not until that case is fully understood, should the more difficult case of oblique incidence from the front of the slab only be tackled.

5.4 The one-dimensional inverse-scattering problem for a lossless dielectric slab

5.4.1 Formulation of the problem

In this section, we consider the inverse-scattering problem of reconstructing the susceptibility profile of an inhomogeneous, lossless slab from known values of the frequency-domain reflection coefficient at normal incidence $r^{+}(s)$. For the case where this coefficient is known for $s = i\omega$ with $\omega \in \mathbb{R}$, it was shown by Kay (1955) that the solution of the inverse-scattering problem is unique. This allows us to investigate the limitations of two types of reconstruction procedures in a situation where any ambiguities in the reconstructions obtained are due to either the ill-posedness of the inverse-scattering problem or the inherent limitations in the methods of solution. In fact, the present problem is, to the best of the author's knowledge, the only one for which such an investigation can be carried out since it is the only one for which a rigorous uniqueness proof is available. The analysis of the present section consists of two parts. In the first part, we will describe how, for the problem at hand, the Born approximation discussed in the previous section can be improved such that it provides an acceptable reconstruction for larger contrasts as well. In the second part, we will use the approximation thus obtained, or some other initial estimate, as a starting point for the iterative procedure proposed in Subsection 5.2.2. In the present subsection, we can refrain from recounting integral and differential equations, since the relevant ones are obtained directly by setting $\sigma(z) = 0$, $\tilde{\sigma}(k) = 0$, and $\cos\theta = 1$ where necessary in the general expressions given in Sections 5.2 and 5.3.

5.4.2 Estimation procedures: formulation

When the susceptibility contrast is small, it can be estimated by applying the Born approximation outlined in Subsection 5.3.2. For the lossless configuration which is presently of interest, this amounts to interchanging the actual reflection coefficient $r^{+}(s)$ and the approximation

$$r_B^+(s) \triangleq -\tfrac{1}{4}s\int_{-\infty}^{\infty}\chi(z)\exp(-2sz)dz, \tag{5.4.1}$$

which is obtained from (5.3.12) by substituting $\sigma(z) = 0$ and the first connecting formula listed in (5.3.8). Starting from (5.4.1), and going through the same steps as in Subsection 5.3.2, we obtain the approximate inversion formula

$$\chi_B(z) \triangleq (2\pi)^{-1}\int_{-\infty}^{\infty}dk\ \exp(ikz)\tilde{\chi}_B(k), \tag{5.4.2}$$

with

$$\tilde{\chi}_B(k) \triangleq (4i/k)r^+(ik/2). \tag{5.4.3}$$

In (5.4.2) and (5.4.3), the approximate spatial Fourier coefficient $\tilde{\chi}_B(k)$ is expressed in terms of the frequency-domain reflection coefficient $r^+(s)$. An alternative way to understand the Born approximation follows by considering, as in Subsections 5.2.3 and 5.2.5, the time-domain signal corresponding to the quantity $s^{-1}r^+(s)$. As remarked in those subsections, the factor of s^{-1} can be identified as being the Laplace transform of the unit-step incident-pulse shape $F(t) = U(t)$. In that context, the quantity $s^{-1}r^+(s)$ is recognized as the reflected field at the front of the slab caused by the incident field given in (2.2.1). With the identification $k = -2is$, the approximate inversion formula (5.4.2) then reduces to the relation

$$E_+^r(0,t) = -\tfrac{1}{4}\chi_B(z_B), \quad \text{with } z_B = t/2, \tag{5.4.4}$$

which holds rigorously for all t. In interpreting the definition of z_B, one should keep in mind that both z_B and t are dimensionless coordinates as introduced on p. 37.

For larger susceptibility contrasts, the estimate obtained from either (5.4.2) or (5.4.4) can no longer be relied upon. Since we do want to handle such contrasts, some correction procedure should be devised to improve these estimates. A first example of such a correction procedure is the one proposed by Tabbara (1976,1979). Rather than copying the exact formulation of that author, we will repeat his argumentation such that it can be connected with the WKB results derived in Subsection 5.2.5. From homogeneous-slab expressions similar to the ones given in (2.4.25), Tabbara argued that the result of applying the Born approximation should be corrected in two ways. In the first place, the nonlinear variation of the reflected field with the local value of $\chi(z)$ must be taken into account. This could be provided for by assigning the value of the reflection coefficient at the interface between a vacuum and a homogeneous, lossless dielectric half-space with the proper susceptibility to the field reflected at a given location in the slab. In the second place, one should correct for the fact that the dimensionless local wave speed is $1/N(z)$, with $N(z) = [1 + \chi(z)]^{\frac{1}{2}}$, and not 1, as assumed in the Born approximation. The combination of both these observations results in the following approximate expression for the time-domain reflected field:

$$E^r_{+T}(0,t) = -R(z_T), \quad \text{with } z_T = z^+(t), \tag{5.4.5}$$

where $z^+(t)$ is the solution of (5.2.38), and where the local reflection coefficient $R(z)$ is given by

$$R(z) \triangleq [N(z) - 1]/[N(z) - 1]. \tag{5.4.6}$$

Now Tabbara's correction procedure follows from neglecting the difference between the approximate expression (5.4.5) and the exact reflected field $E^r_+(0,t)$ in (5.4.4). The level correction is obtained by equating the magnitudes of the reflected fields E^r_{+T} and E^r_+ at each instant. This results in a linear equation for $N(z)$, whose solution yields

$$\chi'_T(z_B) = \left[\frac{4 + \chi_B(z_B)}{4 - \chi_B(z_B)} \right]^2 - 1, \tag{5.4.7}$$

where $\chi_B(z_B)$ is the value found by setting $z = z_B$ in the integral in (5.4.2). Because of the wave speed error mentioned above, z_B is not the position at which the corrected value of the susceptibility is actually observed. Hence, we apply a position correction of the form

$$\chi_T(z_T) = \chi'_T(z_B), \tag{5.4.8}$$

where z_T is obtained by relating the positions associated with a given value of t in (5.4.4) and (5.4.5). Since $\chi'_T(z_B)$ is assumed to be a better approximation than $\chi_B(z_B)$, we arrive, with the aid of (5.2.38), at the approximate relations

$$z_B = \int_0^{z_T} N_T(z'_T)\,dz'_T, \qquad z_T = \int_0^{z_B} N'_T(z'_B)^{-1}\,dz'_B. \tag{5.4.9}$$

In (5.4.9), $N_T(z_T) \triangleq [1 + \chi_T(z_T)]^{\frac{1}{2}}$ and $N'_T(z_B) \triangleq [1 + \chi'_T(z_B)]^{\frac{1}{2}}$ denote the improved values of the refractive index produced by the expression between brackets in (5.4.7). The coordinates $\{z_B, z'_B\}$ and $\{z_T, z'_T\}$ refer to the original and the corrected positions, respectively. Finally, it should be observed that, as $\chi_B \to 0$, the relations (5.4.7) and (5.4.9) can be rewritten as

$$\chi_T = \chi_B[1 + \mathcal{O}(\chi_B)], \qquad z_T = z_B[1 + \mathcal{O}(\chi_B)], \tag{5.4.10}$$

which shows the agreement between the original Born approximation and Tabbara's improved version.

The one weak point of the argument given above seems to be the arbitrariness in the choice of the amplitude of the approximate reflected field in (5.4.5). One way to arrive at a more well-founded choice for this amplitude is to start from the WKB results (5.2.37) and (5.2.41). Combining these results and substituting the definition of T^+_{0a} given in (2.3.21) directly leads to

$$\begin{aligned} \mathcal{E}^r_{+WKB}(0,t) &= \bar{\mathcal{E}}^r_{+WKB}(0,t) - \frac{2\bar{N}_0[N(z^+(t)) - \bar{N}(z^+(t))]}{(\bar{N}_0 + 1)^2 N(z^+(t))} \\ &\quad + \mathcal{O}[(N - \bar{N})^2] \quad \text{for } 0 < t < 2\bar{\tau}. \end{aligned} \tag{5.4.11}$$

In (5.4.11), the barred quantities pertain, as before, to some approximate configuration with a refractive index $\bar{N}(z)$ while $z^+(t)$ is again the solution of (5.2.38). Now the first two terms on the right-hand side of (5.4.11) can be identified as the first two terms of a Taylor expansion of the approximate amplitude E^r_{+WKB} around $N = \bar{N}$. Accordingly, we have, for that amplitude, the differential equation

$$d_N E^r_{+WKB} = -2N_0/(N_0 + 1)^2 N, \tag{5.4.12}$$

which can be integrated in closed form. A suitable starting point for this integration is also available, since, for a homogeneous, lossless slab, we have

$$E^r_{+WKB}(0,t) = E^r(0,t) = -R_0 U(t) \tag{5.4.13}$$

with $R_0 \triangleq (N_0 - 1)/(N_0 + 1)$. After performing the integration, we end up with

$$E^r_{+WKB}(0,t) = -R_0 - \frac{2N_0}{(N_0 + 1)^2} \ln\left(\frac{N(z^+(t))}{N_0}\right), \tag{5.4.14}$$

which holds for $0 < t < 2\tau$.

Like Tabbara's correction procedure, the WKB correction procedure follows from neglecting the difference between the approximate expression (5.4.14) and the exact reflected field $E^r(0,t)$ in (5.4.4). Taking for E^r_{+WKB} the correct principal value at $t = 0$ and interpreting the value $\chi_B(0)$ in the same manner, we first find

$$\begin{aligned} R_0 &= \chi_B(0)/2, \\ N_0 &= [2 + \chi_B(0)]/[2 - \chi_B(0)]. \end{aligned} \tag{5.4.15}$$

Once these auxiliary parameters have been established, the counterpart of the level correction (5.4.8) is found directly by equating the right-hand sides of (5.4.4) and (5.4.14) for $t > 0$. This results in

$$\chi_{WKB} = N_0^2 \exp[(\chi_B/4 - R_0)(N_0 + 1)^2/N_0] - 1. \tag{5.4.16}$$

The corresponding position correction takes the same form as (5.4.9) with N_T and N'_T replaced by the relevant values obtained from χ_{WKB}.

To complete the present formulation, we must indicate the connection between the two estimation procedures formulated above. As with these procedures themselves, their connection can be understood best on the level of the underlying reflected-field expressions (5.4.5) and (5.4.14). Starting from the latter, let us consider the situation where N(z) is approximately equal to N_0 across the whole slab. In that case, we have, for the logarithm in (5.4.14), the estimate

$$\ln(N/N_0) = (N - N_0)/N_0 + O[(N - N_0)^2]. \tag{5.4.17}$$

Substitution of this estimate in (5.4.14) results in

$$E^r_{+WKB}(0,t) = -R_0 - \frac{2[N(z^+(t)) - N_0]}{(N_0 + 1)^2} + O[(N - N_0)^2]. \tag{5.4.18}$$

In (5.4.18), the first two terms on the right-hand side are precisely the first two terms of the Taylor expansion of $R(z^+(t))$ at $N(z^+(t)) = N_0$. Hence, Tabbara's correction procedure can be envisaged as a limiting case of the WKB correction procedure for the case of an almost constant susceptibility profile and vice versa.

5.4.3 Estimation procedures: numerical implementation and results

The first step in the numerical implementation of both the estimation procedures described in Subsection 5.4.2 is the evaluation of the spectral integral in the Born approximation (5.4.2). This integral was evaluated in exactly the same manner as the inversion integrals in (5.3.19), namely by confining the domain of integration to the finite interval $-k_{max} < k < k_{max}$ and applying a (2K + 1) times repeated four-point Gauss-Legendre quadrature rule. Since $r^+(-i\omega) = r^+(i\omega)^*$, the corresponding direct-scattering problem only needs to be solved for positive values of ω. The integration in the position correction (5.4.9) was performed by a repeated trapezoidal rule. In the determination of the auxiliary parameters R_0 and N_0 in the WKB estimation, the feature made use of was that, at the exact location of a discontinuity in $\chi_B(z)$, even a truncated version of the spectral integral in (5.4.2) yields in good approximation the correct principal value. Hence, the relations (5.4.15) remain applicable.

The numerical results show that for most susceptibility profiles, even for high contrast, both estimation procedures provide a fairly good estimate, apart from a readily recognizable spatial fluctuation, the period of which varies with k_{max}. In particular, this fluctuation is observed when the unknown profile exhibits one or more discontinuities inside the slab or at its interfaces. As in Subsection 5.3.3, these observations point to the conclusion that the unwanted oscillating behavior is mainly caused by Gibbs' phenomenon. For larger contrasts, however, a part of this behavior should also be attributed to the approximations made in devising the estimation procedures.

When the unknown susceptibility profile does not differ too much in value across the slab, the results of both procedures appear to be almost indistinguishable. This is in agreement with the connection between the two methods found in the last paragraph of Subsection 5.4.2. A special situation in which the equivalence of both methods is observed occurs when the unknown susceptibility contrast is not too large. When the value of the unknown susceptibility varies over a wider range, the results of both methods do differ. The most significant difference turns out to be that Tabbara's version provides a better level correction while the WKB version provides a better position correction.

For the WKB version, finally, we investigated the effect on the reconstructed profile due to a minor variation in the auxiliary parameters R_0 and N_0. As remarked above, the numerically obtained values for these parameters may be inaccurate due to the truncation of the spectral integral in (5.4.2). However, in spite of the dominant appearance of these parameters in the level correction (5.4.16), a minor variation in them does not cause a significant deviation in the value of χ_{WKB} obtained. Hence, the possible presence of an error in the computed value of $\chi_B(0)$ should not be considered a true objection to the application of the WKB estimation procedure.

Representative examples of the successful application of both methods are shown in Figures 5.4.1 and 5.4.2, where, for increasing contrast, the result of the improved Born approximations are compared with the actual sine-squared susceptibility profiles. An example of an application to a discontinuous susceptibility profile will be observed in Figure 5.4.5, where it is compared with both the actual profile and the result of an iterative reconstruction. The computation time spent in the numerical determination of the results shown in Figures 5.4.1 and 5.4.2 was mainly consumed in solving the direct-scattering problem as explained in Subsection 2.4.2. The complete reconstruction of a single profile, including the determination of the reflection coefficients, took about 4 seconds.

A disadvantage of both estimation procedures is that, for high contrast, the result sometimes locally does not resemble the actual profile at all. Generally, this is due to the inaccurate reconstruction of a discontinuity in $\chi(z)$, e.g. at a slab interface. As a consequence, the estimate obained may not be suitable as an initial guess in the iterative scheme. An example will be shown below.

5.4.4 Iterative procedure: numerical implementation and results

In this last subsection, we discuss the numerical aspects of applying the iterative technique formulated in Subsection 5.2.2. The formulation of that technique for the lossless case presently under consideration follows immediately by taking $\sigma^{(n)}(z) = 0$ and $\alpha = +$ in the discussion concerning Equations (5.2.5) - (5.2.7). Hence, we will not repeat it here. To obtain a good initial guess, the first step of the reconstruction was generally the application of Tabbara's estimation procedure as formulated and implemented in the previous two subsections.

In subsequent steps of the iterative scheme, $\chi^{(n)}(z)$ is determined after $\hat{E}_{+}^{(n)}(z,i\omega)$ has been computed by the procedure discussed in Subsection 2.4.2. As argued in Subsection 5.2.4, in each iteration step of that scheme we can only recover band-limited information on $\chi^{(n)}(z)$. In the problem at hand, the band limitation is caused by both the truncation of the frequency interval at $\omega = k_{max}/2$ and the

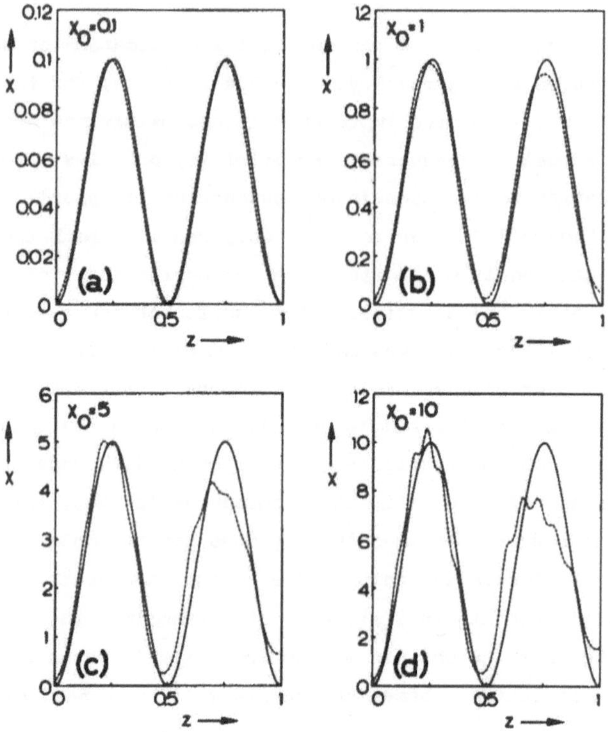

Figure 5.4.1 Results of Tabbara's improved Born approximation (dashed lines) compared with the actual susceptibility profiles (solid lines) $\chi(z) = \chi_0 \sin^2(2\pi z)$ for increasing contrast computed for $k_{max} = 30$ and $K = 6$.

frequency behavior of the Born error specified in (5.2.32). To circumvent this problem, we must, as in Subsection 5.3.3, supplement the band-limited profile information by *a priori* information about the short-range behavior of $\chi(z)$. We restrict our considerations to a continuous susceptibility profile within the slab. Then, the unknown susceptibility profile can be represented by a piecewise-linear expansion of the type (5.3.28), in which the coefficients $\{\chi_m\}$ are taken as unknowns. In such an expansion, the domain of the slab $\mathcal{D}_2$ is subdivided into M cells. The coefficients $\{\chi_m\}$ can be regarded as the values at the cell boundaries, i.e. $\chi(m/M) = \chi_m$; in between these boundaries the profile is interpolated linearly. Next,

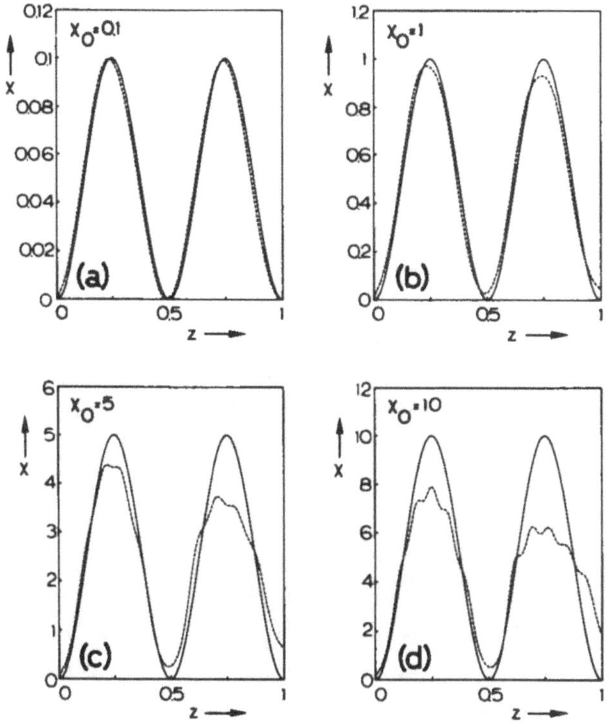

Figure 5.4.2 Results of the WKB version of the improved Born approximation (dashed lines) compared with the actual susceptibility profiles (solid lines) $\chi(z) = \chi_0 \sin^2(2\pi z)$ for increasing contrast computed for $k_{max} = 30$ and $K = 6$.

we discretize the integral on the right-hand side of (5.2.4) by an N_{cell} times refined repeated trapezoidal rule with space step $h = 1/N$, with $N = M\,N_{cell}$. For this choice of the space step, each cell boundary coincides with a boundary of a subinterval in the repeated trapezoidal rule. Thus, we avoid possible problems due to the discontinuities in the z-derivative of the expansion (5.3.28) at the cell boundaries. The relation (5.2.4) then reduces to

$$r^+(i\omega) \simeq -\frac{i\omega}{2}\sum_{m=0}^{M} E_m(i\omega)\chi_m, \tag{5.4.19}$$

with

$$E_m(i\omega) \triangleq (1 - \delta_{mM})\sum_{j=0}^{N_{cell}} \alpha_j \frac{N_{cell} - j}{N_{cell}}$$
$$\times \exp[-i\omega(mN_{cell} + j)h]\hat{E}_+((mN_{cell} + j)h, i\omega)$$
$$+ (1 - \delta_{m0})\sum_{j=0}^{N_{cell}} \alpha_j \frac{N_{cell} - j}{N_{cell}}$$
$$\times \exp[-i\omega(mN_{cell} - j)h]\hat{E}_+((mN_{cell} - j)h, i\omega), \tag{5.4.20}$$

where δ_{mn} denotes the Kronecker symbol and $\alpha_j \triangleq h$ for $0 < m < M$, while $\alpha_0 = \alpha_{N_{cell}} \triangleq h/2$. If we now limit ω to a finite set $\{\omega_\ell\}$ with $1 \le \ell \le L$, e.g. the set of positive frequencies for which $r^+(i\omega)$ was computed in the evaluation of $\chi_B(z)$, we end up with

$$r_\ell \simeq \sum_{m=0}^{M} E_{\ell m}\chi_m, \qquad 1 \le \ell \le L, \tag{5.4.21}$$

where $r_\ell \triangleq r^+(i\omega_\ell)$ and $E_{\ell m} \triangleq -\tfrac{1}{2}i\omega_\ell E_m(i\omega_\ell)$. Since the coefficients $\{\chi_m\}$ are real-valued and both $E_{\ell m}$ and r_ℓ have complex values, (5.4.21) constitutes a system of 2L linear equations with real-valued coefficients for M + 1 unknowns. This system is inverted by a linear-least-squares inversion such that an error of the type obtained from adding (5.2.6) and (5.2.7) is minimized.

Numerical results were obtained for various profiles and for several combinations of a set of frequencies $\{\omega_\ell\}$ and a choice of the squared error of the type indicated above in the inversion of (5.4.21). The best results were obtained if the frequencies were more or less evenly dispersed across the positive half of the confined integration interval used in the numerical evaluation of (5.4.2). Successful computations were carried out for the set of frequencies corresponding to the abcissae that are relevant in that evaluation as well for a set of equally spaced frequencies. In both these cases, the squared error was

$$ERR[\chi^{(n)};\hat{E}_+^{(n)}] \triangleq \sum_{\ell=1}^{L} w(\omega_\ell)^2 \left| r_\ell - \sum_{m=0}^{M} E_{\ell m}^{(n)}\chi_m^{(n)} \right|^2$$
$$+ \delta\sum_{m=1}^{M-1}(\chi_{m-1}^{(n)} - 2\chi_m^{(n)} + \chi_{m+1}^{(n)})^2. \tag{5.4.22}$$

The weighting function $w(\omega_\ell)$ limits the influence of the high frequencies. Numerical experiments showed that the best rate of convergence is obtained when $\lim_{\omega\to\infty} \omega w(\omega) = 1$, while the low-frequency behavior of $w(\omega)$ has hardly any influence. Successful computations were carried out for $w(\omega) = (\omega^2 + \omega_0^2)^{-\frac{1}{2}}$ with $0 \le \omega_0 \le 2$ while the most rapid convergence was obtained for $w(\omega) = 1/\omega$. An alternative way to emphasize the low frequencies is to choose a set of frequencies that are spaced equally on a logarithmic scale. For such a set, convergence was obtained - at a slightly slower rate - even for $w(\omega) = 1$. Although these results were originally obtained by numerical experimentation, they can now be substantiated by the same two arguments that were used in the selection of the cost functions (5.3.31) and (5.3.32). Both the linear increase of the error $\Delta r_B(i\omega)$ as $|\omega|\to\infty$ found in (5.2.32) and the relative decrease of the amount of profile information present in $r^+(i\omega)$ - signaled in connection with (5.2.23a) and (5.2.23b) - indicate that each equation of the type (5.4.19) should, in effect, be weighted by a factor that is proportional to ω^{-1} as $\omega\to\infty$. In the first successful computation reported above, this was achieved by constructing the system (5.4.21) for a more or less equally spaced set of frequencies $\{\omega_\ell\}$ and including a properly behaved weighting function in the squared error (5.4.22). In the second successful computation, a similar effect was attained by introducing a density variation in the set of frequencies $\{\omega_\ell\}$ and keeping the size of the weighting function constant.

As in the low-contrast case discussed in Subsection 5.3.4, the number of cells M should be chosen large enough to avoid a possible suppression of part of the profile information present in the reflection coefficients $\{r^+(i\omega_\ell)\}$ by the a priori information supplied by the piecewise-linear expansion (5.3.28). In view of the approximate results (5.2.22) and (5.2.17), it seems sensible to take M in the same range as in the Born approximation, i.e. $M \approx k_{max}/\pi$. In fact, the form of the path length $\zeta^+(z)$ occurring in (5.2.16), which is the frequency-domain counterpart of (5.2.17), indicates that, for larger contrasts, it may be possible to increase M beyond that value. The presence of the regularization term on the right-hand side of (5.4.22) is necessary because, as in the computations of

Subsection 5.3.4, the square matrix that results from minimizing the first term of (5.4.22) with respect to $\chi_m^{(n)}$ turns out to have an eigenvector $\underline{\nu}$ with components $\nu_m = (-1)^m$, with an almost vanishing eigenvalue. Since in each iteration step, $E_{\ell m}^{(n)}$ is only known approximately, this implies that the approximation error may result in a solution for $\chi_m^{(n)}$ with a large, spurious component along ν_m. The amplification of this component in subsequent iteration steps will then cause instabilities in the iterative procedure. The addition of the regularization term, with δ as a small parameter, annihilates this instability by restraining the variation in the slope of the piecewise-linear approximation $\chi^{(n)}(z)$ between adjacent cells.

The effect of the regularization term is illustrated in Figure 5.4.3, where results are given for a parabolic susceptibility profile. Figure 5.4.3a shows a linearized version of Tabbara's improved Born approximation, and Figures 5.4.3b,c,d offer the result after three, five and nine iteration steps without a regularization term. It is observed that even for a good starting value, the iteration scheme exhibits instabilities. In Figures 5.4.3e,f, the results are presented of two iteration steps with $\delta = 0.01$, starting from the already unstable result of Figure 5.4.3c. Now the iteration scheme does converge. Obviously, the regularization parameter δ should be chosen as small as possible. In all cases that we considered, the choice $\delta = 0.01$ was sufficient to guarantee convergence. For that value of δ, the elements of the square matrix that has to be inverted in order to obtain the $\chi_m^{(n)}$ that minimize (5.4.22) differ about one percent from the corresponding elements for $\delta = 0$. The choice $\delta = 0.01$ was obtained by repeatedly performing, for a slightly inhomogeneous, parabolic profile, a single iteration step starting from a constant susceptibility profile. Both the initial estimate and the value of δ were varied in this process. The parameter δ was determined such that for M between 10 and 20, a change of a factor of 2 in δ did not significantly change the results, while a reduction of δ by a factor of 10 did cause fluctuations in the results similar to the ones shown in Figures 5.4.3b,c.

Compared to the errors defined in (5.2.6), (5.3.31), and (5.3.32), the first term of (5.4.22) lacks the quadrature weighting factors by

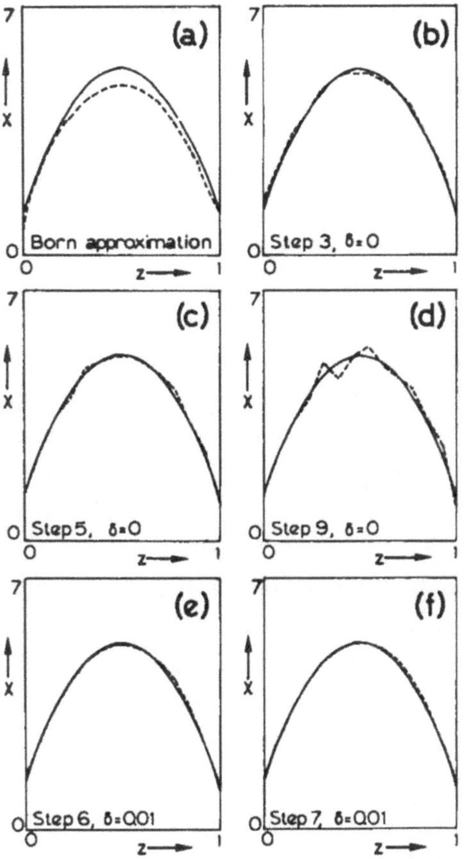

Figure 5.4.3 Results of the iterative reconstruction of $\chi(z) = 5.25 - 16(z - 0.5)^2$ (dashed lines) compared with the actual profile (solid lines) for $L = 26$, $M = 13$, $N_{cell} = 15$, $k_{max} = 30$ and $w(\omega) = 1/\omega$. Figures (e) and (f) show the result of two "regularized" iteration steps starting from Figure (c).

which it would become an approximation of an integral over the domain $0 < \omega < k_{max}/2$. After the inclusion of these quadrature factors, the rate of convergence was observed to be slightly faster for $\chi^{(n)}(z)$ close to the actual profile but considerably slower for $\chi^{(n)}(z)$ further away from it. In order to preserve the general applicability of the method, it therefore seems preferable to omit the quadrature weighting factors from (5.4.22).

For smoothly varying, continuous profiles the regularized scheme described above converges in about five iteration steps even for a "bad" initial guess. However, if either the initial estimate or the unknown profile violates the restriction that they should be smoothly varying, it may be necessary to stress the low-frequency range additionally. This can be achieved by extending the set $\{\omega_\ell\}$ by an equivalent frequency set for a reduced integration interval. An example is shown in Figure 5.4.4, where results are presented for a sinusoidal profile. It is observed that, near the slab's interfaces, the estimate resulting from Tabbara's improved Born approximation differs considerably from the actual profile and, particularly at the end of the slab, varies more rapidly with z than in the remaining part of the interval. In spite of this numerically unfavorable estimate, the iteration scheme turns out to converge in about ten iteration steps. For this profile, the computation was also carried out starting from the initial estimate $\chi(z) = 3$. This computation took about 3 iteration steps less to converge. Comparing the result of both computations after 15 iteration steps, we observed a maximum difference in χ_ℓ of 0.007, which is approximately equal to the difference between the results of steps 14 and 15 for the Born start.

Figure 5.4.4 also illustrates two features of our scheme that show up if the unknown profile cannot be matched exactly. In the first place, the algorithm tends to approximate especially those parts of the profile where the gradient of $\chi(z)$ is largest, i.e. where the susceptibility varies most rapidly. Physically, this can be attributed to the effect that they are exactly the gradients of $\chi(z)$ that cause a reflection. Secondly, the estimate obtained is more accurate at the front of the slab than at the back. This corresponds to the fact that a wave which is reflected at some point $z = z_0$ inside the slab traverses the interval $0 < z < z_0$ twice before it is observed at the front of the slab. As a consequence, the reflected field at the front of the slab depends more strongly on the local value of $\chi(z)$ according to whether z is closer to that front. Since the repeated minimization of the squared error (5.4.22) will produce the best fit where this dependence is strongest, the effect mentioned

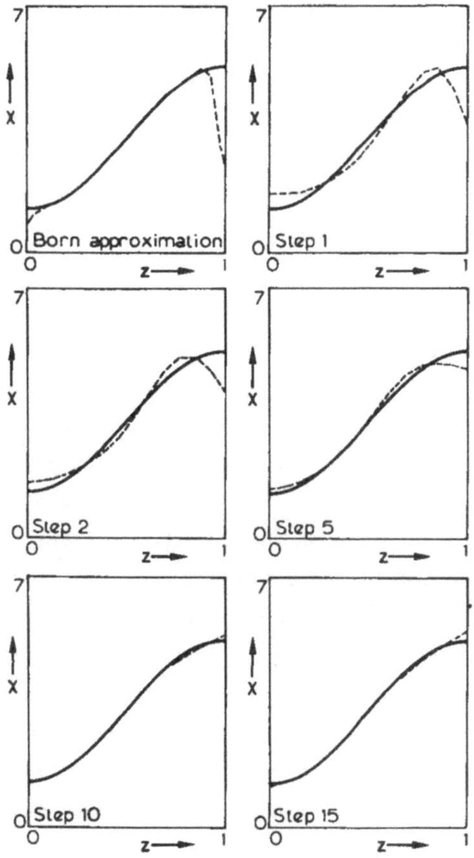

Figure 5.4.4 Results of the iterative reconstruction of $\chi(z) = 1.25 + 4\sin^2(\pi z/2)$ (dashed lines) compared with the actual profile (solid lines) for $M = 13$, $N_{cell} = 15$, $w(\omega) = 1/\omega$ and $\delta = 0.01$. The set of frequencies is obtained by combining two subsets with $L = 26$ and $k_{max} = 3$ and 30, respectively.

above must be observed when the representation (5.3.28) is inadequate. Note that this effect is different from the one that shows up in Figure 5.3.6. As remarked in Subsection 5.3.4, in that case, the deterioration in the estimate obtained is simply due to the error in replacing the unknown actual field by the known incident field. For rapidly varying profiles, the two features explained above prevent an accurate reconstruction of the unknown susceptibility in terms of the continuous piecewise-linear expansion functions. Generally, the results do converge but especially the more slowly varying parts of $\chi(z)$ are reconstructed poorly. Although the estimate can be somewhat improved by slightly increasing the regularization parameter δ, the reconstruction obtained is usually not better than the result of either of the improved Born approximations. As an example, Figure 5.4.5 shows results for a discontinuous profile. For these profiles, a different set of expansion functions should be chosen, possibly by allowing the susceptibility to be discontinuous across the cell boundaries. An attempt to reconstruct the profile specified in Figure 5.4.5 was also made by Coen et al. (1981), who used a quasi-Newton method to minimize a squared error similar to the one figuring in our iterative scheme. The result obtained by these authors resembles the one shown in the lower part of Figure 5.4.5. As the only other resemblance between their method and the one employed in the present subsection is the restriction of $\chi(z)$ to a smoothly varying profile, we arrive at the conclusion that it is precisely this restriction which impedes a more accurate reconstruction. The maximum difference between the actual reflection coefficients and the reflection coefficients for the approximated profile turns out to be a good indication of the quality of the estimate obtained. For the results of Figures 5.4.4 and 5.4.5, the magnitude of this difference was 0.02 and 0.5, respectively.

For the computations shown in Figures 5.4.3 - 5.4.5, most of the computational effort was spent in determining the approximate fields $\{\hat{E}_+^{(n)}(z,i\omega_\ell)\}$. For the set of frequencies used in Figure 5.4.3, i.e. $k_{max} = 30$, $K = 6$ or, equivalently, $L = 26$, a single iteration step took about 9 seconds. For the extended sets of frequencies used in Figures 5.4.4 and 5.4.5, a single step took about 17 seconds. Compared to the time-domain case, which will be discussed in Section 6.2, a single iteration step consumes considerably more computation time due to the repeated solution of the direct-scattering problem for each frequency ω_ℓ. For large contrasts, however, the time-domain scheme requires more iteration steps. As a result, the total computation time of both procedures is then of the same order of magnitude.

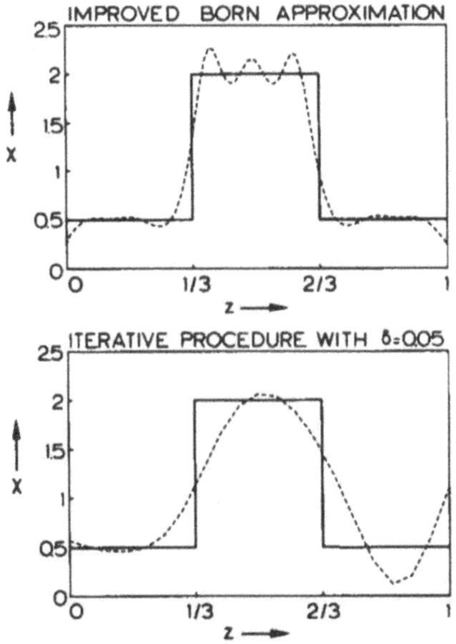

Figure 5.4.5 Upper part: results of Tabbara's improved Born approximation (dashed line) compared with the actual profile (solid line) for $\chi(z) = 0.5 + 1.5\text{rect}(z - 0.5;1/3)$ for $k_{max} = 30$ and $L = 26$. Lower part: result of the iterative reconstruction of this profile for the same frequencies as in Figure 5.4.4 and for $M = 20$, $N_{cell} = 10$, $w(\omega) = 1/\omega$, $\delta = 0.05$ and $\chi^{(0)}(z) = 1$.

5.5 Conclusions

In this chapter, we have formulated and analyzed a Born-type iterative technique for the solution of the one-dimensional inverse-scattering problem of reconstructing one or both constitutive parameters of an inhomogeneous, lossy dielectric slab from reflected-field data. In Tijhuis (1981), where this solution technique was first proposed, it was developed mainly by numerical experimentation, supplemented by some physical and mathematical insight. In the present chapter, a more theoretical analysis has been presented which provides at least a good qualitative explanation of the numerical results obtained. In particular, it was argued that, for a unique determination of both unknown constitutive parameters by our technique, reflected-field data for at least two independent directions of incidence are needed. In addition, the inherent band limitation in the reconstruction observed in previous applications of the technique (Tijhuis (1981), Tijhuis and Van der Worm (1984)) was elucidated. The theoretical analysis can also serve as a guideline in the actual numerical implementation of the iterative procedure. Particularly the inclusion of a priori information and the selection of the cost function can now be justified more rigorously. Finally, the analysis has presented us with an improved method of solution, in which the inherent band limitation is overcome as the iterative scheme converges.

There is one problem in considering the results of our theory as being rigorous, namely that we have approximated the unknown field in an inhomogeneous slab configuration by its WKB counterpart. In some respects, however, this need not be a true objection. For example, the non-uniqueness results obtained for the iterative scheme are even exact, since, for the special example of a homogeneous, lossless dielectric slab, the WKB expressions employed reduce to the actual solutions. In particular, this applies to the result that our scheme may not be able to reconstruct more than one unknown constitutive parameter from the reflected field for a single direction of incidence. Furthermore, the conclusions about the inherent band limitation must be reliable since they pertain to the high-frequency behavior of the fields involved, which is described accurately by the relevant WKB expressions. Finally, some confidence in our conclusions can be obtained from the good qualitative agreement between the actual and WKB results observed in Subsections 2.3.5 and 2.4.3.

In this chapter, we also considered the numerical implementation of our Born-type scheme for two special frequency-domain problems, namely the case of low contrast for oblique incidence, and that of a lossless slab for normal incidence. In both cases, the original version of the technique was employed. The numerical results turn out to agree well with the theory. Moreover, the inherent band limitation in the reconstruction does not seem to prevent the recovery of at least smoothly varying profiles. One aspect which was not addressed is the feature that, in using frequency-domain data, we have forfeited the possibility of considering only the time intervals specified in connection with (5.2.17) and (5.2.37). In doing so, we may have spoiled the almost linear relation between the known reflected field and the unknown constitutive profiles without adding any new information about these profiles. The only way to investigate this particular aspect is to apply our solution procedure to time-domain data.

A second aspect that remains to be considered is the comparison of the original Born-type iterative procedure with the modified version proposed in Subsection 5.2.5. Also, both iterative procedures should be compared with minimizing the cost function figuring in the approximate inverse problem solved in each iteration step by a conventional optimization technique. Obviously, in such a minimization the unknown profiles should be approximated by the same piecewise-linear expansions as in the Born-type scheme to ensure a fair comparison.

Both aspects signaled above will be discussed in the next chapter, where we deal with the reconstruction of constitutive slab parameters from time-domain data. Consequently, we will postpone our final assessment of the Born-type iterative technique until the end of that chapter.

References

Berryman, J.G., and Greene, R.R. (1980), Discrete inverse methods for elastic waves in layered media, *Geophysics 45*, 213-233.

Bojarski, N.N. (1980), One-dimensional direct and inverse scattering in causal space, *Wave Motion 2*, 115-124.

Bolomey, J.-Ch., Durix, Ch., and Lesselier, D. (1979), Determination of conductivity profiles by time-domain reflectometry, *IEEE Trans. Antennas Propagat. 27*, 244-248.

Burridge, R. (1980), The Gelfand-Levitan, the Marchenko, and the Gopinath-Sondhi integral equation of inverse scattering theory, regarded in the context of inverse impulse-response problems, *Wave Motion 2*, 305-323.

Chadan, K., and Sabatier, P.C. (1977), *Inverse problems in quantum scattering theory*, Springer-Verlag, New York, Chap. 17.

Coen, Sh. (1981), Inverse scattering of a layered and dispersionless dielectric half-space, Part I: reflection data from plane waves at normal incidence, *IEEE Trans. Antennas Propagat. 29*, 726-732.

Coen, Sh., Mei, K.K., and Angelakos, D.J. (1981), Inverse scattering technique applied to remote sensing of layered media, *IEEE Trans. Antennas Propagat. 29*, 298-306.

Corones, J.P., Davison, M.E., and Krueger, R.J. (1985), Dissipative problems in the time domain. In: *Inverse methods in electromagnetic imaging Part I.*, Boerner, W.-M (Ed.), Reidel Publishing Company, Dordrecht, 121-130.

Corones, J., and Krueger, R.J. (1983), Obtaining scattering kernels using invariant imbedding, *J. Math. Anal. Appl. 95*, 393-415.

Devaney, A.J., and Oristaglio, M.L. (1983), Inversion procedure for inverse scattering within the distorted-wave Born approximation, *Phys. Rev. Lett. 51*, 237-240.

Fawcett, J. (1984), On the stability of inverse scattering problems, *Wave Motion 6*, 489-499.

Goupillaud, P.L. (1961), An approach to inverse filtering of near-surface layer effects from seismic records, *Geophysics 26*, 754-760.

Jaggard, D.L., and Kim, Y. (1985), Accurate one-dimensional inverse scattering using a nonlinear renormalization technique, *J. Opt. Soc. Am. A 2*, 1922-1930.

Jaggard, D.L., and Olson, K.E. (1985), Numerical reconstruction for dispersionless refractive profiles, *J. Opt. Soc. Am. A 2*, 1931-1936.

Jaulent, M. (1976), Inverse scattering problems for absorbing media, *J. Math. Phys. 17*, 1351-1360.

Jaulent, M. (1982), The inverse scattering problem for LCRG transmission lines, *J. Math. Phys. 23*, 2286-2290.

Kaiser, H., and Kaiser, H.Ch. (1983), Identification of stratified media based on the Bremmer series representation of the reflection coefficient, *Applied Optics 22*, 1337-1345.

Kay, I. (1955), The inverse scattering problem, *Research Report EM-74*, New York University, New York.

Kristensson, G., and Krueger, R.J. (1986a), Direct and inverse scattering in the time domain for a dissipative wave equation. Part I: scattering operators, *J. Math. Phys. 27*, 1667-1682.

Kristensson, G., and Krueger, R.J. (1986b), Direct and inverse scattering in the time domain for a dissipative wave equation. Part II: simultaneous reconstruction of dissipation and phase velocity profiles, *J. Math. Phys.* *27*, 1683-1693.

Krueger, R.J. (1976), An inverse problem for a dissipative hyperbolic equation with discontinuous coefficients, *Quart. Appl. Math.* *34*, 129-147.

Krueger, R.J. (1978), An inverse problem for an absorbing medium with multiple discontinuities, *Quart. Appl. Math.* *36*, 235-253.

Krueger, R.J. (1981), Numerical aspects of a dissipative inverse problem, *IEEE Trans. Antennas Propagat.* *29*, 253-261.

Ladouceur, H.D., and Jordan, A.K. (1985), Renormalization of an inverse-scattering theory for inhomogeneous dielectrics, *J. Opt. Soc. Am. A* *2*, 1916-1921.

Lee, C.Q. (1982), Wave propagation and profile inversion in lossy inhomogeneous media, *Proc. IEEE* *70*, 219-228.

Lesselier, D. (1978), Determination of index profiles by time domain reflectometry, *J. Optics* *9*, 349-358.

Lesselier, D. (1982a), Optimization techniques and inverse problems: reconstruction of conductivity profiles in the time domain, *IEEE Trans. Antennas Propagat.* *30*, 59-65.

Lesselier, D. (1982b), *Diagnostic optimal de la lame inhomogène en régime temporel. Applications à l'électromagnétisme et à l'acoustique*, Ph.D. Thesis, l'Université Pierre et Marie Curie, Paris.

Mendel, J.M., and Habibi-Ashrafi, F. (1980), A survey of approaches to solving inverse problems for lossless layered media systems, *IEEE Trans. Geoscience Remote Sensing* *18*, 320-330.

Mostafavi, M., and Mittra, R. (1972), Remote probing of inhomogeneous media using parameter optimization techniques, *Radio Sci.* *7*, 1105-1111.

NAG Fortran Library Mark 11 (1984), Numerical Algorithms Group, Oxford.

Newton, R.G. (1980), Inverse scattering. I. One dimension, *J. Math. Phys.* *21*, 493-505.

Roger, A. (1981), *Problèmes inverses de diffraction en électromagnétisme, Théorie, traitement numérique, et applications à l'optique*, Ph.D. Thesis, l'Université d'Aix-Marseille III, Marseille.

Roger, A., Maystre, D., and Cadilhac, M. (1978), On a problem of inverse scattering in optics: the dielectric inhomogeneous medium, *J. Optics* *9*, 83-90.

Sabatier, P.C. (1983), Theoretical considerations for inverse scattering, *Radio Sci.* *18*, 1-18.

Sarkar, T.K., Weiner, D.D., and Jain, V.K. (1981), Some mathematical considerations in dealing with the inverse problem, *IEEE Trans. Antennas Propagat.* *29*, 373-379.

Schaubert, D.H., and Mittra, R. (1977), A spectral domain method for remotely probing stratified media, *IEEE Trans. Antennas Propagat.* *25*, 261-265.

Tabbara, W. (1976), *Etude de problèmes de diffraction inverse électromagnétique par analyse multifréquentielle des champs diffractés*, Ph.D. Thesis, l'Université Pierre et Marie Curie, Paris.

Tabbara, W. (1979), Reconstruction of permittivity profiles from a spectral analysis of the reflection coefficient, *IEEE Trans. Antennas Propagat.* *27*, 241-244.

Tijhuis, A.G. (1981), Iterative determination of permittivity and conductivity profiles of a dielectric slab in the time domain, *IEEE Trans. Antennas Propagat. 29*, 239-245.

Tijhuis, A.G., and Van der Worm, C. (1984), Iterative approach to the frequency-domain solution of the inverse-scattering problem for an inhomogeneous lossless dielectric slab, *IEEE Trans. Antennas Propagat. 32*, 711-716.

Titchmarsh, E.C. (1950), *The theory of functions*, Oxford University Press, London, second edition, Chap. 13.

Uno, T., and Adachi, S. (1985), Electromagnetic inverse scattering method for one-dimensional inhomogeneous layered media, *Proceedings of the International Symposium on Antennas and Propagation, Kyoto, August 20-22*, pp. 887-890.

Weston, V.H. (1972), On the inverse problem for a hyperbolic dispersive partial differential equation, *J. Math. Phys. 13*, 1952-1956.

Weston, V.H. (1974), On inverse scattering, *J. Math. Phys. 15*, 209-213.

Weston, V.H., and Krueger, R.J. (1973), On the inverse problem for a hyperbolic dispersive partial differential equation. II, *J. Math. Phys. 14*, 406-408.

Widder, D.V. (1946), *The Laplace transform*, Princeton University Press, Princeton, pp. 61-63.

6. TIME-DOMAIN TECHNIQUES

6.1 Introduction

In this chapter, we continue the analysis of the one-dimensional inverse-scattering problem of reconstructing the susceptibility profile and/or the conductivity profile of a dielectric slab in vacuo from the reflected fields caused by one or two plane electromagnetic waves, incident from one or both sides of the slab. In contradistinction to Chapter 5, we consider the situation where the relevant incident and reflected fields are known in the *time domain*. For an extensive introduction to the subject, the reader is referred to the introduction of the previous chapter, i.e. Section 5.1.

In the present introduction, we will confine ourselves to describing the organization of this chapter, which runs as follows. In Section 6.2, we investigate the time-domain application of the Born-type iterative procedure with a vacuum as the background medium, as formulated in Subsection 5.2.2. As remarked in Section 5.5, this enables us to consider only reflected-field data in the time interval specified in connection with (5.2.17). This may restore the almost linear relation between the known reflected field and the unknown constitutive profiles observed from that equation. Consequently, a better reconstruction may be obtained. In Section 6.2, we will consider the situation investigated previously in Tijhuis (1981), where one of the constitutive parameters is unknown, as well as the case where both constitutive parameters are unknown.

In Section 6.3, we consider solving the inverse-profiling problem by applying a conventional optimization approach to minimize a cost function involving *all* the reflected-field information available. A similar approach was followed previously by Lesselier (1982a,b). However, by using the band-limitation arguments developed in Section 5.2, we are able to obtain a much better performance. In Section 6.3, we have included a review of some common optimization methods as well as a derivation of a closed-form expression for the so-called profile gradient of the cost function. In view of the computational effort required, we consider only the situation where a single unknown constitutive parameter is to be reconstructed from a single time-domain reflected field.

In Section 6.4, we investigate the application of the modified version of the Born-type iterative procedure proposed in Subsection 5.2.5. Special attention will be devoted to the construction of the time-domain field matrix whose inversion yields the expansion coefficients in the piecewise-linear approximation, and to comparing the results obtained with corresponding ones presented in previous sections of Chapters 5 and 6. In addition, we will compare the natural frequencies of the actual and the reconstructed slab configurations. Finally, the conclusions are stated in Section 6.5.

6.2 Born-type iterative procedure with a vacuum as the background medium

6.2.1 Formulation of the problem

In this chapter, we consider the time-domain equivalent of the inverse-scattering problem discussed in Chapter 5. In this problem, we wish to reconstruct the dielectric susceptibility $\chi(z)$ and/or the conductivity $\sigma(z)$ of an inhomogeneous, lossy dielectric slab from known incident and reflected fields at one or both of the slab's interfaces. For the incident fields, we take one or both of the electromagnetic pulses of finite duration given by

$$\underline{E}^i_+(\underline{r},t) = E^i_+(z,t)\underline{i}_y = F(t - z)\underline{i}_y, \qquad (6.2.1a)$$

$$\underline{E}^i_-(\underline{r},t) = E^i_-(z,t)\underline{i}_y = F(t - (1 - z))\underline{i}_y, \qquad (6.2.1b)$$

and by corresponding expressions for the magnetic-field strength. Specifically, we will solve this inverse-scattering problem for the case where $F(t)$ represents the sine-squared incident pulse

$$F(t) = \sin^2(\pi t/T)\mathrm{rect}(t - T/2;T), \qquad (6.2.2)$$

where T is the pulse duration. In (6.2.1), the subscripts + and - refer to the direction of propagation of the incident field (see Figure 6.2.1). The sine-squared incident pulse was selected primarily because it is one of the pulses amenable to the marching-on-in-time method which allows the frequency range over which the slab is excited to be adjusted. Thus, we can carry over the ambiguity arguments given in Chapter 5 for the frequency-domain problem. Moreover, we will be able to use the analysis given in Section 4.3 both to estimate and to filter the frequency content of our incident pulse.

To each of the incident fields specified in (6.2.1), there corresponds a total field. In line with the notation used until now, these fields will be denoted as $E_+(z,t)$ and $E_-(z,t)$, respectively. As remarked earlier, for a general, inhomogeneous slab, these fields must be computed numerically. One way to achieve this is to solve

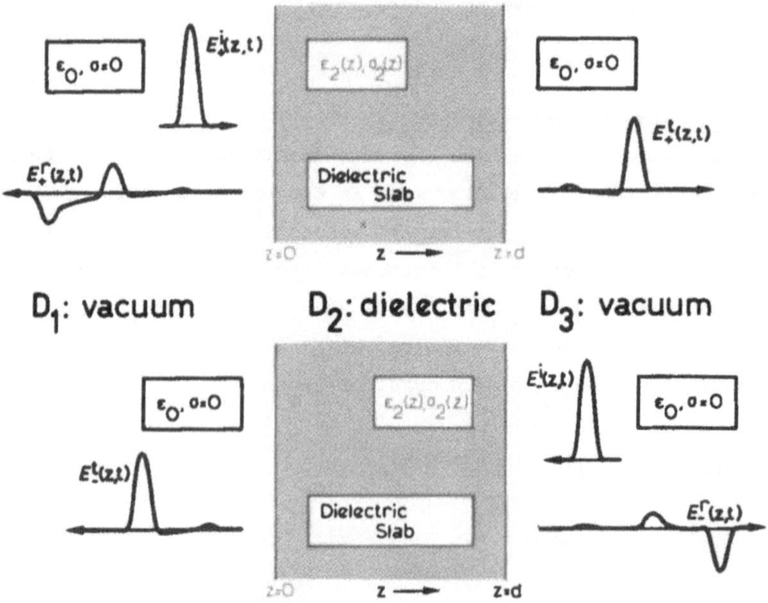

Figure 6.2.1 Pulsed plane waves normally incident from the left and from the right on an inhomogeneous, lossy dielectric slab embedded in vacuum.

the relevant integral equation (3.3.2), i.e.

$$E(z,t) = E^i(z,t) - \tfrac{1}{2}\int_0^1[\chi(z')\partial_t + \sigma(z')]E(z',t')dz', \tag{6.2.3}$$

with $t' \triangleq t - |z - z'|$, by applying the marching-on-in-time method as described in Section 3.3. Alternatively, we can follow the procedure outlined in Section 2.4, which amounts to repeatedly resolving the corresponding frequency-domain problem for real-valued angular frequencies, and subjecting the result to a numerical Fourier inversion.

In the present section, we attempt to resolve the inverse-scattering problem formulated above by means of a time-domain version of the Born-type iterative procedure proposed in Subsection 5.2.2. To this end, we need the time-domain counterpart of the relation (5.2.4). By setting z = 0 and z = 1 in (6.2.3), we directly obtain

$$E^r_+(0,t) = -\tfrac{1}{2}\int_0^1[\chi(z)\partial_t + \sigma(z)]E_+(z,t-z)dz, \tag{6.2.4a}$$

$$E^r_-(1,t) = -\tfrac{1}{2}\int_0^1[\chi(z)\partial_t + \sigma(z)]E_-(z,t-(1-z))dz. \tag{6.2.4b}$$

As noted in Section 5.2, the integral relations in (6.2.4) can be envisaged as source-type field representations with respect to a vacuum background medium. In that respect, both (6.2.4a) and (6.2.4b) may be regarded as special cases of more general source-type representations with respect to an arbitrary background medium. The choice of a vacuum background medium leading to (6.2.4) was made primarily with the simplicity of the numerical implementation in mind.

In this section, we will consider two situations. In the first situation, either the susceptibility profile $\chi(z)$ or the conductivity profile $\sigma(z)$ is determined from the known values of $E^i_+(z,t)$ and $E^r_+(0,t)$, while the remaining constitutive parameter is assumed to be known. This problem will be discussed in Subsections 6.2.2 and 6.2.3. In Subsections 6.2.4 and 6.2.5, we address the full inverse-scattering problem where both $\chi(z)$ and $\sigma(z)$ are reconstructed from the known values of $E^i_+(z,t)$, $E^r_+(0,t)$, $E^i_-(z,t)$ and $E^r_-(1,t)$.

6.2.2 Either $\chi(z)$ or $\sigma(z)$ unknown: implementation

As mentioned above, we wish to apply the time-domain equivalent of the Born-type iterative procedure outlined in Subsection 5.2.2. Each iteration step of that procedure consists of solving two approximate problems. First we solve the approximate direct-scattering problem of determining the field $E^{(n)}_+(z,t)$ that results from the known incident field $E^i_+(z,t)$ if the unknown profile equals the approximation found in the previous step. From the direct-scattering results obtained in Subsections 2.4.3 and 3.3.3, it turns out that the most efficient way of obtaining a numerical solution to this subproblem is to solve either of the time-domain integral equations

$$E^{(n)}_+(z,t) = E^i_+(z,t) - \tfrac{1}{2}\int_0^1[\chi^{(n-1)}(z')\partial_t + \sigma(z')]E^{(n)}_+(z',t')dz', \tag{6.2.5a}$$

$$E_+^{(n)}(z,t) = E_+^i(z,t) - \tfrac{1}{2}\int_0^1 [\chi(z')\partial_t + \sigma^{(n-1)}(z')]E_+^{(n)}(z',t')dz', \qquad (6.2.5b)$$

where $t' \triangleq t - |z - z'|$. Both these equations are solved directly in the time domain by means of the marching-on-in-time method explained in Subsection 3.3.2. Equations (6.2.5a) and (6.2.5b) pertain to the case of unknown $\chi(z)$ and $\sigma(z)$, respectively. Since this particular application of the marching-on-in-time method has been discussed extensively in Section 3.3, we need not dwell on it here.

The second part of each iteration step consists of determining a new approximation to the unknown profile from either of the approximate integral relations

$$E_+^r(0,t) + \tfrac{1}{2}\int_0^1 \sigma(z)E_+^{(n)}(z,t-z)dz = -\tfrac{1}{2}\int_0^1 \chi^{(n)}(z)\partial_t E_+^{(n)}(z,t-z)dz, \qquad (6.2.6a)$$

$$E_+^r(0,t) + \tfrac{1}{2}\int_0^1 \chi(z)\partial_t E_+^{(n)}(z,t-z)dz = -\tfrac{1}{2}\int_0^1 \sigma^{(n)}(z)E_+^{(n)}(z,t-z)dz. \qquad (6.2.6b)$$

As in (6.2.5), Equations (6.2.6a) and (6.2.6b) pertain to the case of unknown $\chi(z)$ and $\sigma(z)$, respectively. Each equation in (6.2.6) can be regarded as an integral equation of the first kind for the next approximation to the unknown profile, i.e. $\chi^{(n)}(z)$ or $\sigma^{(n)}(z)$. As argued in Section 5.2, these equations may not suffice to establish this approximation in full detail. In particular, the band limitations associated with (6.2.6) may prevent the reconstruction of the short-range behavior of the unknown profile.

In the present time-domain problem, we have the same two types of band limitation that we encountered in the corresponding frequency-domain case. The first type now originates from the choice of the sine-squared incident pulse specified in (6.2.2). From the discussion in Subsection 4.3.2 and from the frequency spectrum displayed in Figure 4.3.2b, it is apparent that the Fourier transform $F(i\omega)$

of that pulse may at least be considered as negligible beyond its second zero on the real-frequency axis, i.e. for $|\omega| > \omega_{max} = 6\pi/T$. In view of the linearity of the underlying direct-scattering problem, this observation also applies to the reflected field it excites, and, in turn, to the corresponding spectral components of the unknown profile figuring in (5.2.16). The second band limitation is the one inherent in the Born-type approximation of replacing the exact field $E_+(z,t)$ by its approximate counterpart $E_+^{(n)}(z,t)$, as analyzed in Subsection 5.2.4.

The band limitations mentioned above are dealt with in the same manner as in Subsections 5.3.4 and 5.4.4, namely by representing the unknown profile by a piecewise-linear expansion as specified in (5.3.28) and (5.3.29). Substitution of such an expansion reduces (6.2.6) to

$$E_+^r(0,t) + \tfrac{1}{2}\int_0^1 \sigma(z) E_+^{(n)}(z,t-z)\,dz = -\tfrac{1}{2}\sum_{m=0}^{M} \chi_m^{(n)} \int_0^1 \phi_m(z)\,\partial_t E_+^{(n)}(z,t-z)\,dz, \tag{6.2.7a}$$

$$E_+^r(0,t) + \tfrac{1}{2}\int_0^1 \chi(z)\,\partial_t E_+^{(n)}(z,t-z)\,dz = -\tfrac{1}{2}\sum_{m=0}^{M} \sigma_m^{(n)} \int_0^1 \phi_m(z) E_+^{(n)}(z,t-z)\,dz, \tag{6.2.7b}$$

where $\{\chi_m^{(n)}\}$ and $\{\sigma_m^{(n)}\}$ denote the expansion coefficients obtained in the nth step of reconstructing $\chi(z)$ and $\sigma(z)$, respectively. Since, in the second part of each iteration step, the approximate field $E_+^{(n)}(z,t)$ is known, both equations in (6.2.7) constitute, for each fixed t, an approximate linear equation for these expansion coefficients.

In order to obtain a system of such equations allowing a unique determination of the unknown coefficients $\{\chi_m^{(n)}\}$ or $\{\sigma_m^{(n)}\}$, we have to resolve two more problems. First, we must choose an appropriate range for the time t. In accordance with the results obtained in Subsection 5.2.3 (particularly Equations (5.2.17), (5.2.18) and (5.2.22)), we select the time interval covering the first traverse of the pulse across the slab. In view of the finite duration of this pulse, we define this traverse as the course of the field inside the slab in between the situation in which precisely half of the incident pulse has reached the slab and that in which precisely half of the directly transmitted pulse has left its rear end. Accounting for the time delay in the integrands in (6.2.7), we end up with the time interval

$$T/2 < t < \tau^{(n-1)} + 1 + T/2, \tag{6.2.8}$$

where $\tau^{(n)}$ denotes the one-way travel time corresponding to $\chi^{(n)}(z)$ as defined in the fourth line of (2.3.21).

Next, we have to devise a procedure to convert the equations of the type (6.2.7) with t in the range given in (6.2.8) to a square matrix equation for the approximate expansion coefficients $\{\chi_m^{(n)}\}$ or $\{\sigma_m^{(n)}\}$. If (6.2.7) were exact, this could simply be achieved by selecting M + 1 equally spaced values of t covering the time interval specified in (6.2.8). However, in view of both the approximations made in drawing up (6.2.7) and the possible presence of noise on the available reflected-field data, it seems more sensible to attach equal importance to each value of t in the interval. We found two ways to do this. In the first place, we can subdivide the time interval in (6.2.8) into M + 1 subintervals of equal length and integrate either equation in (6.2.7) over each subinterval. This leads to an almost diagonal, square matrix equation for the unknown expansion coefficients in the piecewise-linear approximation. In effect, we establish in this manner an almost one-to-one correspondence between the reflected field in a given subinterval and the unknown profile in a single cell in the approximation employed. On the other hand, we cannot include a regularization term to prevent the resulting field matrix from having unwanted eigenvectors. This necessitates applying a smoothing procedure to the result of each iteration step. The smoothing procedure is similar to the one specified in (3.2.30) and will be justified in more detail in Appendix 6.2.A. Since, in the present method, each row of the new, square matrix equation is the sum of a number of equations of the type (6.2.7), we will designate this method as the *summation method*.

In the second place, we can arrive at a square matrix equation by defining $\{\chi_m^{(n)}\}$ as that set of coefficients that minimize the integrated squared error

$$\int_{T/2}^{\tau^{(n-1)}+1+T/2}\{E_+^r(0,t) + \tfrac{1}{2}\int_0^1\sigma(z)E_+^{(n)}(z,t-z)dz \\ + \tfrac{1}{2}\sum_{m=0}^{M}\chi_m^{(n)}\int_0^1\phi_m(z)\partial_t E_+^{(n)}(z,t-z)dz\}^2 dt, \tag{6.2.9}$$

and by introducing a similar definition for $\{\sigma_m^{(n)}\}$. In principle, such a definition does allow unwanted eigenvectors to be avoided by augmenting the relevant integrated squared errors by a suitably chosen regularization term. In practice, we chose to omit such a term from the squared errors to facilitate the comparison between both methods of composing a square matrix equation.

This subsection would not be complete without paying some attention to the actual discretization of the integral relations in (6.2.7). Let us start with the selection of the number of cells in the piecewise-linear approximation, i.e. M. As in Subsections 5.3.4 and 5.4.4, our choice of M is guided by the highest angular frequency for which the frequency-domain reflection coefficient would be obtainable. By the same argument as in Subsection 5.4.4, we arrive at

$$M \approx 2\omega_{max}/\pi = 12/T, \tag{6.2.10}$$

where we have substituted the maximum frequency estimated above. Alternatively, M can be chosen by estimating the number of independent samples that can be taken from the band-limited time-domain reflected field in the time interval specified in (6.2.8). According to Shannon's sampling theorem, the shortest sampling period for which the samples are independent is $\Delta t = \pi/\omega_{max} = T/6$. This means that we can recover at most

$$M \approx (\tau + 1)/\Delta t = 6(\tau + 1)/T \tag{6.2.11}$$

independent parameters from the relevant reflected field. Note that this estimate confirms the conjecture made in Subsection 5.4.4, that, for larger contrasts, it may be possible to increase M beyond the value found in the Born approximation.

Finally, we should indicate how the discretized version of (6.2.7) is obtained from the marching-on-in-time solution $\tilde{E}_+^{(n)}(m',n')$ of the discretized equation (3.3.4). Since Equations (6.2.7a) and (6.2.7b) are completely similar in structure, we restrict ourselves to discretizing the latter. Going through the same steps as in Subsection 5.4.4, we arrive at a system of equations of the form

$$r_\ell + p_\ell^{(n)} = \sum_{m=0}^{M} E_{\ell m}^{(n)} \sigma_m^{(n)}, \tag{6.2.12}$$

where ℓ is an integer with $T/2 < 2\ell\Delta t < \tau^{(n-1)} + 1 + T/2$, with $\Delta t = h = 1/N$ as specified in Subsection 3.3.2. In (6.2.12), the vector components on the left-hand side are defined by

$$\begin{aligned} r_\ell &\triangleq E_+^r(0,2\ell\Delta t) \\ p_\ell^{(n)} &\triangleq \tfrac{1}{4}\sum_{k=0}^{N} w_k \frac{\tilde{\chi}(k)}{2h}\left[\tfrac{3}{2}\tilde{E}_+^{(n)}(k,2\ell-k)\right. \\ &\qquad \left. - 2\tilde{E}_+^{(n)}(k,2\ell-k-2) + \tfrac{1}{2}\tilde{E}_+^{(n)}(k,2\ell-k-4)\right], \end{aligned} \tag{6.2.13}$$

with w_k and $\tilde{\chi}(k)$ as defined in Subsection 3.3.2. The elements of the field matrix in (6.2.12) are given by

$$\begin{aligned} E_{\ell m}^{(n)} \triangleq\ & (1-\delta_{mM})\sum_{j=0}^{N_{cell}} \alpha_j \frac{N_{cell}-j}{N_{cell}} \\ &\times \tilde{E}_+^{(n)}(mN_{cell}+j, 2\ell - mN_{cell} - j) \\ &+ (1-\delta_{m0})\sum_{j=0}^{N_{cell}} \alpha_j \frac{N_{cell}-j}{N_{cell}} \\ &\times \tilde{E}_+^{(n)}(mN_{cell}-j, 2\ell - mN_{cell} + j), \end{aligned} \tag{6.2.14}$$

where the same notations and definitions as in (5.4.20) have been used for $\delta_{mm'}$, α_j and N_{cell}. From the right-hand sides of (6.2.13) and (6.2.14), it is observed that, in the evaluation of the vector component $p_\ell^{(n)}$ and the matrix elements $E_{\ell m}^{(n)}$ for a fixed value of ℓ, we need only the marching-on-in-time results $\tilde{E}_+^{(n)}(m',n')$ with $m' + n' = 2\ell$. Hence, it seems advantageous to organize the numerical computation of these results such that the relevant field values are computed along lines where m' + n' is fixed, i.e. by the recurrent schemes employed in Bolomey et al. (1979) and in Tijhuis (1981). As remarked in connection with (6.2.7), the system of equations (6.2.14) is overdetermined and may be inconsistent. Hence, we can only solve it approximately by composing a square matrix equation according to one of the procedures outlined above.

6.2.3 Either $\chi(z)$ or $\sigma(z)$ unknown: numerical results

We have solved the inverse-scattering problem numerically for various permittivity and conductivity profiles. The reflected fields were computed for the exact profiles which do not satisfy (5.3.28) exactly. Generally, the result obtained in the first iteration step already gives a reasonable indication of the unknown profile. If the difference between the unknown profile and the initial estimate is small, a single iteration step may even suffice to determine the unknown profile exactly. For greater differences, the first approximation can be used to select a set of parameters, suitable for the determination of the unknown profile. For example, N_{cell} varies typically from five to ten, while M may vary from ten to 40. The duration of the incident pulse is usually chosen such that its spatial width is of the order of the expected minimum distance between two subsequent extrema in the unknown constitutive parameter. As argued above, this choice allows us, in principle, to take as much as 12 cells in the part of the piecewise-linear expansion covering the z-interval in between these extrema. In practice, however, the band limitation inherent in the Born-type approximation may force us to reduce the number of cells below this estimate.

For a number of profiles, we also investigated the stability of the iterative procedure with respect to a variation in the initial estimate. It shows that the procedure converges for a wide range of guesses. An example is presented in Figure 6.2.2, which displays the results $\chi^{(1)}(z)$ of repeatedly carrying out a single iteration

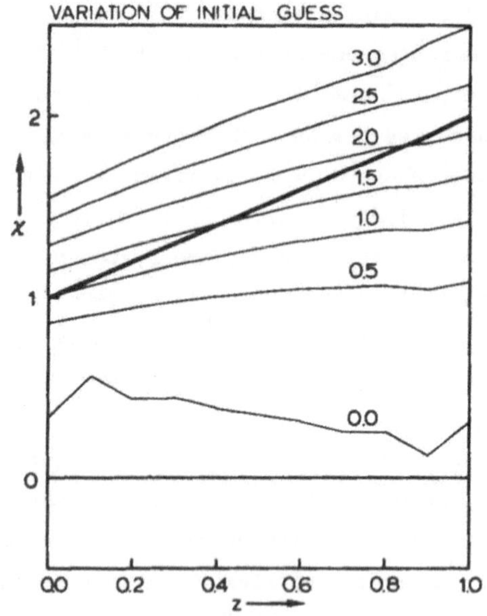

Figure 6.2.2 Influence of initial estimate. The thin lines show the result obtained in the first iteration step of reconstructing the susceptibility profile $\chi(z) = 1 + z$ (thick line) for $\sigma(z) = 0$, $M = 10$, $N = 100$ and $T = 0.5$. The initial estimate was $\chi(z) = \chi^{(0)}$ with $\chi^{(0)}$ as indicated.

step, starting from the initial estimate of a homogeneous, lossless dielectric slab with a varying susceptibility $\chi(z) = \chi^{(0)}$.

Next, we investigated the rate of convergence. For most profiles, convergence is obtained within three to five iteration steps. A representative result can be found in Table 6.2.1. Especially for complicated profiles, the accuracy of the reconstruction gradually decreases towards the end of the slab. This is the same effect that was also observed in the frequency-domain application of the iterative procedure in Subsection 5.4.4, particularly in Figures 5.4.4 and 5.4.5. Figure 6.2.3 shows a time-domain application where this effect is encountered.

Generally, the accuracy of the reconstruction is not influenced by the choice of the method employed to construct a square-matrix equation for the expansion coefficients $\{\chi_m^{(n)}\}$ or $\{\sigma_m^{(n)}\}$. However, when

Table 6.2.1 Actual susceptibility profile $\chi(z) = 5(z - 0.5)^2$ compared to the results after successive iteration steps for $\sigma(z) = 1$, $M = 10$, $N = 100$, $T = 0.5$, and initial estimate $\chi^{(0)}(z) = 1$.

z	Actual	Step 1	Step 2	Step 3	Step 4	Step 5
0.0	1.25	1.20	1.16	1.21	1.22	1.22
0.1	0.80	0.93	0.81	0.81	0.81	0.81
0.2	0.45	0.69	0.45	0.45	0.45	0.45
0.3	0.20	0.48	0.19	0.19	0.20	0.20
0.4	0.05	0.36	0.02	0.04	0.05	0.06
0.5	0.00	0.39	0.00	0.01	0.01	0.01
0.6	0.05	0.57	0.11	0.09	0.08	0.08
0.7	0.20	0.84	0.35	0.27	0.25	0.25
0.8	0.45	1.05	0.67	0.57	0.53	0.52
0.9	0.80	1.50	1.09	0.90	0.88	0.87
1.0	1.25	1.08	1.23	1.33	1.29	1.27

these coefficients are solved by linear-least-squares inversion, it may happen that parts of the retrieved profile are inaccurate. If, for example, the unknown profile has a relatively high maximum, the value of the parameter at points behind the maximum can only be retrieved accurately by employing the summation method. In that case, the reflected field in the time interval given by (6.2.8) is mainly determined by the value of the constitutive parameter in the front part of the slab. This is demonstrated in Figures 6.2.4 and 6.2.5.

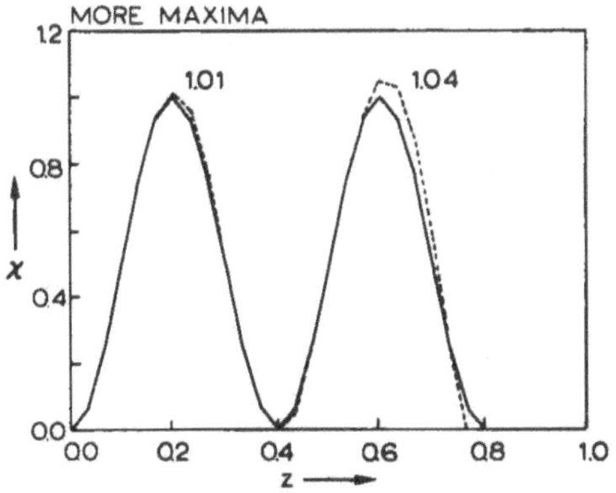

Figure 6.2.3 Actual susceptibility profile $\chi(z) = \sin^2(2.5\pi z) \times U(0.8 - z)$ compared to the approximation found after five iteration steps for $\sigma(z) = 0$, $M = 30$, $N = 300$, $T = 0.15$ and initial estimate $\chi^{(0)}(z) = 0.5$. Solid line: linearized version of actual profile; dashed line: reconstructed profile.

Figure 6.2.4 contains the final results of reconstructing a parabolic susceptibility profile with an increasing maximum by applying a linear least-squares inversion. Figure 6.2.5 presents the results $\chi^{(n)}(z)$ of all iteration steps leading to a reconstruction of profile C as specified in Figure 6.2.4 by employing the summation method. A similar effect occurs if, for unknown conductivity profiles, the known permittivity increases in magnitude, and vice versa. The reflected field is then mainly determined by the known constitutive coefficient. In these cases, too, only the summation method provides the correct reconstruction.

For some types of profiles, special correction procedures have to be applied after each iteration step. Discontinuous profiles, for

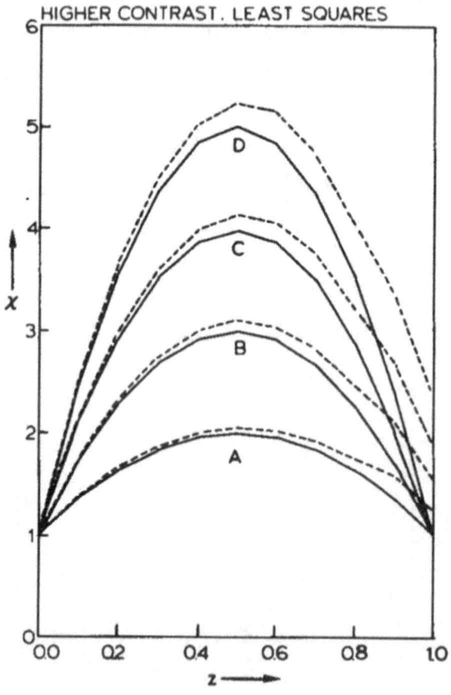

Figure 6.2.4 Actual susceptibility profile $1 + \Delta\chi[1 - 4(z - 0.5)^2]$ compared to an approximation found after 25 iteration steps by employing the linear-least-squares inversion. Contrasts: (A): $\Delta\chi = 1$; (B): $\Delta\chi = 2$; (C): $\Delta\chi = 3$; (D): $\Delta\chi = 4$. Remaining data: $\sigma(z) = 0$, $M = 10$, $N = 100$, $T = 0.5$. Solid lines: actual profiles; dashed lines: reconstructed profiles.

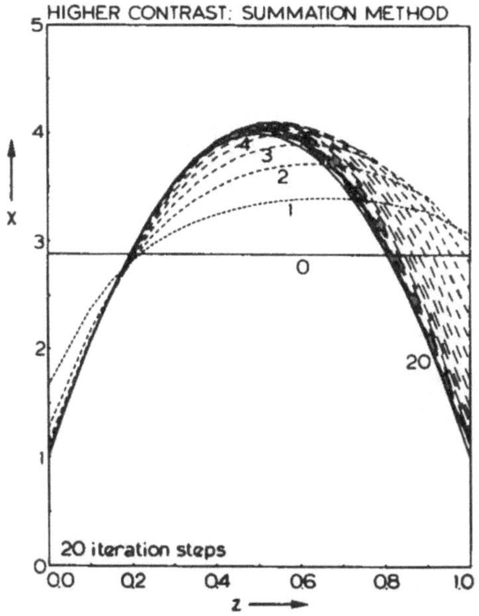

Figure 6.2.5 Results of reconstructing profile C as specified in Figure 6.2.4 by employing the summation method. Remaining data: $\sigma(z) = 0$, $M = 20$, $N = 100$, $T = 0.5$. Solid line: actual profile. Dashed lines: results of subsequent iteration steps. The dash length increases with the step number.

instance, do not satisfy the assumptions that have led to the system of equations (6.2.12). Although this may cause relatively large errors near the discontinuity, they can be determined by applying a modified version of the smoothing procedure. In devising our discontinuity correction, we used the feature that the initial guess already indicates the presence of discontinuities in the unknown profile. Some illustrative results have been plotted in Figure 6.2.6. Note that in this case as well the inversion is better for a lower susceptibility at the front of the slab.

The result obtained after each iteration step is also corrected for two types of behavior that may cause unnecessary computational problems in the direct-scattering part of the next iteration step. First, if the value of the coefficients $\{\chi_m^{(n)}\}$ or $\{\sigma_m^{(n)}\}$ should be-

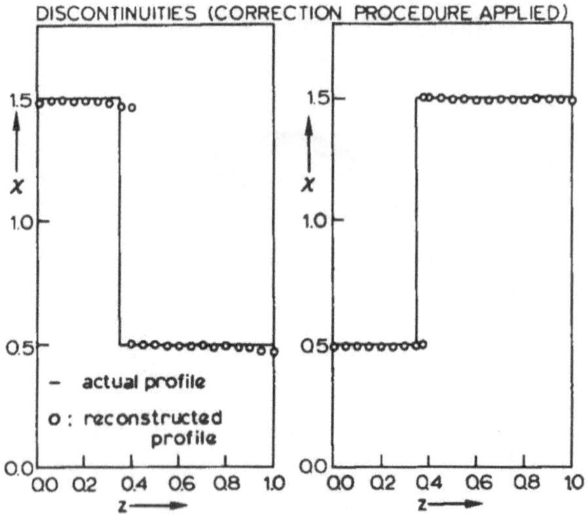

Figure 6.2.6 Actual discontinuous susceptibility profiles compared to results obtained after five iteration steps for $\sigma(z) = 0$, $M = 40$, $N = 400$, $T = 0.3$, and initial estimate $\chi^{(0)}(z) = 1.25$.

come negative for some m, this value is set to zero. Second, in the initial iteration steps, rather large fluctuations may occur in these coefficients. These fluctuations can be removed by a repeated application of the smoothing procedure (see Appendix 6.2.A).

Finally, we investigated the influence of random noise on the known sampled fields. For a typical example, superimposing ten percent multiplicative noise as specified in Subsection 4.2.3 on both the incident and the reflected fields had an effect that was well within the accuracy of the original determination from the noiseless fields. Obviously, the noise on both fields was uncorrelated since it is precisely the correlation between the incident and the reflected field which allows an unknown profile to be reconstructed. In terms of the interpretation given in Subsection 4.2.3, this means that we only simulate sampling errors in the independent detection of both fields. All external effects are disregarded. The influence reported above is considerably smaller than the one observed in comparable studies that have appeared in the literature (Bolomey et al. (1979), Lesselier (1978,1982a)). Most likely, this is due to the feature in our scheme in which at least five to ten sampled field values correspond to each unknown profile parameter. In the references cited above, there was almost a one-to-one correspondence between the sampled field values and the unknown profile parameters.

The computation times for the Born-type iterative procedure depend on the complexity of the unknown profile and the desired numerical accuracy. As in the frequency-domain application, most of the computation time is taken up in the solution of the approximate direct-scattering problems. For the results displayed in this subsection, the total computation time varied from five seconds to about one minute.

6.2.4 Both $\chi(z)$ and $\sigma(z)$ unknown: implementation

Next, we turn our attention to the case where both constitutive parameters of the slab are unknown. To verify the conclusions of Subsection 5.2.3, we first attempted to solve this complete inversion problem by determining in each step the expansion coefficients $\{\chi_m^{(n)}\}$ and $\{\sigma_m^{(n)}\}$ that minimize the integrated squared error

$$\int_{T/2}^{\tau^{(n-1)}+1+T/2}\{E_+^r(0,t) + \tfrac{1}{2}\sum_{m=0}^{M}\chi_m^{(n)}\int_0^1\phi_m(z)\partial_t E_+^{(n)}(z,t-z)dz + \tfrac{1}{2}\sum_{m=0}^{M}\sigma_m^{(n)}\int_0^1\phi_m(z)E_+^{(n)}(z,t-z)dz\}^2\,dt + \delta\sum_{m=1}^{M-1}\{(\chi_{m-1}^{(n)} - 2\chi_m^{(n)} + \chi_{m+1}^{(n)})^2 + (\sigma_{m-1}^{(n)} - 2\sigma_m^{(n)} + \sigma_{m+1}^{(n)})^2\}. \tag{6.2.15}$$

In (6.2.15), the functions $\{\phi_m(z)\}$ are, as before, the triangular expansion functions defined in (5.3.29), and δ is a small regularization parameter. As predicted in Subsection 5.2.3, this attempt was totally unsuccessful. This confirms the non-uniqueness result obtained in that subsection.

In accordance with that result, we then included reflection data for incidence from the right in the consideration. To this end, we augmented the squared error in (6.2.15) by the integral

$$\int_{T/2}^{\tau^{(n-1)}+1+T/2}\{E_-^r(1,t) + \tfrac{1}{2}\sum_{m=0}^{M}\chi_m^{(n)}\int_0^1\phi_m(z)\partial_t E_-^{(n)}(z,t-1+z)dz + \tfrac{1}{2}\sum_{m=0}^{M}\sigma_m^{(n)}\int_0^1\phi_m(z)E_-^{(n)}(z,t-1+z)dz\}^2 dt. \tag{6.2.16}$$

With this extension the iterative procedure does converge.

As a result of including the integral in (6.2.16) in the error to be minimized we must, in the present scheme, solve two approximate direct-scattering problems in each iteration step. This is effectuated in the same manner as described in Subsection 6.2.2, i.e. by the marching-on-in-time method specified in Subsection 3.3.2. This leaves us with constructing the field matrix corresponding to the terms in braces in (6.2.15) and (6.2.16). For the reflected-field data, the same approach was followed as in deriving the system of equations (6.2.14). The regularization terms were accounted for in the same manner as in Subsection 5.3.4, namely by converting them into properly normalized, extra equations for the coefficients $\{\chi_m^{(n)}\}$ and $\{\sigma_m^{(n)}\}$. Both these equations and the ones pertaining to the sampled known fields in the time interval (6.2.8) were combined into one large, overdetermined matrix equation. From this equation, both sets of expansion coefficients were obtained simultaneously with the aid of the linear-least-squares inversion procedure based on Householder transformations that we applied in Subsections 4.2.2 and 5.3.4 (NAG (1984), subroutine F04AMF).

In order to monitor the progression of the iterative procedure, two types of errors were computed in the course of each iteration step. These errors are best formulated in terms of the notation introduced in Subsection 5.2.3, where $\{\bar{\chi}(z),\bar{\sigma}(z)\}$ denote the constitutive parameters in the current approximate configuration, and $\bar{E}_+(z,t)$ and $\bar{E}_-(z,t)$ the fields in that configuration resulting from the incident pulses specified in (6.2.1) and (6.2.2). The first error that we evaluated was the deviation between the approximate reflected fields in the time interval (6.2.8) and the corresponding actual reflected fields. This error was defined as

$$err(E) \triangleq \left\{ \frac{\int_{T/2}^{\bar{\tau}+1+T/2}[\Delta E_+(t)^2 + \Delta E_-(t)^2]dt}{\int_{T/2}^{\bar{\tau}+1+T/2}[E_+^r(0,t)^2 + E_-^r(1,t)^2]dt} \right\}^{\frac{1}{2}}, \tag{6.2.17}$$

where $\bar{\tau}$ denotes the one-way travel time corresponding to the approximate susceptibility $\bar{\chi}(z)$, and where

$$\begin{aligned} \Delta E_+(t) &\triangleq \bar{E}_+^r(0,t) - E_+^r(0,t), \\ \Delta E_-(t) &\triangleq \bar{E}_-^r(1,t) - E_-^r(1,t). \end{aligned} \tag{6.2.18}$$

The computation of this error primarily serves to estimate the rate of convergence of the iterative procedure, and to assess the necessity of carrying out more iteration steps.

The second error that we considered was the global error in the reconstructed profiles. This error was defined as

$$err(\chi,\sigma) \triangleq \left\{ \frac{\int_0^1[\Delta\chi(z)^2 + \Delta\sigma(z)^2]dz}{\int_0^1[\chi(z)^2 + \sigma(z)^2]dz} \right\}^{\frac{1}{2}}, \tag{6.2.19}$$

with

$$\begin{aligned} \Delta\chi(z) &= \bar{\chi}(z) - \chi(z), \\ \Delta\sigma(z) &= \bar{\sigma}(z) - \sigma(z), \end{aligned} \tag{6.2.20}$$

and expresses the accuracy of the reconstruction in terms of a single quantity.

6.2.5 Both $\chi(z)$ and $\sigma(z)$ unknown: numerical results

As in the case of a single unknown constitutive parameter, we have tested our scheme for several slab configurations. It turns out that, when the unknown profiles can be matched by the piecewise-linear expansions (5.3.28), a fairly good reconstruction can be obtained in 10 to 15 iteration steps. Since we now have twice the amount of reflected-field data, the number of cells in the piecewise-linear expansion can still be chosen according to (6.2.10) and

(6.2.11). This was confirmed in our numerical experiments. As initial estimate, we took a homogeneous, lossless slab with the correct one-way travel-time. This travel time was estimated from the arrival times of the secondary reflected waves at $z = 0$ and $z = 1$, respectively. Generally, the unknown susceptibility was reconstructed more accurately than the unknown conductivity. The only possible explanation for this phenomenon seems to be that, apparently, the reflected fields $E^r_+(0,t)$ and $E^r_-(1,t)$ depend more strongly on the local value of the susceptibility than on that of the conductivity. This is the same effect that showed up in Subsection 6.2.3, where it caused a reduction in the accuracy of part of the reconstructed profile when either the known or the unknown constitutive parameter increased in magnitude.

An an illustration we present, in Figure 6.2.7, results of the reconstruction of a parabolic susceptibility profile and a linear conductivity profile. Figures 6.2.7a and 6.2.7b compare the successive approximations $\chi^{(n)}(z)$ and $\sigma^{(n)}(z)$ to the corresponding actual profiles. In Figures 6.2.7c and 6.2.7d, the root-mean-square errors $err(E)$ and $err(\chi,\sigma)$ defined in (6.2.17)-(6.2.20) have been plotted as a function of the step number n. The dashed line in Figure 6.2.7d indicates the smallest possible value of $err(\chi,\sigma)$ for $\bar{\chi}(z)$ and $\bar{\sigma}(z)$ being piecewise-linear approximations as employed in the iterative scheme. Figure 6.2.7c illustrates the monotonic decrease of the field error during the procedure. From Figure 6.2.7d, however, it is observed that the profile error starts to increase somewhat as soon as the field error reaches its final value. This phenomenon points to the possible presence of a minor ambiguity in the reconstructed profiles that has not been eliminated by the regularization applied. In fact, Figures 6.2.7a and 6.2.7b indicate that the regularization terms in (6.2.15) may cause a small deviation between the reconstructed and the actual susceptibility profile. Because of the relatively strong dependence of the reflected fields on the susceptibility, this deviation may, in turn, be associated with a larger error in the reconstructed conductivity.

Nevertheless, the regularization term cannot be omitted or reduced. This was verified by repeating the computation of Figure 6.2.7

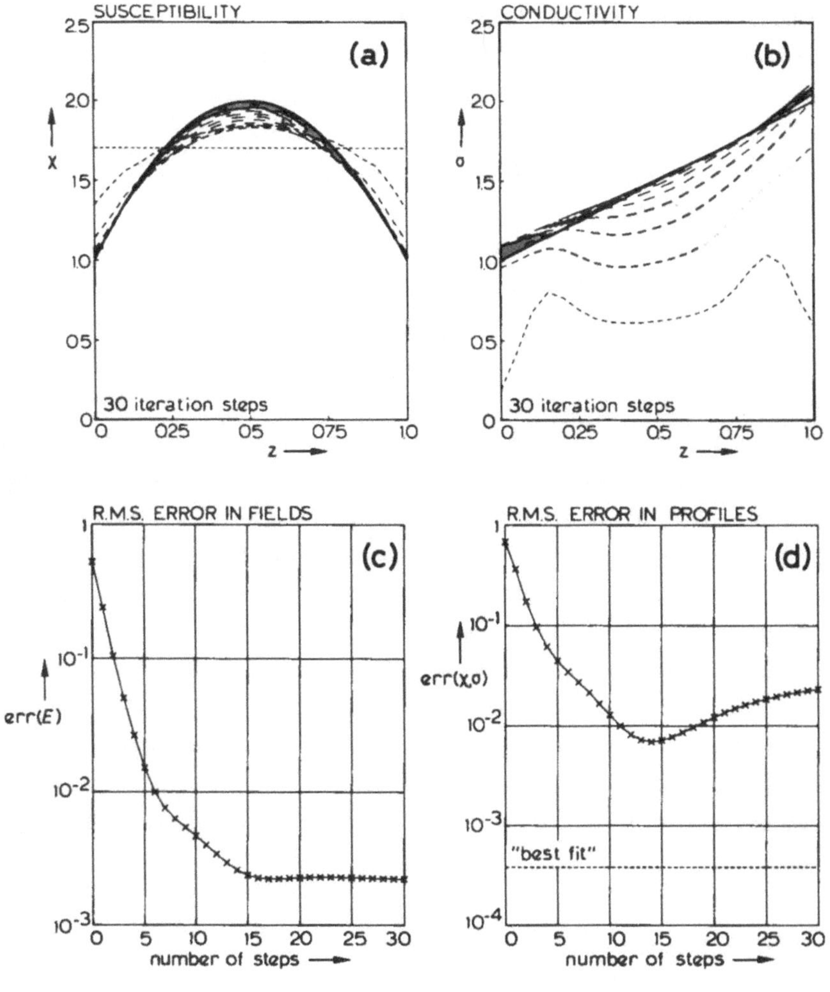

Figure 6.2.7 (a), (b): Results of 30 iteration steps in the simultaneous reconstruction of $\chi(z) = 2 - 4(z - 0.5)^2$ and $\sigma(z) = 1 + z$ as outlined in Subsection 6.2.4 with $M = 20$, $N = 100$, $\delta = 2 \times 10^{-4}$ and $T = 0.5$. The dash length increases with the step number. (c), (d): r.m.s. errors as defined in (6.2.17) and (6.2.19) for the results shown in Figures (a), (b). In Figure (d), the dashed line indicates the smallest possible value.

for a ten times smaller value of the regularization parameter δ. In that circumstance, the iterative procedure did not converge at all. An additional confirmation of our interpretation of the role of the regularization term may be derived from the observation that the minimum value of the profile error found during the iterative procedure is an order of magnitude larger than the smallest possible value. This observation, too, suggests that the regularization terms may serve to relieve a possible excess in flexibility in the piecewise-linear approximation employed.

The ambiguity effect observed in Figure 6.2.7 shows up even stronger when one of the unknown profiles cannot be matched by the piecewise-linear expansion (5.3.28). An example is given in Figure 6.2.8, which contains results of repeating the computation of Figure 6.2.7 for the same susceptibility profile and a discontinuous conductivity profile. Apparently, we now have a range of configurations for which approximately the same field error is observed. Note that Figure 6.2.8d shows an increase similar to the one in Figure 6.2.7d. This seems to justify the conjecture that the best reconstruction will be obtained by terminating the iterative procedure as soon as the global field error no longer changes significantly. Terminating the procedure like that at least prevents the regularization terms from exerting undue influence.

The computation times for our iterative procedure turned out to be quite acceptable. The total computation of the results displayed in Figures 6.2.7 and 6.2.8 took approximately 60 and 40 seconds, respectively.

Appendix 6.2.A Smoothing procedure

As discussed in Subsection 6.2.2, one of the methods for obtaining a subsystem of M + 1 equations from which the expansion coefficients $\{\chi_m^{(n)}\}$ or $\{\sigma_m^{(n)}\}$ can be solved in step number n is the summation method. If this method is applied in the numerical computation of an unknown profile, the resulting solution for $\{\chi_m^{(n)}\}$ and $\{\sigma_m^{(n)}\}$ shows an alternating behavior. The amplification of this behavior in subsequent iteration steps will cause instabilities in the iterative procedure.

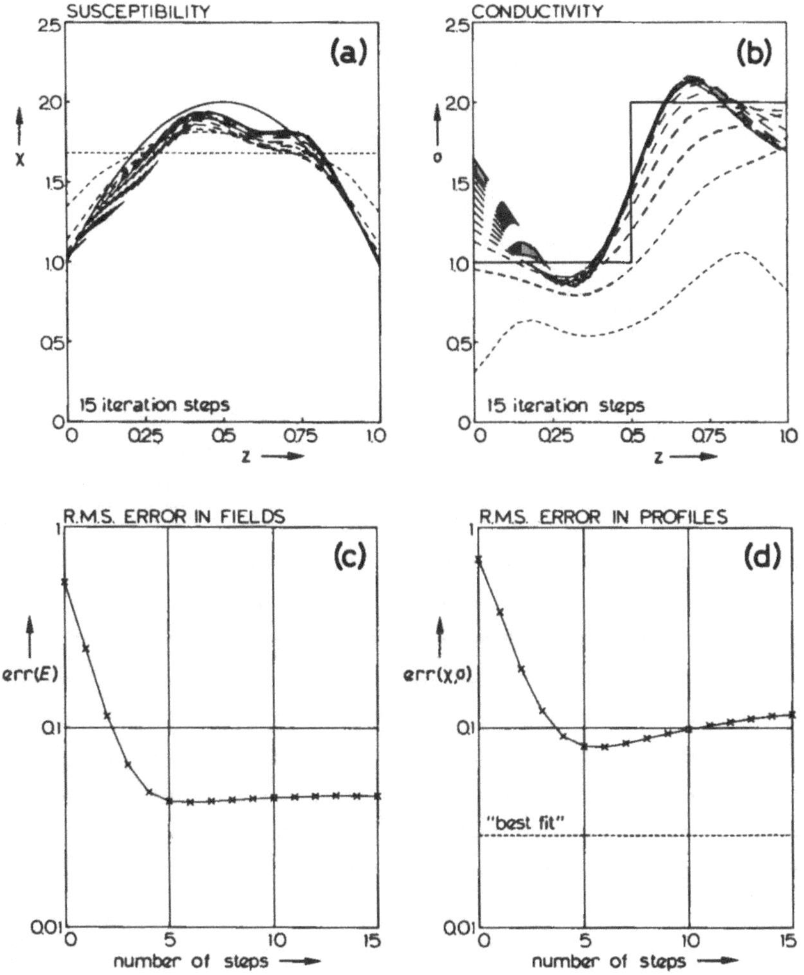

Figure 6.2.8 Results of 15 iteration steps in the simultaneous reconstruction of $\chi(z) = 2 - 4(z - 0.5)^2$ and $\sigma(z) = 1 + U(z - 0.5)$ as outlined in Subsection 6.2.4 with $M = 22$, $N = 110$, $\delta = 10^{-3}$ and $T = 0.5$. Lines as given in Figure 6.2.7.

As argued in Subsections 5.3.4, 5.4.4 and 6.2.2, this alternating behavior can be understood as being a Gibbs-type phenomenon associated with the band limitations in the known reflected-field data and in the method of solution. For the case of the summation method, the alternating behavior can also be understood from the special form of the system matrix. Numerical investigation showed that, for $\chi(z)$ unknown and for M as specified in (6.2.10) or (6.2.11), the resulting system matrix locally assumes the form

$$\begin{bmatrix} 1-a & 1 & a & & & & \bigcirc \\ -1+a & 0 & 1 & & & & \\ -a & -1 & 0 & a & & & \\ & -a & -1 & 1 & a & & \\ & & -a & 0 & 1 & a & \\ \bigcirc & & & -1 & 0 & 1 & a \\ & & & -a & -1 & 0 & 1-a \\ & & & & -a & -1 & -1+a \end{bmatrix} \tag{6.2.21}$$

with a being a small, nonnegative parameter. This special form can be understood by studying the structure of the problem. At the space-time points under consideration (i.e. points covering the first traverse of the pulse across the slab), the electric field behaves locally as a pulse propagating in the positive z direction. As a consequence, from looking at the time derivative of that pulse, we can expect that the system matrix will have a form resembling (6.2.21). Matrices of the type (6.2.21) have an eigenvector $\underline{\nu}$ with a zero eigenvalue, where the components of $\underline{\nu}$ are given by

$$\nu_m = \text{constant} \cdot (-1)^m, \quad m = 0,1,\dots,M. \tag{6.2.22}$$

Hence, it does seem plausible that matrices resembling (6.2.21) will have a similar eigenvector with a small eigenvalue.

A similar argument can be given for the case of unknown conductivity, also leading to an eigenvector with a zero eigenvalue of the form (6.2.22). As a consequence, $\chi_m^{(n)} + \lambda\nu_m$ or $\sigma_m^{(n)} + \lambda\nu_m$, where λ is some constant, is also an approximate solution of the square-matrix equation constructed with the summation method in iteration step n. This implies that the solutions $\{\chi_m^{(n)}\}$ or $\{\sigma_m^{(n)}\}$ found by using this method may indeed have relatively large errors, which locally resemble the components $\{\upsilon_m\}$.

These errors can be suppressed by a smoothing procedure where the value of the coefficients $\{\chi_m^{(n)}\}$ or $\{\sigma_m^{(n)}\}$ is replaced by the average of their local value and a value found from the values of $\chi_m^{(n)}$ or $\sigma_m^{(n)}$ at the two nearest points where the errors have opposite signs. We have

$$\begin{aligned}
\bar{\chi}_m^{(n)} &= (\chi_{m-1}^{(n)} + 2\chi_m^{(n)} + \chi_{m+1}^{(n)})/4 \qquad \text{for } 0 < m < M, \\
\bar{\chi}_0^{(n)} &= (2\chi_0^{(n)} + 3\chi_1^{(n)} - \chi_3^{(n)})/4, \\
\bar{\chi}_M^{(n)} &= (2\chi_M^{(n)} + 3\chi_{M-1}^{(n)} - \chi_{M-3}^{(n)})/4,
\end{aligned} \tag{6.2.23}$$

and a similar definition for $\{\sigma_m^{(n)}\}$. For inner points $(0 < m < M)$, this value is found via linear interpolation and for end points $(m = 0,M)$ via linear extrapolation. Thus, at least the linear variation of the estimate obtained is unaffected. Finally, it should be noted that, for inner points, the smoothing operation specified above is identical to the one specified in (3.2.30).

6.3 Optimization approach

6.3.1 Formulation of the problem

Although the Born-type iterative procedure with a vacuum as the background medium apparently produces an acceptable reconstruction of one or both of the constitutive parameters $\chi(z)$ and $\sigma(z)$ of a slab in vacuo, the results obtained in Subsections 5.4.4, 6.2.3 and 6.2.5 indicate that there may be some room for improvement. Particularly the ambiguity effects observed in the time-domain results displayed in Figures 6.2.3, 6.2.4, 6.2.7 and 6.2.8 confirm this suspicion. Moreover, the analysis of Section 5.2 already suggested that using only the reflected field in the time interval (6.2.8) may cause the reconstruction of the unknown $\chi(z)$ and/or $\sigma(z)$ to be incomplete. In fact, in Subsection 5.2.5 an improved Born-type method of solution was proposed. Like the procedure applied up to now, this improved method as well uses only a prescribed part of the available reflected-field information.

Before tackling the implementation of this second variant of the Born-type iterative procedure, which will be postponed until Section 6.4, we investigate, in the present subsection, the possibility of using *all* the reflected-field information available. To be more specific, we consider the possible reconstruction of one unknown constitutive parameter from the known reflected field $E_+^r(0,t)$ resulting from the incident pulse specified in (6.2.1a) and (6.2.2) for $0 < t < \infty$. Obviously, the remaining constitutive parameter is assumed to be known. Throughout this section we will use the notation introduced in Subsection 5.2.3, i.e. $\{\bar{\chi}(z),\bar{\sigma}(z)\}$ denote a given approximate configuration with, in this case, either $\bar{\chi}(z) = \chi(z)$ or $\bar{\sigma}(z) = \sigma(z)$, and $\bar{E}_+(z,t)$ denotes the field in that configuration resulting from the incident pulse.

In line with the approach followed until now, we attempt to minimize one of the functionals

$$ERR[\bar{\chi}(z)] \stackrel{\Delta}{=} \int_0^\infty [\bar{E}_+^r(0,t) - E_+^r(0,t)]^2\, dt + \delta\sum_{m=1}^{M-1}(\chi_{m-1} - 2\chi_m + \chi_{m+1})^2, \tag{6.3.1a}$$

$$ERR[\bar{\sigma}(z)] = \int_0^\infty [\bar{E}_+^r(0,t) - E_+^r(0,t)]^2\,dt + \delta\sum_{m=1}^{M-1}(\sigma_{m-1} - 2\sigma_m + \sigma_{m+1})^2, \tag{6.3.1b}$$

according to whether $\chi(z)$ or $\sigma(z)$ is to be determined. In (6.3.1), $E_+^r(0,t)$ denotes the actual reflected field, δ is a small, nonnegative regularization or tuning parameter, and $\bar{\chi}(z)$ and $\bar{\sigma}(z)$ have been restricted to the class of piecewise-linear functions

$$\bar{\chi}(z) = \sum_{m=0}^{M}\chi_m\phi_m(z), \tag{6.3.2a}$$

$$\bar{\sigma}(z) = \sum_{m=0}^{M}\sigma_m\phi_m(z), \tag{6.3.2b}$$

with $\phi_m(z)$ being a triangular expansion function as defined in (5.3.29).

As remarked above, the results obtained in Subsections 5.2.3 and 5.2.5 indicate that the values of the expansion coefficients $\{\chi_m\}$ or $\{\sigma_m\}$ for which the cost functions defined in (6.3.1) are minimal cannot be found by applying a Born-type iterative technique. In fact, in the numerical experimentation leading to the time-domain version proposed in Tijhuis (1981) and reformulated in Subsections 6.2.2 and 6.2.3, it was attempted to extend the time interval (6.2.8) towards infinity. This modification had the effect of an almost complete loss of convergence for any configuration with a sizable susceptibility and/or conductivity contrast. The most likely explanation of this effect is that, by extending the time interval under consideration, we have enhanced the non-linearity of the cost function. This leaves us no option but to apply one of the conventional optimization routines for minimizing or maximizing a function. In particular, we focused our attention on the standard routines available in the NAG Fortran Library (1984).

An approach related to the one proposed above has previously been followed by Lesselier (1982a,b). There is, however, one fundamental difference. In the above formulation, we have purposely chosen the expansion functions in conformity with the maximum amount of information that could be available from the known reflected field. In

Lesselier's approach, each sampled value of the unknown profile occurring in the discretized integral equation for the underlying direct-scattering problem was treated as an unknown parameter. This must lead to the type of unstable behavior that was signaled at the end of Subsection 6.2.3. Consequently, we may expect our variant to perform much better.

It should be stressed that the choice of determining only one unknown profile is not essential. The only motivation for our choice was that conventional optimization routines may require a rather large number of evaluations of the cost function. As discussed in Subsection 6.2.4, determining both $\chi(z)$ and $\sigma(z)$ from two-sided time-domain reflection data would double the number of unknown expansion coefficients as well as the numerical effort required to evaluate the cost function. Moreover, the results displayed in Figures 6.2.3 and 6.2.4 show that we should already be able to assess the reliability of the Born-type procedure applied up to now by reconstructing a single unknown profile.

Finally, we wish to remark that it suffices to carry out the present analysis in the time domain only. From Parseval's theorem, we have

$$\int_0^{\infty}[\bar{E}_+^r(0,t) - E_+^r(0,t)]^2\, dt = \frac{1}{2\pi}\int_{-\infty}^{\infty}|F(i\omega)|^2|\bar{r}^+(i\omega) - r^+(i\omega)|^2\, d\omega, \tag{6.3.3}$$

where $F(s)$, $r^+(s)$ and $\bar{r}^+(s)$ are the same frequency-domain quantities that were used in Section 2.4 and Chapter 5. By associating $|F(i\omega)|$ with the weighting function $w(\omega)$, we can therefore identify either of the functionals defined in (6.3.1) as being essentially identical to a square error of the type (5.4.22). Hence, there is no need to carry out a separate minimization for such a frequency-domain error.

The remainder of this section is organized as follows. We will start off by listing the types of optimization methods available along with some of their general properties. This will be done in Subsection 6.3.2. In Subsection 6.3.3, we will then look into the principal problem in these approaches, namely the evaluation of the partial derivatives of the cost function with respect to the unknown expansion coefficients. For the so-called gradient vector containing these derivatives, a closed-form expression will be derived. Finally, in Subsection 6.3.4, some representative results will be presented and discussed.

6.3.2 Optimization methods

For the numerical solution of the minimization problem formulated in Subsection 6.3.1, several methods are available. It lies outside the scope of the present manuscript to discuss these methods and their specific properties exhaustively. For such a discussion, the reader is, for instance, referred to the surveys authored by Brodlie (1977), and by Gill and Murray (1977). The aim of this subsection is merely to enumerate some common optimization methods and to summarize those of their features that we should be aware of in order to understand the numerical results that will be given in Subsection 6.3.4.

Let us first introduce some terminology. As remarked in Subsection 6.3.1, we wish to minimize a function of M + 1 variables. In the present subsection, we will consider these variables as being components $\{\xi_m\}$ of a vector $\underline{\xi}$. These components represent either the unknown expansion coefficients $\{\chi_m\}$ or their counterparts $\{\sigma_m\}$. In the first place, we distinguish between *constrained* and *unconstrained* minimization. When the unknown variables are not limited in any way, we are dealing with "unconstrained" minimization. When the variables are subject to restrictions, we have a "constrained" minimization problem. In the problem formulated above, the variables $\{\xi_m\}$ are subject to the bound constraints $\xi_m \geq 0$ for all m. When one or more of the $\{\xi_m\}$ are considerably smaller than zero, we must expect computational difficulties in evaluating the relevant approximate field $\bar{E}(z,t)$ (see also Subsection 3.3.4). Hence, we should preferably apply that version of an optimization method which allows for bound constraints. When only an unconstrained version is available, we should carefully select our initial estimate such that negative values of the expansion coefficients are avoided during the minimization process.

Since methods for constrained and unconstrained problems are closely related, we restrict the present discussion to the latter. In line with the notation used in Subsection 6.3.1, we will denote the *cost function* by $ERR(\underline{\xi})$. In addition, we define the *gradient vector* of that cost function as the vector whose components are

$$g_m(\underline{\xi}) \triangleq \frac{\partial}{\partial \xi_m} ERR(\underline{\xi}). \tag{6.3.4}$$

In the discussion, we will further encounter the *Hessian matrix* $h(\underline{\xi})$ associated with our cost function, which consists of the second derivatives

$$h_{mm'}(\underline{\xi}) \triangleq \frac{\partial^2}{\partial \xi_m \partial \xi_{m'}} ERR(\underline{\xi}). \tag{6.3.5}$$

It should be pointed out that, for the problem at hand, a closed-form expression for the gradient vector $\underline{g}(\underline{\xi})$ is available. With the aid of this expression, which will be discussed in more detail in Subsection 6.3.3, $\underline{g}(\underline{\xi})$ can be evaluated numerically up to the same accuracy as the cost function $ERR(\underline{\xi})$. For the Hessian matrix, such an expression has not been derived.

The optimization methods under consideration are all iterative, yielding successive approximations $\underline{\xi}^{(n)}$ to the solution of the problem. Each iteration step amounts to analyzing the behavior of $ERR(\underline{\xi})$ along a *line*

$$\underline{\xi} = \underline{\xi}^{(n-1)} + \eta \underline{p}^{(n)}, \tag{6.3.6}$$

where $\underline{\xi}^{(n-1)}$ represents the current estimate, $\underline{p}^{(n)}$ a direction vector, and η a real-valued parameter. The new approximation $\underline{\xi}^{(n)}$ is obtained by searching for an appropriate value of η, i.e.

$$\underline{\xi}^{(n)} = \underline{\xi}^{(n-1)} + \eta^{(n)} \underline{p}^{(n)}. \tag{6.3.7}$$

The various methods differ in the choice of $\eta^{(n)}$ and $\underline{p}^{(n)}$ and depend on the availability of first and second derivatives. In some methods, $\eta^{(n)}$ is chosen such that $ERR(\underline{\xi}^{(n)})$ is minimal (*exact line search*). In others, $\eta^{(n)}$ is determined such that $ERR(\underline{\xi}^{(n)})$ is reduced by a sufficient amount (*approximate line search*).

Now that we have established the terminology, we will review some frequently-used minimization methods. In *Newton-type* methods, the Hessian matrix $h(\underline{\xi}^{(n-1)})$ (or some finite-difference approximation) is used to define the search direction $\underline{p}^{(n)}$. The choice of $\underline{p}^{(n)}$ is based on the approximation of $ERR(\underline{\xi})$ around $\underline{\xi} = \underline{\xi}^{(n-1)}$ by the first three terms of its Taylor expansion. In the literature, various Newton-type methods have been described in which both the exact and the approximate line search occur. In *quasi-Newton* methods, an approximation of the Hessian matrix $h(\underline{\xi})$ (or its inverse, as in early methods) is gradually built up as the algorithm proceeds. The approximation of $h(\underline{\xi})$ is based upon the assumption that, near the minimum, $ERR(\underline{\xi})$ is quadratic, i.e. that the contours of constant $ERR(\underline{\xi})$ are described well by hyperellipsoids. In recent years, much research has been done on the updating of the Hessian, the choice of the search direction, and the line search itself. In modern quasi-Newton methods, an approximate line search is carried out. In *conjugate-gradient methods*, the Hessian matrix is not used. In these methods, the initial search direction is given by the gradient:

$$\underline{p}^{(1)} = -\underline{g}(\underline{\xi}^{(0)}), \tag{6.3.8}$$

with $\underline{\xi}^{(0)}$ being an initial estimate. In the next M iteration steps, the search direction is given by

$$\underline{p}^{(n)} = -\underline{g}(\xi^{(n-1)}) + \beta^{(n)}\underline{p}^{(n-1)}, \tag{6.3.9}$$

where n = 2,3,...,M + 1, and

$$\beta^{(n)} = g^2(\underline{\xi}^{(n-1)})/g^2(\underline{\xi}^{(n-2)}). \tag{6.3.10}$$

In each step, an exact line search is carried out. The choice of $\beta^{(n)}$ ensures that, for a quadratic cost function, the successive directions $\underline{p}^{(n)}$ are conjugate. For such a cost function, termination is thus achieved in at most M + 1 iterations. For general, nonlinear functions, the iterations are applied in cycles, with $\underline{p}^{(n)}$ being re-set to $-\underline{g}(\underline{\xi}^{(n-1)})$ after every M + 1 iterations.

As remarked in Subsection 6.3.1, we restricted ourselves, in the numerical experimentation, to applying ready-to-hand standard routines available from the NAG Fortran Library. This subroutine library supplies a wide range of procedures for minimizing or maximizing a function, using both quasi-Newton and conjugate-gradient methods. The quasi-Newton routines are based on the work of Gill and Murray (1974,1977), and the conjugate-gradient routines on that of Fletcher and Reeves (1964). For the solution of the problem at hand, four of these routines would appear to be eligible. Their properties are summarized in Table 6.3.1. Only the typification in the last column of that table may require further clarification. In the NAG Library, two types of routines are distinguished. In the first place, one has *easy-to-use* routines; these include in the calling sequence only those parameters that are absolutely essential in the definition of the problem, as opposed to parameters relevant to the method of solution. In the second place, one can invoke *comprehensive* routines, which have additional parameters that allow the user to improve the computational efficiency by tuning the method to his or her particular problem. In the easy-to-use routines, these parameters are generated by the subroutine itself.

Table 6.3.1 Subroutines available from the NAG Fortran Library applicable to the problem formulated in Subsection 6.3.1, and their main properties.

Name of routine	Solution method	Constraints included?	Uses gradient?	Type of routine
E04JAF	quasi Newton	yes	no	easy to use
E04KAF	quasi Newton	yes	yes	easy to use
E04KBF	quasi Newton	optional	yes	comprehensive
E04DBF	conjugate gradient	no	yes	comprehensive

6.3.3 Evaluation of the gradient

In this subsection, we consider the evaluation of the gradients of the cost functions $ERR[\bar{\chi}(z)]$ and $ERR[\bar{\sigma}(z)]$ defined in (6.3.1) with respect to the relevant expansion coefficients $\{\chi_m\}$ and $\{\sigma_m\}$. From (6.3.1), it is observed that there is no problem in differentiating the regularization terms. In the present subsection, we will therefore restrict ourselves to the situation where these terms are absent, i.e. $\delta = 0$. In principle, it seems possible to evaluate the gradient numerically by applying a suitable finite-difference formula. However, the evaluation of a single gradient would then require the solution of at least M + 2 approximate direct-scattering problems. Moreover, these problems would have to be resolved very accurately in order to end up with an acceptable accuracy in the gradient thus obtained. Hence, such an evaluation should be avoided. In this subsection, we will describe one way to do this.

From (6.3.1) and (6.3.4), it follows that the components of the relevant gradient vectors are given by

$$\frac{\partial}{\partial\chi_m} ERR[\bar{\chi}(z)] = 2\int_0^\infty[\bar{E}_+^r(0,t) - E_+^r(0,t)]\frac{\partial}{\partial\chi_m}\bar{E}_+^r(0,t)dt, \tag{6.3.11a}$$

$$\frac{\partial}{\partial\sigma_m} ERR[\bar{\sigma}(z)] = 2\int_0^\infty[\bar{E}_+^r(0,t) - E_+^r(0,t)]\frac{\partial}{\partial\sigma_m}\bar{E}_+^r(0,t)dt, \tag{6.3.11b}$$

respectively. This moves the problem to the determination of the derivatives of $\bar{E}_+^r(0,t)$ with respect to χ_m and σ_m. Formally, these derivatives can be obtained from the direct-scattering equation (6.2.3). Differentiating that equation results in

$$\begin{aligned}\frac{\partial}{\partial\chi_m}\bar{E}_+(z,t) = &-\tfrac{1}{2}\int_0^1\phi_m(z')\partial_t\bar{E}_+(z',t')dz' \\ &-\tfrac{1}{2}\int_0^1[\sigma(z') + \bar{\chi}(z')\partial_t]\frac{\partial}{\partial\chi_m}\bar{E}_+(z',t')dz',\end{aligned} \tag{6.3.12a}$$

$$\begin{aligned}\frac{\partial}{\partial\sigma_m}\bar{E}_+(z,t) = &-\tfrac{1}{2}\int_0^1\phi_m(z')\bar{E}_+(z',t')dz' \\ &-\tfrac{1}{2}\int_0^1[\bar{\sigma}(z') + \chi(z')\partial_t]\frac{\partial}{\partial\sigma_m}\bar{E}_+(z',t')dz',\end{aligned} \tag{6.3.12b}$$

with $t' \triangleq t - |z - z'|$. The equations listed in (6.3.12) have the same form as the direct-scattering equation (6.2.3), with the known first integral on their right-hand sides playing the role of the incident field. Hence, we could resolve them numerically by applying the marching-on-in-time method as described in Subsection 3.3.2. Even though this procedure would resolve the accuracy problem, because no significant accuracy is lost in the evaluation of the time integrals on the right-hand sides of (6.3.11), it would still leave us with the solution of M + 1 integral equations of the type (6.3.12). For larger values of M, in particular, this would still give rise to unacceptably long computation times.

This problem can be circumvented by a procedure described by Lesselier (1982a,b). The basic idea is to use the integral equations (6.3.12) to express the right-hand sides of (6.3.11) in terms of the known field $\bar{E}_+(z,t)$ rather than in its unknown derivatives with respect to χ_m and σ_m. To this end, we organize (6.3.12) into

$$\begin{aligned}\frac{\partial}{\partial\chi_m}\bar{E}_+(z,t) + \tfrac{1}{2}\int_0^1[\sigma(z') + \bar{\chi}(z')\partial_t]\frac{\partial}{\partial\chi_m}\bar{E}_+(z',t')dz' &= \\ -\tfrac{1}{2}\int_0^1\phi_m(z')\partial_t\bar{E}_+(z',t')dz',&\end{aligned} \tag{6.3.13a}$$

$$\begin{aligned}\frac{\partial}{\partial\sigma_m}\bar{E}_+(z,t) + \tfrac{1}{2}\int_0^1[\bar{\sigma}(z') + \chi(z')\partial_t]\frac{\partial}{\partial\sigma_m}\bar{E}_+(z',t')dz' &= \\ -\tfrac{1}{2}\int_0^1\phi_m(z')\bar{E}_+(z',t')dz',&\end{aligned} \tag{6.3.13b}$$

where again $t' \triangleq t - |z - z'|$. In either equation, we have related a linear functional of one of the unknown derivatives of $\bar{E}_+(z,t)$ to an integral involving only the known field $\bar{E}_+(z,t)$. This means that it suffices to express the right-hand sides of (6.3.11) in terms of the functionals occurring on the left-hand sides of (6.3.13). Moreover, the similarity in structure between the relevant integrals indicates that there may be a systematic way to obtain the desired expressions.

Now *suppose* that for each combination of two profiles $\{\bar{\chi}(z),\bar{\sigma}(z)\}$, we can find an *adjoint state* $A(z,t)$ of $\bar{E}_+(z,t)$ such that

$$2\int_0^\infty[\bar{E}_+^r(0,t) - E_+^r(0,t)]\psi(0,t)dt =$$
$$\int_0^\infty\int_0^1 A(z,t)\{\psi(z,t) + \tfrac{1}{2}\int_0^1[\bar{\sigma}(z') + \bar{\chi}(z')\partial_t]\psi(z',t')dz'\}dzdt, \tag{6.3.14}$$

for all properly behaved functions $\psi(z,t)$. Then, our problem is solved since this relation also holds for either of the unknown derivatives. Combining (6.3.11), (6.3.13) and (6.3.14) directly yields

$$\frac{\partial}{\partial\chi_m} ERR[\bar{\chi}(z)] = \int_0^1\phi_m(z)g_\chi(z)dz, \tag{6.3.15a}$$

$$\frac{\partial}{\partial\sigma_m} ERR[\bar{\sigma}(z)] = \int_0^1\phi_m(z)g_\sigma(z)dz, \tag{6.3.15b}$$

where the profile-gradients $g_\chi(z)$ and $g_\sigma(z)$ are given by

$$g_\chi(z) \triangleq -\tfrac{1}{2}\int_0^\infty\partial_t\bar{E}_+(z,t)\int_0^1 A(z',t + |z - z'|)dz'dt, \tag{6.3.16a}$$

$$g_\sigma(z) \triangleq -\tfrac{1}{2}\int_0^\infty\bar{E}_+(z,t)\int_0^1 A(z',t + |z - z'|)dz'dt. \tag{6.3.16b}$$

This result then provides a way to determine *all* components of a gradient vector simultaneously. Depending on the problem at hand, either of the profiles $\bar{\chi}(z)$ and $\bar{\sigma}(z)$ occurring in (6.3.14) should, obviously, be taken identical to its known value.

Now, let us consider whether the condition (6.3.14) can be satisfied at all. Integrating by parts in the time integral and rearranging the order of the integrations reduces this condition to the equivalent form

$$\int_{-\infty}^\infty\int_0^1\psi(z,t)\{A(z,t) - 2[\bar{E}_+^r(0,t) - E_+^r(0,t)]\delta(z)$$
$$+ \tfrac{1}{2}[\bar{\sigma}(z) - \bar{\chi}(z)\partial_t]\int_0^1 A(z',t'')dz'\}dzdt = 0, \tag{6.3.17}$$

where $t'' \triangleq t + |z - z'|$ and where the support of $\delta(z)$ is understood to lie entirely inside the integration interval. From (6.3.17), it is observed that this condition and, equivalently, (6.3.14) indeed hold for all $\psi(z,t)$ if and only if $A(z,t)$ satisfies the *adjoint integral equation*

$$A(z,t) = 2[\bar{E}_+^r(0,t) - E_+^r(0,t)]\delta(z) - \tfrac{1}{2}[\bar{\sigma}(z) - \bar{\chi}(z)\partial_t]\int_0^1 A(z',t + |z - z'|)dz'. \tag{6.3.18}$$

This leaves us with just one problem, namely that, in its present form, this adjoint integral equation is unfit for numerical implementation because of the occurrence of the delta function $\delta(z)$. This problem is readily resolved by substituting

$$A(z,t) = 2[\bar{E}_+^r(0,t) - E_+^r(0,t)]\delta(z) + A'(z,t), \tag{6.3.19}$$

which simplifies (6.3.18) to

$$A'(z,t) = [\bar{\chi}(z)\partial_t - \bar{\sigma}(z)][\bar{E}_+^r(0,t + z) - E_+^r(0,t + z)] + \tfrac{1}{2}[\bar{\chi}(z)\partial_t - \bar{\sigma}(z)]\int_0^1 A'(z',t + |z - z'|)dz'. \tag{6.3.20}$$

This last equation can be solved numerically by a marching-back-in-time method based on a space-time discretization analogous to the one employed in Section 3.3.

Finally, we should still express the profile gradients defined in (6.3.16) in terms of the reduced adjoint state $A'(z,t)$. We end up with

$$g_\chi(z) = -\int_0^\infty \partial_t \bar{E}_+(z,t)[\bar{E}_+^r(0,t + z) - E_+^r(0,t + z)]dt - \tfrac{1}{2}\int_0^\infty \partial_t \bar{E}_+(z,t)\int_0^1 A'(z',t + |z - z'|)dz'dt, \tag{6.3.21a}$$

$$g_\sigma(z) = -\int_0^\infty \bar{E}_+(z,t)[\bar{E}_+^r(0,t + z) - E_+^r(0,t + z)]dt - \tfrac{1}{2}\int_0^\infty \bar{E}_+(z,t)\int_0^1 A'(z',t + |z - z'|)dz'dt. \tag{6.3.21b}$$

With these formulas and (6.3.15), the required gradient can be computed directly from the numerically obtained $\bar{E}(z,t)$ and $A'(z,t)$ by using trapezoidal rules. Thus, the numerical effort required to determine a single gradient has been reduced to the equivalent of solving one extra direct-scattering problem.

6.3.4 Numerical results

With the optimization method outlined in Subsections 6.3.1-6.3.3, a wide range of numerical experiments was carried out. First, we investigated which of the subroutines listed in Table 6.3.1 would be most suitable for solving the problem at hand. To this end, we applied these routines to the reconstruction of some simple susceptibility and conductivity profiles, which could be recovered by carrying out at most five iteration steps of the Born-type procedure discussed in Section 6.2. As in Subsections 6.2.4 and 6.2.5, we judged the quality of the results by computing the root-mean-square errors in the reflected field and in the reconstructed profile. In the present problem, these are defined as

$$err(E) \triangleq \left\{ \frac{\int_0^\infty [\bar{E}_+^r(0,t) - E_+^r(0,t)]^2 dt}{\int_0^\infty E_+^r(0,t)^2 dt} \right\}^{\frac{1}{2}}, \tag{6.3.22}$$

and

$$err(\chi) \triangleq \left\{ \frac{\int_0^1 [\bar{\chi}(z) - \chi(z)]^2 dz}{\int_0^1 \chi(z)^2 dz} \right\}^{\frac{1}{2}}, \tag{6.3.23a}$$

$$err(\sigma) \triangleq \left\{ \frac{\int_0^1 [\bar{\sigma}(z) - \sigma(z)]^2 dz}{\int_0^1 \sigma(z)^2 dz} \right\}^{\frac{1}{2}}, \tag{6.3.23b}$$

respectively.

In Table 6.3.2 and Figure 6.3.1, we consider the representative example of reconstructing the parabolic conductivity profile $\sigma(z) = 1 - 4(z - \frac{1}{2})^2$ in the presence of the known, constant susceptibility profile $\chi(z) = 1$. All computations are carried out for an incident-pulse duration $T = 1$, an initial estimate $\sigma^{(0)}(z) = 0.5$, and for ten subintervals in the piecewise-linear expansion. Note that this choice $M = 10$ is near the limit specified in (6.2.10). Computational data on those computations that were terminated successfully are summarized in Table 6.3.2. The most striking point in this table is that it contains no results of applying quasi-Newton routines to

Table 6.3.2 Computational data of successful reconstructions of the conductivity profile $\sigma(z) = 1 - 4(z - \frac{1}{2})^2$ for $\chi(z) = 1$, $T = 1$, $M = 10$, $N = 50$, with the aid of the optimization routines listed in Table 6.3.1. The parameter δ is the regularization parameter and the number of calls refers to the subroutine that evaluates the cost function and its gradient.

Routine	δ	$err(E)$	$err(\sigma)$	# of calls	CPU time
E04KAF	10^{-4}	5.7×10^{-3}	3.0×10^{-2}	379	157s
E04KBF	10^{-4}	5.8×10^{-3}	3.1×10^{-2}	189	78s
E04DBF	0	9.2×10^{-5}	5.1×10^{-3}	79	33s
E04DBF	10^{-4}	5.7×10^{-3}	3.0×10^{-2}	33	14s

minimizing the cost function for $\delta = 0$, i.e. in the absence of a regularization term. The reason is that for $\delta = 0$ all the quasi-Newton routines failed to converge. This failure can be understood from Figure 6.3.1, which shows results obtained upon the unsuccessful termination of the computations with E04KAF and E04KBF for $\delta = 0$. Apparently, we still have a small, Gibbs-type ambiguity in the direction of the "problematic" vector $v_m = (-1)^m$. This ambiguity can be explained from the discretization errors in the evaluation of the cost function and the gradient; for $N = 50$, the relative power-balance error as considered in Table 3.3.1 is about 6×10^{-3}, which

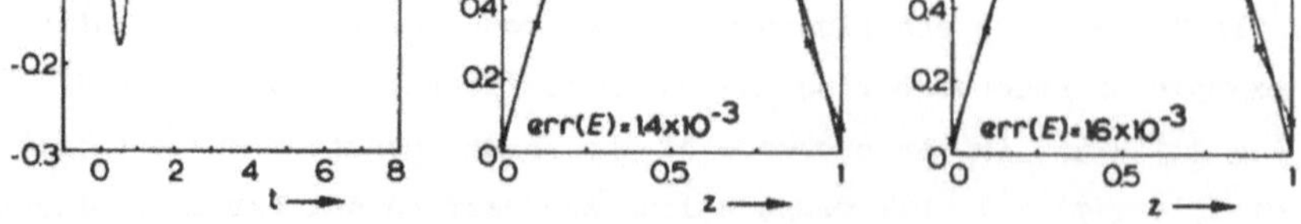

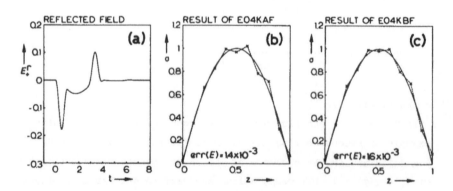

Figure 6.3.1 Results of reconstructing the conductivity profile $\sigma(z) = 1 - 4(z - 0.5)^2$ for $\chi(z) = 1$, $T = 1$, $M = 10$, $N = 50$ and $\delta = 0$ and initial estimate $\sigma^{(0)}(z) = 0.5$ by employing the quasi-Newton routines E04KAF and E04KBF. (a) Actual reflected field (solid line) and field reflected by the approximate configuration reconstructed by E04KAF (dashed line, indistinguishable). (b),(c) Exact profile (solid lines) compared with reconstructions as indicated (dashed lines).

exceeds the field error remaining in the results shown in Figure 6.3.1b,c. The difference in behavior between quasi-Newton and conjugate-gradient methods stems from the fact that in the former, the inaccuracies are amplified in the estimation of the Hessian matrix. As a result, the search direction may be incorrect. For $\delta = 10^{-4}$, the regularization term is apparently precisely large enough to compensate the discretization errors when the gradient is computed explicitly. However, it still does not suffice for an acceptable estimation of the Hessian matrix from values of the cost function only, which explains the lack of success in applying E04JAF. This observation was verified by reducing the number of subintervals in the slab to M = 5, which did lead to a successful application of that subroutine.

In the conjugate-gradient method, carrying out a single line search takes at least two calls to the subroutines evaluating the cost function and its gradient. Hence, it follows from the last line of Table 6.3.2 that the optimization for $\delta = 10^{-4}$ took at most 16 iteration steps. Note that this is only half as much again as the M + 1 steps that would be required to minimize a quadratic cost function. This confirms the conclusion reached in Section 5.2, that the reflected field is an almost linear functional of the corresponding profiles $\{\bar{\chi}(z),\bar{\sigma}(z)\}$. Moreover, this suggests that, for the problem under consideration, we can hardly do better than to minimize the regularized cost function by the conjugate-gradient method. The regularization parameter δ should be chosen such that the root-mean-square error $err(E)$ for the field in the reconstructed configuration just exceeds the errors expected in the determination of the relevant known and approximate fields. One way to achieve this is to compare $err(E)$ with the relative error in the power balance observed for those fields.

In further numerical experimentation, we restricted ourselves to applying the conjugate-gradient routine E04DBF. In particular, we attempted to reconstruct some more complicated profiles of the type considered in Figures 6.2.3-6.2.6. The final results of the optimization are shown in Figures 6.3.2-6.3.4. For each profile, we compare the actual field $E_+^r(0,t)$ and the actual susceptibility profile

$\chi(z)$ with their counterparts in the reconstructed configuration. Some additional computational data have been summarized in Table 6.3.3. In Figure 6.3.2, results are given for the same sine-squared susceptibility problem as in Figure 6.2.3. Clearly, the optimization approach leads to a better reconstruction, even for a considerably longer incident pulse duration T. As in Subsection 6.2.3, those expansion coefficients $\{\chi_m\}$ that came out negative were set to zero before Figure 6.3.2 was plotted. For the uncorrected values, we had $err(E) = 1.3 \times 10^{-2}$ and $err(\chi) = 3.0 \times 10^{-2}$, respectively, i.e. a slightly smaller field error at the cost of a slightly larger profile error. Figure 6.3.2a confirms once more that, for this range of susceptibility contrasts and for $\sigma = 0$, the transient reflected field assumes approximately the form of $d_z\chi(z)$. Finally, we observe that the field error $err(E)$ indeed exceeds the power-balance error found from Table 6.3.3. Hence, no difficulties need be expected due to the discretization errors in the evaluation of the cost function and its gradient.

In Figure 6.3.3, we reconsider the parabolic profile C as specified in Figure 6.2.4 and further reconstructed in Figure 6.2.5. The results indicate that, for the cost function defined in (6.3.1a), we no longer have the ambiguity effect observed in Figure 6.2.4. This confirms our suspicion that using only the reflected field in the time interval (6.2.8) may prevent a complete determination of the unknown profile. In Figure 6.3.4, lastly, we consider two discontinuous susceptibility profiles. It is observed that, even though no discontinuity correction was applied and the number of subintervals was reduced, an acceptable reconstruction has been obtained. In fact, the deviations between the actual and the reconstructed sus-

Table 6.3.3 Additional computational data regarding the numerical experiments leading to Figures 6.3.2 - 6.3.4.

Figure	Regularization parameter δ	Power-balance error	Number of function calls	Computation time
6.3.2	5×10^{-4}	1.6×10^{-3}	24	29s
6.3.3	5×10^{-4}	2.7×10^{-3}	43	69s
6.3.4a,b	10^{-3}	1.2×10^{-3}	28	27s
6.3.4c,d	10^{-3}	1.3×10^{-3}	32	30s

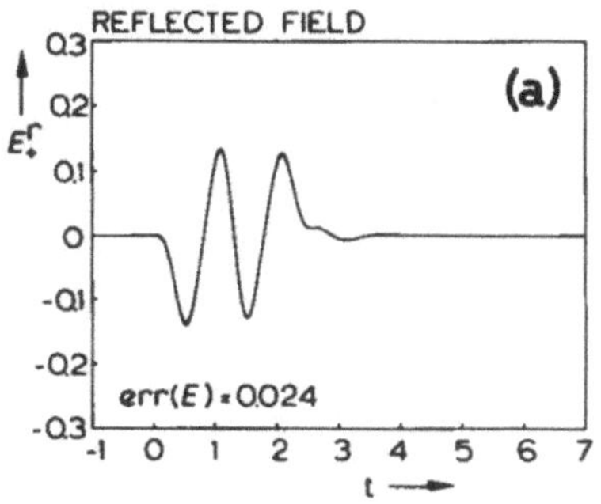

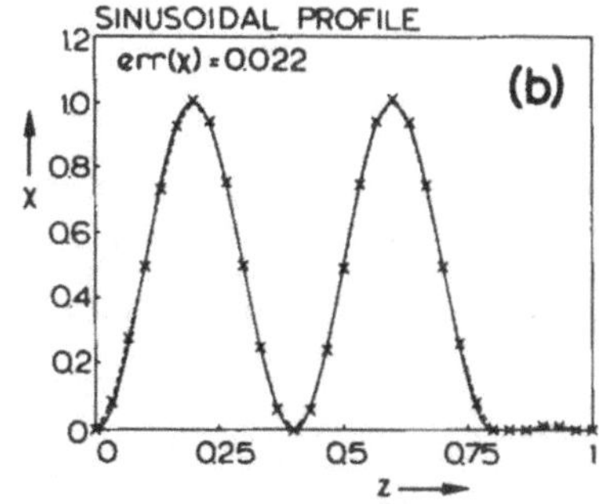

Figure 6.3.2 Results of reconstructing the susceptibility profile $\chi(z) = \sin^2(2.5\pi z)U(0.8 - z)$ for $\sigma(z) = 0$, $M = 30$, $N = 90$, $T = 0.7$ and initial estimate $\chi^{(0)}(z) = 0.5$ by employing the conjugate-gradient subroutine E04DBF. Notation as in Figure 6.3.1.

ceptibility profiles can be envisaged as low-pass filtering effects.

It should be remarked that, in the computations leading to Figures 6.3.2-6.3.4, the number of cells in the piecewise-linear expansion was systematically taken somewhat larger than the maximum estimated in (6.2.11). As before, this choice was made to ensure that the expansion employed is flexible enough to handle all the profile information available from the known reflected field. Any possible excess in flexibility is handled by the regularization terms.

As in Subsection 6.2.3, we investigated the influence of random noise on the sampled fields. Superimposing such noise has an effect similar to the one the discretization error has for quasi-Newton methods. It causes a minor variation in the cost function and its gradient which may prevent the minimization procedure from converging.

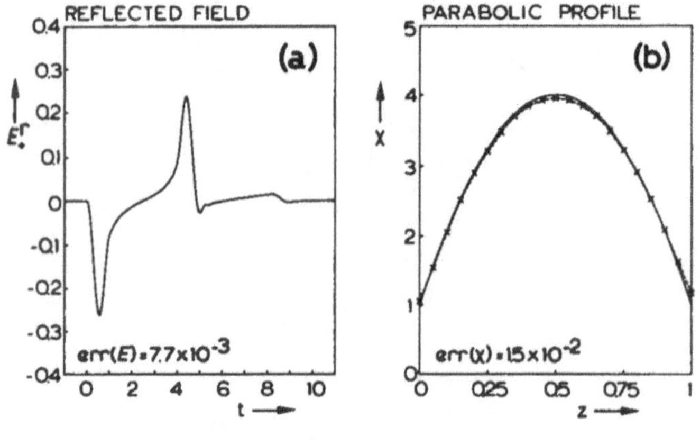

Figure 6.3.3 Results of reconstructing the susceptibility profile $\chi(z) = 4 - 12(z - 0.5)^2$ for $\sigma(z) = 0$, M = 20, N = 80, T = 1 and initial estimate $\chi^{(0)}(z) = 1.5$ by employing the conjugate-gradient subroutine E04DBF. Notation as in Figure 6.3.1.

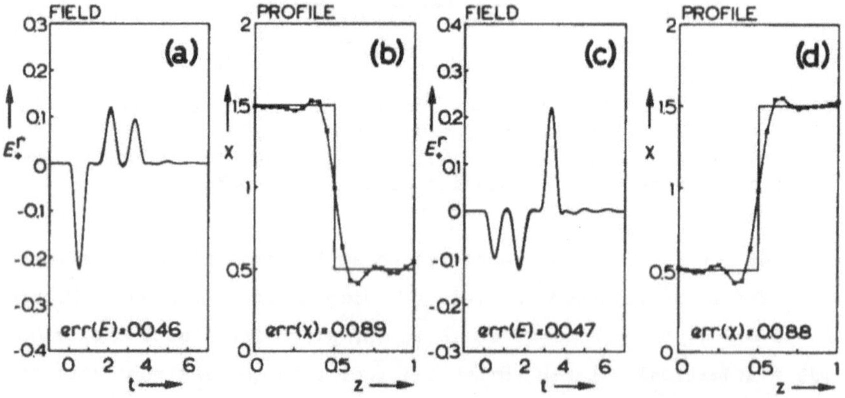

Figure 6.3.4 Results of reconstructing two discontinuous susceptibility profiles as indicated for $\sigma(z) = 0$, M = 20, N = 80, T = 1 and initial estimate $\chi^{(0)}(z) = 1$ by employing the conjugate-gradient subroutine E04DBF. Notation as in Figure 6.3.1.

As before, the resulting ambiguity can be removed by enhancing the regularization term. In fact, for the results displayed in Figures 6.3.2–6.3.4, the regularization terms employed already suffice to deal with up to five percent multiplicative noise as defined in the last part of Subsection 4.2.3.

Finally, it should be remarked that the results presented in this and the previous section should not be compared with respect to their computation time. In the present section, the more efficient recurrence scheme proposed in Subsection 3.3.2 was used as proposed to the less efficient one described in Tijhuis (1981). A more realistic comparison can be made by realizing that each iteration step in Section 6.2 requires the solution of a single approximate direct-scattering problem, while each evaluation of the cost function and its gradient requires an effort equivalent to solving two such problems. Moreover, we should keep in mind that this effort is proportional to N^2 with N being the number of steps in the repeated trapezoidal rule, and with the length of the time interval under consideration. Judging by these criteria, the computation leading to Figure 6.3.2, for instance, takes about ten times as much effort as the one leading to Figure 6.2.3.

In conclusion, it can be stated that the results of the Born-type iterative scheme with a vacuum background medium can indeed be improved. One way of obtaining better results, albeit an inefficient one, is to apply an optimization approach. In such an approach, we are free to use all of the reflected-field information instead of a prescribed subset.

6.4 Born-type iterative procedure with the reference medium as the background medium

6.4.1 Formulation of the problem

Comparing the results presented in Sections 6.2 and 6.3, we observe that it is indeed possible to obtain a better solution to the inverse-scattering problem formulated in Subsection 6.2.1 than the one obtained by applying the Born-type iterative procedure with a vacuum as the background medium. In fact, the results confirm the conclusion reached in Subsection 5.2.4 that this procedure suffers from an inherent band limitation. On the other hand, the optimization approach applied in the previous section, which produces a better reconstruction, turns out to be computationally expensive. Hence, it does seem warranted to try out the improved method of solution proposed in Subsection 5.2.5.

To start with, let us recapitulate the equations that are relevant in the n-th step of the iterative procedure. In this step, we first consider solving the approximate direct-scattering problem of determining the fields $\mathcal{E}^{(n)}(z,t)$, or their frequency-domain counterparts $E^{(n)}(z,\omega)$, that result from the known incident fields in a reference medium with susceptibility $\chi^{(n-1)}(z)$ and dimensionless conductivity $\sigma^{(n-1)}(z)$. For the time-domain fields, this can be achieved by solving an integral equation of the form (6.2.3) by the marching-on-in-time method as implemented in Subsection 3.3.2; for their frequency-domain counterparts, we can integrate the system of first-order differential equations (5.2.5) (with $s = i\omega$) by the Runge-Kutta-Verner algorithm explained in Subsection 2.4.2. The corresponding time-domain response is then found directly by a Fast Fourier Transformation as employed in Subsection 2.4.3. This part of the iteration step is essentially the same as in the scheme employed in Sections 5.4 and 6.2.

The essential difference is in the second part of the iteration step, where we now take the background medium equal to the reference medium. Using (5.2.3) or (5.2.34), taking into account the specification of the incident pulses given in (6.2.1) and (6.2.2), then leads to the approximate equation

$$F(i\omega)[\hat{E}^{r}_{\alpha}(z_{\alpha},i\omega) - \hat{E}^{r(n)}_{\alpha}(z_{\alpha},i\omega)] = \frac{-i\omega}{2} F(i\omega)\int_0^1 C(z,i\omega)\hat{E}^{(n)}_{\alpha}(z,i\omega)^2 dz, \tag{6.4.1}$$

where the label α stands for either of the labels + of -, where $z_+ = 0$, $z_- = 1$, and where

$$C(z,s) = \chi(z) - \chi^{(n-1)}(z) + [\sigma(z) - \sigma^{(n-1)}(z)]/s. \tag{6.4.2}$$

The definitions of $\hat{E}_{\alpha}(z,s)$, $\hat{E}^{r}_{\alpha}(z,s)$ and $F(s)$ are the same as given before (see e.g. Section 2.4).

As argued in Subsection 5.2.5, we should *not* solve the approximate equation (6.4.1) in the frequency domain. Instead, we should transform it back to the time domain and consider those of the resulting equations with a time parameter t lying in the time interval during which the field directly reflected at the inhomogeneities inside the slab is arriving at the interface $z = z_{\alpha}$. Hence, it seems worthwile to consider the time-domain equivalent of (6.4.1). Using Lerch's theorem (Widder (1946)), we directly identify this equivalent as

$$\begin{aligned} E^{r}_{\alpha}(z_{\alpha},t) - E^{r(n)}_{\alpha}(z_{\alpha},t) = \\ -\tfrac{1}{2}\int_0^1 dz[\chi(z)-\chi^{(n-1)}(z)]\int_0^{t-\tau_{\alpha}^{(n-1)}(z)} dt'\hat{E}^{(n)}_{\alpha}(z,t')\partial_t E^{(n)}_{\alpha}(z,t-t') \\ -\tfrac{1}{2}\int_0^1 dz[\sigma(z)-\sigma^{(n-1)}(z)]\int_0^{t-\tau_{\alpha}^{(n-1)}(z)} dt'\hat{E}^{(n)}_{\alpha}(z,t')E^{(n)}_{\alpha}(z,t-t'). \end{aligned} \tag{6.4.3}$$

In (6.4.3) we have, in accordance with (2.3.38), the arrival times

$$\begin{aligned} \tau_+^{(n)}(z) &= \int_0^z [1 + \chi^{(n)}(z')]^{\frac{1}{2}} dz', \\ \tau_-^{(n)}(z) &= \int_z^1 [1 + \chi^{(n)}(z')]^{\frac{1}{2}} dz', \end{aligned} \tag{6.4.4}$$

while $\hat{E}_{\alpha}(z,t)$ denotes a delta-function response, i.e. the response to one of the incident fields specified in (6.2.1), with $F(t) =$

$\delta(t)$. The remaining field quantities figuring in (6.4.3) correspond to this delta-function response via the relations

$$E_\alpha(z,t) = \int_0^T F(t')\hat{E}_\alpha(z,t - t')dt',$$
$$\partial_t E_\alpha(z,t) = \int_0^T \partial_{t'} F(t')\hat{E}_\alpha(z,t - t')dt', \qquad (6.4.5)$$

where $F(t)$ now denotes the sine-squared incident pulse specified in (6.2.2). With (6.4.5), we have, in principle, reduced the approximate direct-scattering computation to the evaluation of one or both of the delta-function responses $\hat{E}_\alpha(z,t)$.

To conclude our formulation, we should incorporate the finite duration of the incident pulse. This can be done in the same manner as in Subsection 6.2.2. In the present case, we select the time interval in between the situation where precisely half of the incident pulse has reached the slab and that in which precisely half the secondary reflection has emerged from it. This changes (6.2.8) into

$$T/2 < t < 2\tau^{(n-1)} + T/2, \qquad (6.4.6)$$

where $\tau^{(n-1)}$ again denotes the one-way travel time in the current reference medium.

The remainder of this section is organized in almost the same way as in Section 6.2. In Subsections 6.4.2 and 6.4.3, we discuss the problem of reconstructing either $\chi(z)$ or $\sigma(z)$ from the known values of $E_+^i(z,t)$ and $E_+^r(0,t)$, provided that the remaining constitutive parameter is known. In Subsection 6.4.4, we consider the full inverse-scattering problem of reconstructing both profiles from two-sided reflection data. Finally, in Subsection 6.4.5, we add a new element to the discussion by comparing the natural frequencies of the actual and the reconstructed slab configurations.

6.4.2 Either $\chi(z)$ or $\sigma(z)$ unknown: implementation

In this and the next subsection, we first consider the situation where one unknown profile is to be reconstructed from known fields at the left-hand interface of the slab. In that case, the approximate frequency-domain equation (6.4.1) reduces to

$$F(i\omega)[\hat{E}_{+}^{r}(0,i\omega) - \hat{E}_{+}^{r(n)}(0,i\omega)] = \frac{-i\omega}{2} F(i\omega)\int_0^1[\chi(z) - \chi^{(n-1)}(z)]\hat{E}_{+}^{(n)}(z,i\omega)^2 dz, \tag{6.4.7a}$$

$$F(i\omega)[\hat{E}_{+}^{r}(0,i\omega) - \hat{E}_{+}^{r(n)}(0,i\omega)] = -\tfrac{1}{2}F(i\omega)\int_0^1[\sigma(z) - \sigma^{(n-1)}(z)]\hat{E}_{+}^{(n)}(z,i\omega)^2 dz, \tag{6.4.7b}$$

for $\chi(z)$ and $\sigma(z)$ unknown, respectively. The corresponding time-domain equivalent (6.4.3) reduces to

$$E_{+}^{r}(0,t) - E_{+}^{r(n)}(0,t) = -\tfrac{1}{2}\int_0^1 dz[\chi(z)-\chi^{(n-1)}(z)]\int_0^{t-\tau_{+}^{(n-1)}(z)} dt'\hat{E}_{+}^{(n)}(z,t')\partial_t E_{+}^{(n)}(z,t-t'), \tag{6.4.8a}$$

and

$$E_{+}^{r}(0,t) - E_{+}^{r(n)}(0,t) = -\tfrac{1}{2}\int_0^1 dz[\sigma(z)-\sigma^{(n-1)}(z)]\int_0^{t-\tau_{+}^{(n-1)}(z)} dt'\hat{E}_{+}^{(n)}(z,t')E_{+}^{(n)}(z,t-t'), \tag{6.4.8b}$$

respectively. Comparing these equations with their counterparts for a vacuum background medium, i.e. (5.2.10) and (6.2.6), we observe a number of differences. In the first place, (6.4.7) and (6.4.8) are equations for a *profile update* rather than for the unknown profile itself. In the second place, the known profile $\sigma(z)$ or $\chi(z)$ no longer occurs explicitly in the equations. Finally, we encounter, on the left-hand sides, the known reflected field as excited in the approximate slab configuration found in the previous iteration step. Apparently, the presence of this field suffices to account for the influence of both the known and the estimated unknown profile. Note that the evaluation of this field does not require additional computational effort, since the corresponding total field occurs in the integrals on the right-hand sides.

As usual, the equations (6.4.7) and (6.4.8) may not suffice to establish the new approximate profile $\chi^{(n)}(z)$ or $\sigma^{(n)}(z)$ in full detail. Hence, we follow the same approach as in the reconstruction methods discussed before, and assume the unknown profile to be piecewise-linear as specified in (5.3.28). This reduces the approximate relations in (6.4.7) to

$$F(i\omega)[\hat{E}_+^r(0,i\omega) - \hat{E}_+^{r(n)}(0,i\omega)] = \frac{-i\omega}{2}F(i\omega)\sum_{m=0}^{M}[\chi_m^{(n)} - \chi_m^{(n-1)}]\int_0^1\phi_m(z)\hat{E}_+^{(n)}(z,i\omega)^2dz, \tag{6.4.9a}$$

and

$$F(i\omega)[\hat{E}_+^r(0,i\omega) - \hat{E}_+^{r(n)}(0,i\omega)] = -\tfrac{1}{2}F(i\omega)\sum_{m=0}^{M}[\sigma_m^{(n)} - \sigma_m^{(n-1)}]\int_0^1\phi_m(z)\hat{E}_+^{(n)}(z,i\omega)^2dz. \tag{6.4.9b}$$

Thus, we have obtained approximate linear equations for the unknown expansion coefficients $\{\chi_m^{(n)} - \chi_m^{(n-1)}\}$ and $\{\sigma_m^{(n)} - \sigma_m^{(n-1)}\}$ with $0 \leq m \leq M$. The approximate time-domain relations in (6.4.8) reduce to similar linear equations.

From (6.4.7)-(6.4.9), it is observed that there are essentially two different ways to actually arrive at a system of equations for the unknown expansion coefficients. In the first place, we can start in the time domain by computing a discrete version of the delta-function response $\hat{\mathcal{E}}_+^{(n)}(z,t)$. Once this field is obtained, we have to evaluate the convolution integrals appearing in (6.4.5) and (6.4.8). It seems most efficient to carry out both integrations simultaneously by transforming to the real-frequency domain and back with the aid of an FFT algorithm. In that case, we have the additional advantage that we can reduce the number of inverse Fourier transformations by evaluating the space integration in (6.4.8) already for the frequency-domain fields, as prescribed by (6.4.9). That equation also shows that the high-frequency errors made in computing and transforming the discretized delta-function response are suppressed by the factor of $F(i\omega)$, which represents the spectrum of the incident pulse. The one objection against this procedure is that, in the numerical computation, we must simultaneously store the fields $\hat{\mathcal{E}}_{+}^{(n)}(z,t)$ and $\hat{E}_{+}^{(n)}(z,i\omega)$ for all relevant z, t and ω. This may give rise to storage problems.

In the second place, we can start directly in the frequency domain and evaluate the harmonic plane-wave response $\hat{E}_{+}^{(n)}(z,i\omega)$ by the Runge-Kutta-Verner procedure discussed in Subsection 2.4.2. This has the advantage that we can construct the linear equations in (6.4.9) one at a time for each new frequency ω. As a result, the computer storage required decreases by a factor of $N_{cell} = N/M$, where N is the number of subintervals in the space discretization for the electric field. A second advantage is that we have already implemented the necessary field computation in the course of the numerical experiments described in Subsection 5.4.4. The principal disanvantage of following this procedure is that the necessary frequency-domain computations are more time-consuming than the corresponding time-domain computations (see also Subsection 3.3.5).

The specific advantages and disadvantages of both of the procedures outlined above seem to outweigh each other. In our numerical computations, we therefore followed the one that took the least amount of programming work, i.e. the one starting from frequency-domain results. The disadvantage of the relatively large consumption of computation time was somewhat alleviated by truncating the frequency spectrum of the incident pulse at its first zero, as discussed in Subsection 4.3.2. With this truncation and the Schwarz reflection principle, the solution of the approximate direct-scattering problem yielding $\hat{E}_{+}^{(n)}(z,i\omega)$ could be restricted to the frequency interval $0 \le \omega < 4\pi/T$, with T being the duration of the sine-squared incident pulse (6.2.2). The truncation has the additional advantage that we now have a truly band-limited signal. This facilitates estimating the number of independent expansion coefficients that can be retrieved from the known reflected field. That field, in turn, was computed by the marching-on-in-time method described in Subsection 3.3.2. The truncation of the frequency spectrum was incorporated by employing the same low-pass filtering as in Subsection 4.3.2. By invoking that procedure, we have automatically included the possibility of contaminating the marching-on-in-time result with multiplicative or additive noise as defined in the last part of Subsection 4.2.3. Obviously, the noise is superimposed before the actual filtering takes place.

Regardless of the procedure by which we obtain the approximate plane-wave responses figuring on the right-hand sides of (6.4.9a) and (6.4.9b), we now have to carry out the following steps to obtain the new estimates for the expansion coefficients $\{\chi_m\}$ and $\{\sigma_m\}$. First we must evaluate the space integral that occurs on the right-hand sides of (6.4.9a) and (6.4.9b). This can be carried out by using the same repeated trapezoidal rule as in the derivation of (5.4.20). We arrive at

$$\hat{E}_m^{(n)}(i\omega) \triangleq \int_0^1 \phi_m(z)\hat{E}_+^{(n)}(z,i\omega)^2 dz = \tag{6.4.10}$$
$$(1-\delta_{mM})\sum_{j=0}^{N_{cell}} \alpha_j \frac{N_{cell}-j}{N_{cell}} \hat{E}_+^{(n)}((mN_{cell}+j)h,i\omega)^2$$
$$+ (1-\delta_{m0})\sum_{j=0}^{N_{cell}} \alpha_j \frac{N_{cell}-j}{N_{cell}} \hat{E}_+^{(n)}((mN_{cell}-j)h,i\omega)^2,$$

where the same notations and definitions have been used for $\delta_{mm'}$, α_j, h and N_{cell} as in (5.4.20). As before, the hat on $\hat{E}_m^{(n)}(i\omega)$ denotes the fact that a factor of $F(i\omega)$ has been taken out. Next, we define the time-domain matrix coefficients $E_m^{(n)}(t)$ via the correspondence

$$E_m^{(n)}(t) \leftrightarrow F(i\omega)\hat{E}_m^{(n)}(i\omega). \tag{6.4.11}$$

Combining (6.4.9)-(6.4.11) then immediately leads to the approximate time-domain equations

$$E_+^r(0,t) - E_+^{r(n)}(0,t) = -\tfrac{1}{2}\sum_{m=0}^{M}[\chi_m^{(n)} - \chi_m^{(n-1)}]\partial_t E_m^{(n)}(t), \tag{6.4.12a}$$

$$E_+^r(0,t) - E_+^{r(n)}(0,t) = -\tfrac{1}{2}\sum_{m=0}^{M}[\sigma_m^{(n)} - \sigma_m^{(n-1)}]E_m^{(n)}(t), \tag{6.4.12b}$$

which, like the ones in (6.2.7), should be converted to a square matrix equation. This can be accomplished by rewriting the expansion coefficients $\{\chi_m^{(n)}\}$ and $\{\sigma_m^{(n)}\}$ as

$$\chi_m^{(n)} = \chi_m^{(n-1)} + \Delta\chi_m^{(n)}, \quad \sigma_m^{(n)} = \sigma_m^{(n-1)} + \Delta\sigma_m^{(n)}, \tag{6.4.13}$$

with $0 \le m \le M$, where the coefficients $\{\Delta\chi_m^{(n)}\}$ and $\{\Delta\sigma_m^{(n)}\}$ minimize the cost functions

$$\int_{T/2}^{2\tau^{(n-1)}+T/2}\left\{E_+^r(0,t) - E_+^{r(n)}(0,t) + \tfrac{1}{2}\sum_{m=0}^{M}\Delta\chi_m^{(n)}\partial_t E_m^{(n)}(t)\right\}^2 dt$$
$$+ \delta\sum_{m=1}^{M-1}(\Delta\chi_{m-1}^{(n)} - 2\Delta\chi_m^{(n)} + \Delta\chi_{m+1}^{(n)})^2, \tag{6.4.14a}$$

and

$$\int_{T/2}^{2\tau^{(n-1)}+T/2}\left\{E_+^r(0,t) - E_+^{r(n)}(0,t) + \tfrac{1}{2}\sum_{m=0}^{M}\Delta\sigma_m^{(n)}E_m^{(n)}(t)\right\}^2 dt$$
$$+ \delta\sum_{m=1}^{M-1}(\Delta\sigma_{m-1}^{(n)} - 2\Delta\sigma_m^{(n)} + \Delta\sigma_{m+1}^{(n)})^2, \tag{6.4.14b}$$

respectively. These cost functions have the same structure as the ones defined in (6.2.15) and (6.3.1) and, hence, require no further explanation. The only new element is that the time integration now runs over the interval specified in (6.4.6), which is in accordance with the conclusions obtained in Subsection 5.2.5. In the numerical implementation, we followed the same procedure as in Subsections 5.3.4 and 6.2.4. The equations pertaining to the sampled known field values in the time interval (6.4.6) and the ones corresponding to the $M - 1$ regularization terms were properly normalized and collected into one large, overdetermined matrix equation. This equation was solved in a least-squares sense using an inversion procedure based on Householder transformations (NAG (1984), subroutine F04AMF).

The parameter M, which represents the number of cells in the piecewise-linear expansion (5.3.28), should obviously be adjusted to the cut-off frequency in the low-pass filtering as well as to the length of the time interval in (6.4.6). Using Shannon's sampling theorem in the same manner as in deriving (6.2.11), we find that we can at most recover $M \approx 8\tau/T$ independent parameters from the filtered reflected field. In practice, we will choose M slightly larger than this estimate to ensure that the maximum amount of profile information is drawn from that field. As argued before, we can rely on the regularization terms to eliminate any ambiguity from the piecewise-linear expansion employed.

During the course of the iterative procedure, we continuously monitored a profile error as well as a field error. For the profile errors, we used the definitions given in (6.3.23). The field error was defined as the root-mean-square error in the approximate field resulting from the incident field in a slab with $\chi(z) = \chi^{(n)}(z)$ or $\sigma(z) = \sigma^{(n)}(z)$ over the time interval (6.4.6). In analogy with (6.2.17) and (6.3.22), this error was defined as

$$err^{(n)}(E) \triangleq \left\{ \frac{\int_{T/2}^{2\tau^{(n)}+T/2} [E_+^{r\,(n+1)}(0,t) - E_+^r(0,t)]^2 dt}{\int_{T/2}^{2\tau^{(n)}+T/2} E_+^r(0,t)^2 dt} \right\}^{\frac{1}{2}} . \quad (6.4.15)$$

In this definition, we have kept in mind that $E_+^{(n)}(z,t)$, which is the field computed in the n-th iteration step, has been defined as the field excited in the previous approximate configuration, i.e. the one with $\chi(z) = \chi^{(n-1)}(z)$ or $\sigma(z) = \sigma^{(n-1)}(z)$. As (6.4.15) indicates, this error is computed in the iteration step subsequent to the one in which the relevant approximate profile is determined.

Finally, it should be remarked that, by generating the actual reflected field by the marching-on-in-time method and the approximate fields excited in the successive reference configurations by direct Fourier inversion, we have automatically excluded the possibility that the solution of the inverse-profiling problem is biased by using the same direct-scattering algorithm in both computations. Only when the respective computations are both carried out with sufficient accuracy may we obtain an accurate reconstruction of the unknown profile. Indirectly, this feature may also serve to verify the validity of the inverse-profiling results obtained until now, where the same algorithm was used in solving the corresponding actual and approximate direct-scattering problems.

6.4.3 Either $\chi(z)$ or $\sigma(z)$ unknown: numerical results

In order to obtain insight into the potentialities of the computational scheme described in Subsections 6.4.1 and 6.4.2 we carried out a number of numerical experiments. In particular, we attempted to reconstruct some of the profiles that were previously considered in Subsections 5.4.4, 6.2.3 and 6.3.4. In this manner, we can compare the reconstructive potentialities of the present iterative procedure with those of the solution schemes described previously.

Figures 6.4.1-6.4.4 contain results of reconstructing the same types of profile that were considered before in Figures 6.2.3-6.2.6 and 6.3.2-6.3.4. To make the comparison as complete as possible, we employed the same piecewise-linear expansions, regularization parameters and initial estimates as in the computations leading to Figures 6.3.2-6.3.4. In Figures 6.4.1 and 6.4.2, results are given for the same sine-squared susceptibility profile as in Figures 6.2.3 and 6.3.2. As in Subsection 6.2.3, negative values of the expansion coefficients $\{\chi_m^{(n)}\}$ were set to zero after each iteration step. Apparently, the reconstruction obtained after six iteration steps is of about the same quality as the one obtained by applying the conjugate-gradient method. This is hardly surprising since it is observed from Figure 6.3.2a that, for this special configuration, the reflected field virtually vanishes outside the time interval specified in (6.4.6). It should be noted, however, that the computation leading to Figure 6.3.2b took a numerical effort equivalent to solving 48 approximate direct-scattering problems (see also Table 6.3.3). In terms of the efficiency criteria introduced towards the end of Subsection 6.3.4, this means that we have gained a factor of eight in efficiency.

In Figure 6.4.3, we show the result of reconstructing the parabolic susceptibility profile specified as profile C in Figure 6.2.4 and further considered in Figures 6.2.5 and 6.3.3. Compared with the reconstructions displayed in the latter two figures, the present reconstruction appears to be less accurate in the last stretch of the piecewise-linear expansion. This can be explained from the fact that, with the discretization employed, the marching-on-in-time method generates the secondary reflection with less accuracy than

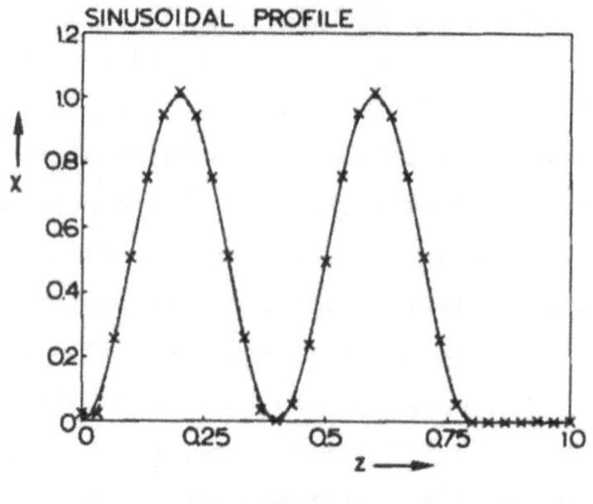

Figure 6.4.1 Actual susceptibility profile $\chi(z) = \sin^2(2.5\pi z) \times U(0.8 - z)$ compared with the reconstruction obtained after six iteration steps of the improved Born-type procedure for $\sigma(z) = 0$, $M = 30$, $N = 120$, $T = 0.7$, $\delta = 5 \times 10^{-4}$ and initial estimate $\chi^{(0)}(z) = 0.5$. Solid line: actual profile; dashed line: reconstructed profile; x: values of expansion coefficients $\{\chi_m^{(n)}\}$.

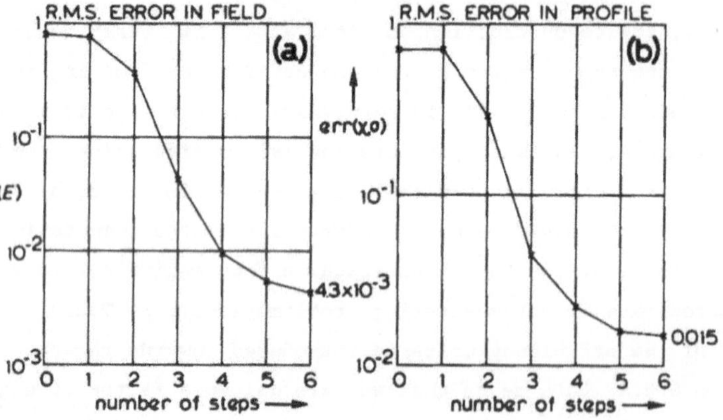

Figure 6.4.2 Root-mean-square errors as defined in (6.4.15) and (6.3.23) as a function of the number of iteration steps for the computation leading to Figure 6.4.1.

the field that emerges from the slab before this reflection arrives. This fact was verified both by refining the discretization in the marching-on-in-time computation and by comparing the time-marching results with results of the Fourier-inversion scheme described in

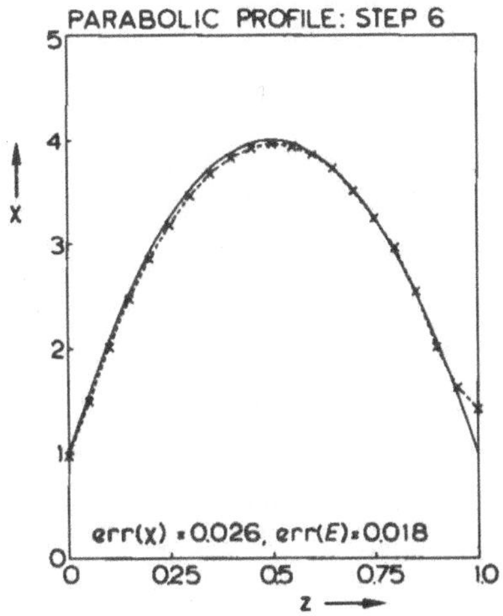

Figure 6.4.3 Actual susceptibility profile $\chi(z) = 4 - 12(z - 0.5)^2$ compared with the reconstruction obtained after six iteration steps of the improved Born-type procedure for $\sigma(z) = 0$, $M = 20$, $N = 80$, $T = 1$, $\delta = 5 \times 10^{-4}$ and initial estimate $\chi^{(0)}(z) = 2.5$. Notation as in Figure 6.4.1, and r.m.s. errors referring to the reconstruction indicated.

Section 2.4. In the present reconstruction procedure, the behavior of the unknown profile at the rear end of the slab is determined entirely from the amplitude of the secondary reflection. Hence, the relatively large deviation in the reconstruction. In order to confirm our interpretation, we repeated the computation for the improved time-marching result mentioned above. The resulting profile estimate did show good agreement with its actual counterpart.

The deviation observed in Figure 6.4.3 demonstrates the validity of the remark made towards the end of Subsection 6.4.2 that, in a numerical simulation, it seems preferable to generate the known reflected field by a different direct-scattering algorithm than the approximate fields in the successive reference configurations. In fact, the same inaccurate time-marching result was used in computing the accurate reconstruction shown in Figure 6.3.3.

In Figure 6.4.4, results are presented of reconstructing the same two discontinuous susceptibility profiles that were recovered in

Figure 6.3.4. For both profiles, ten iteration steps were carried out. Unlike the Born-type iterative procedure with a vacuum background medium, the present reconstruction scheme did not require the application of a discontinuity correction. Moreover, the profile errors for the reconstructions obtained are comparable to the ones listed in Figure 6.3.4. The field errors appearing in Figures 6.3.4 and 6.4.4 should be compared with some care. In Figure 6.3.4 these errors pertain to the exact reflected fields and in Figure 6.4.4 to the filtered versions. The only conclusion that can be drawn is that the deviation between the reflected fields excited in the actual and in the reconstructed slab configurations shows up mainly in those high-frequency components of those fields that are removed in the low-pass filtering.

To conclude this subsection, we carry out a different type of comparison. By applying the low-pass filtering procedure to the known time-domain reflected field, we have in effect restricted our knowledge of the frequency-domain reflection coefficient to a finite

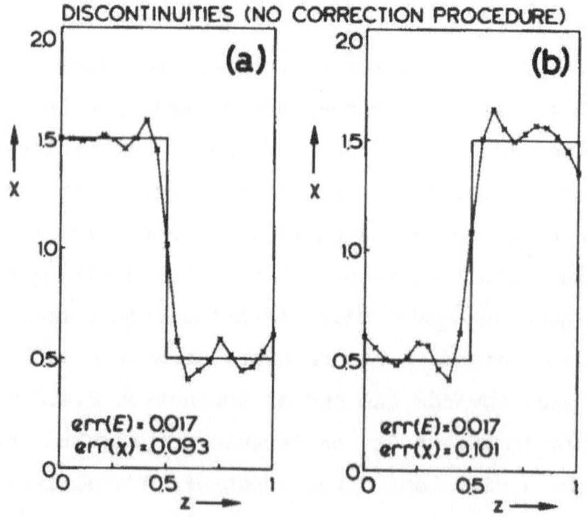

Figure 6.4.4 Actual susceptibility profiles $\chi(z) = 1.5 - U(z - 0.5)$ and $\chi(z) = 0.5 + U(z - 0.5)$ compared with the reconstructions obtained after ten iteration steps of the improved Born-type procedure for $\sigma(z) = 0$, $M = 20$, $N = 80$, $T = 1$, $\delta = 10^{-3}$, and initial estimate $\chi^{(0)}(z) = 1.0$. Notation as in Figure 6.4.1, and r.m.s. errors referring to reconstructions indicated.

range of real-valued frequencies. Moreover, we use the frequency-domain solutions for the approximate configuration over that range to construct the field matrix whose inversion yields the next estimate of this configuration. In both respects, we have the same situation as in Subsection 5.4.4, where we applied the Born-type iterative procedure with a vacuum background medium to the frequency-domain inverse-profiling problem for a lossless slab. It seems worthwile, therefore, to apply the present method in the same circumstances as the method employed in Subsection 5.4.4. To this end we reconsider in Figure 6.4.5 the most complicated profile reconstructed in that subsection, namely the piecewise-homogeneous three-layer medium considered in the paper by Coen et al. (1981) and in Figure 5.4.5. The duration of the incident pulse was chosen such that the range of frequencies remaining after the low-pass filtering was comparable to the one selected in Figure 5.4.5. Figure 6.4.5 gives the results after every second iteration step. From the results it is observed that, in the first few iteration steps, the average behavior of the unknown profile is recovered. In subsequent iteration steps, we obtain a more detailed impression of its local behavior. This observation agrees well with the conclusion reached in Subsection 5.2.5 that, in the present scheme, the inherent band limitation due to the error in the Born-type approximation is gradually removed as the iterative procedure progresses. The relatively large oscillations in the eventual reconstruction can be explained in the same manner as in Subsection 5.4.4, namely from the fact that they are exactly the gradients of $\chi(z)$ which cause a reflection. As in that subsection, we further observe a gradual deterioration of the reconstruction as z increases. As argued in connection with Figure 5.4.4, this corresponds to the fact that a wave which is directly reflected at $z = z_0$ only traverses the interval $0 < z < z_0$ before leaving the slab.

Finally, some computation times should be reported. For the configurations investigated in this subsection, a single iteration step typically took between 5 and 15 seconds. Of this time, about half a second was consumed by the matrix inversion. The remainder was spent in evaluating the field in the current reference configuration. Comparing these times with the ones listed in Table 3.3.1, we see that,

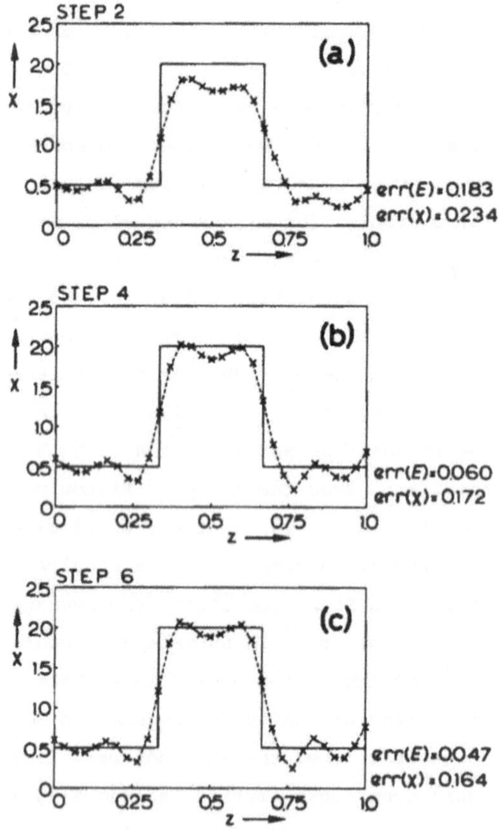

Figure 6.4.5 Actual susceptibility profile $\chi(z) = 0.5 + 1.5\text{rect}(z - 0.5; \frac{1}{3})$ compared with the reconstruction obtained after 2, 4 and 6 iteration steps of the improved Born-type procedure for $\sigma(z) = 0$, $M = 30$, $N = 120$, $T = 0.7$, $\delta = 10^{-3}$ and initial estimate $\chi^{(0)}(z) = 1$. Notation as in Figure 6.4.1 and r.m.s. errors referring to reconstructions indicated.

in principle, the computational efficiency of the procedure can be improved by employing an alternative algorithm in the field computations.

6.4.4 Both $X(z)$ and $\sigma(z)$ unknown

As in Section 6.2, we have also considered the case where both an unknown susceptibility profile and an unknown conductivity profile must be reconstructed from the known time-domain reflected fields at both the slab's interfaces, i.e. $E^r_+(0,t)$ and $E^r_-(1,t)$. The numerical implementation was carried out in the same manner as described in Subsection 6.4.2. The only difference was the duplication in the number of approximate frequency-domain direct-scattering problems solved per iteration step and in the dimension of the overdetermined matrix equation for the expansion coefficients $\{\chi_m^{(n)}\}$ and $\{\sigma_m^{(n)}\}$. As in the Born-type iterative procedure with a vacuum background medium, these two sets of expansion coefficients are obtained simultaneously by inverting this overdetermined matrix equation in a least-squares sense. In the matrix equation, both the available reflected-field information and the a priori information brought in via the regularization terms are collected.

In the course of each iteration step, we again computed a field error and a profile error. For the profile error, we used the same definition as given in (6.2.19) and (6.2.20). In the definition of the field error, we took into account the lengthened time interval specified in (6.4.6). This changes (6.2.17) into

$$err(E) \triangleq \left\{ \frac{\int_{T/2}^{2\bar{\tau}+T/2}[\Delta E_+(t)^2 + \Delta E_-(t)^2]dt}{\int_{T/2}^{2\bar{\tau}+T/2}[E^r_+(0,t)^2 + E^r_-(0,t)^2]dt} \right\}^{\frac{1}{2}}, \qquad (6.4.16)$$

where the same notations and definitions have been used.

To compare the efficacy of the present method to that of the one described in Subsection 6.2.4 and employed in Subsection 6.2.5, we attempted to reconstruct the same slab configurations as considered in the latter subsection. With the present method, an acceptable reconstruction is usually obtained in 5 to 7 iteration steps, which is about twice as efficient as reported in Subsection 6.2.5. Moreover, ambiguity effects as encountered in Figures 6.2.7 and 6.2.8 are no longer observed. This is in agreement with the conclusion reached in Subsection 5.2.5, that only the knowledge of both the reflected fields $E^r_+(0,t)$ and $E^r_-(1,t)$ over a time interval of length 2τ suffices to establish the unknown slab configuration completely.

In Figure 6.4.6, we present results of reconstructing the same

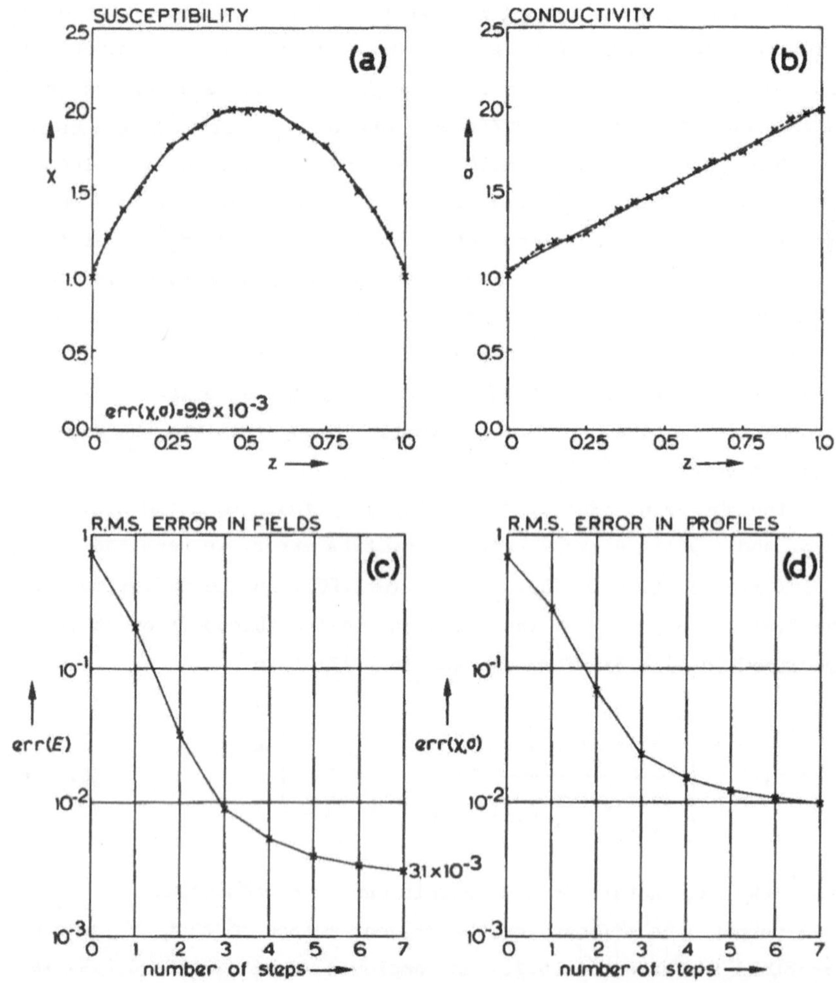

Figure 6.4.6 (a),(b): Results obtained after seven iteration steps in the simultaneous reconstruction of $\chi(z) = 2 - 4(z - 0.5)^2$ and $\sigma(z) = 1 + z$ by the improved Born-type procedure with $M = 20$, $N = 80$, $T = 1$ and initial estimate generated as explained in Subsection 6.2.5. Notation as in Figure 6.4.1. (c),(d): r.m.s. errors as defined in (6.4.16) and (6.2.19) as a function of the number of the iteration step. The computation of the actual reflected field was carried out with $N = 160$.

parabolic susceptibility profile and the same linear conductivity profile considered previously in Figure 6.2.7. Comparing the error plots in the lower half of both figures, we observe that the present reconstruction procedure indeed converges faster than the one with a vacuum background medium. From the upper half of Figure 6.4.6, we observe that the only noticeable deviation between the actual and the reconstructed profiles is a readily recognizable, oscillating difference, whose presence may be due to the low-pass filtering employed. The results obtained in iteration steps 5 to 7 indicate that the amplitude of this oscillation could be reduced even further by carrying on with the computation.

Figure 6.4.7 contains results of reconstructing the same quadratic susceptibility profile and a discontinuous conductivity profile. This configuration was also investigated in Figure 6.2.8. The iteration was kept up somewhat longer than in Figure 6.4.6 in order to look for the same type of ambiguity effect that showed up in Figure 6.2.8. As Figure 6.4.7 indicates, no such effect was observed. On the contrary, even in the final iteration steps, a minor improvement of the reconstruction was attained. For the discontinuous conductivity profile, we obtain a "filtered" reconstruction similar to the ones obtained for the discontinuous susceptibility profiles analyzed in Figures 6.4.4 and 6.4.5. As Figure 6.4.7a shows, the imperfections in this reconstruction apparently have no significant effect on the quality of the reconstructed susceptibility profile. In fact, the reconstruction displayed in Figure 6.4.7a would appear to be slightly better than the one presented in Figure 6.4.6a. This again demonstrates the relatively strong dependence of the reflected fields on the susceptibility profile.

The ultimate test of our improved Born-type iterative method was the reconstruction of a similar piecewise-homogeneous three-layer medium as considered by Coen et al. (1981). Figure 6.4.8 shows the results obtained in every second iteration step. For the unknown susceptibility profile, we obtain a reconstruction of at least the same quality as the one obtained in the lossless case studied in Figure 6.4.5. Towards the end of the slab, the present reconstruction is even better. This can be understood from the fact that we

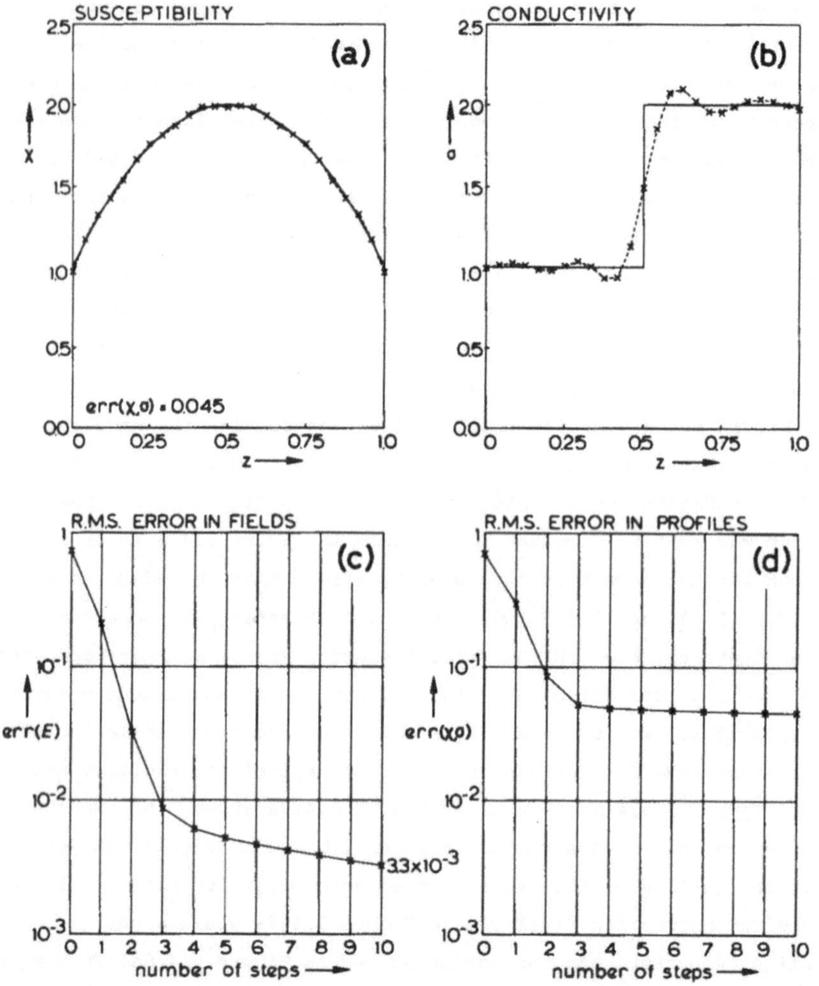

Figure 6.4.7 Results of carrying out ten iteration steps in simultaneously reconstructing the profiles $\chi(z) = 2 - 4(z - 0.5)^2$ and $\sigma(z) = 1 + U(z - 0.5)$ by the improved Born-type method with $M = 24$, $N = 96$, $T = 1$ and initial estimates generated as explained in Subsection 6.2.5. Notation as in Figure 6.4.6. The computation of the actual reflected fields was carried out with $N = 160$.

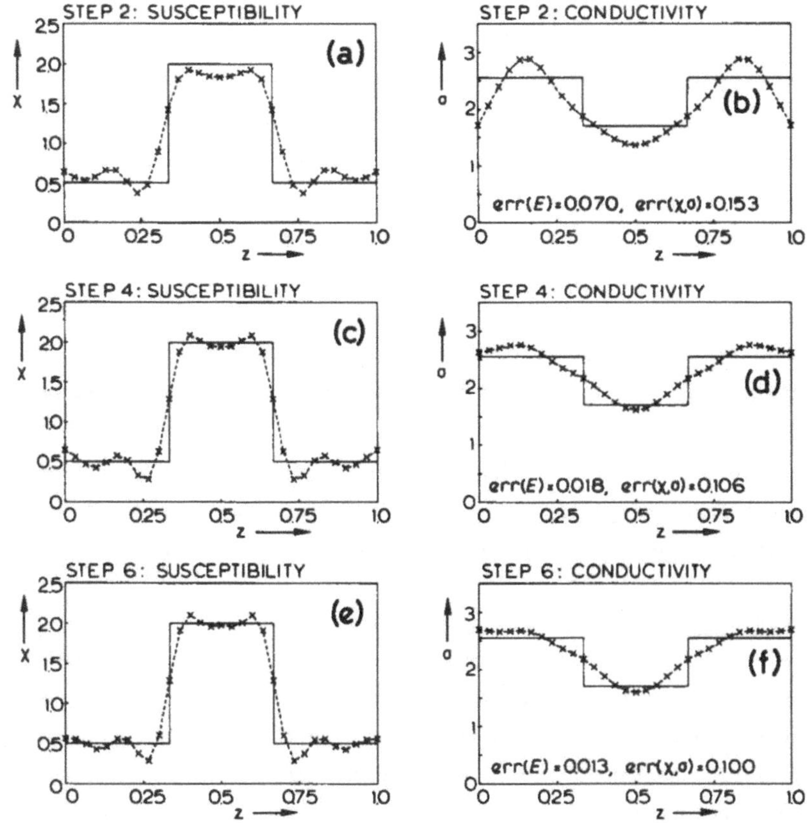

Figure 6.4.8 Actual profiles $\chi(z) = 0.5 + 1.5\text{rect}(z - 0.5;\frac{1}{3})$ and $\sigma(z) = 2.55 - 0.85\text{rect}(z - 0.5;\frac{1}{3})$ compared with the reconstructions obtained after 2, 4 and 6 iteration steps of the improved Born-type procedure with $M = 30$, $N = 120$, $\delta = 10^{-4}$ and initial estimates $\chi^{(0)}(z) = 1$ and $\sigma^{(0)}(z) = 0$. Notation as given in Figure 6.4.1 and r.m.s. errors referring to the reconstructions indicated.

now use two-sided reflection data. Apparently, the imperfections in the recovered susceptibility profile do degrade the reconstruction of the conductivity profile. Nevertheless, the correct trend is obtained for this profile as well. Moreover, the size of the root-mean-square field errors listed in Figure 6.4.8 indicates that the reflection coefficients of the exact configuration and its reconstruction given in Figures 6.4.8e,f may be almost indistinguishable over the frequency range excited by the filtered incident pulse. This would mean that a better reconstruction may be obtained from the reflected fields corresponding to a shorter incident pulse. We will come back to this subject in the next subsection.

As in Subsection 6.4.3, we wish to complete the presentation of the numerical results by reporting some computation times. For the computations leading to Figures 6.4.6-6.4.8, a typical iteration step took between 20 and 35 seconds. The linear-least-squares inversion required up to one and a half seconds. The remainder of the computation time was mainly consumed in solving the subsequent approximate direct-scattering problems. The amount of computation time required was dictated mainly by the complexity of the unknown slab configuration. Owing to the adaptive nature of the Runge-Kutta-Verner integration employed, some minor variations were also observed in the times taken up in successive iteration steps.

6.4.5 Using the natural frequencies

The analysis of the one-dimensional inverse-scattering problem discussed in this and the previous chapter would not be complete without reconsidering the natural frequencies that were analyzed extensively in Chapters 2 and 4. Now that we have solved the inverse-profiling problem, we have two ways of finding the natural frequencies of a slab configuration from time-domain reflected-field data. In the first place, we can find these natural frequencies directly by subsequently applying the preprocessing procedure outlined in Subsection 4.3.2, and Prony's method as formulated in Subsection 4.2.1. In the second place, we can determine the natural frequencies indirectly by reconstructing the unknown susceptibility and/or conductivity profile with the aid of one of the schemes employed in the present chapter. For the configuration thus recovered, we can then compute the natural frequencies as explained in Subsection 2.3.4. If we employ the improved Born-type iterative procedure investigated in this section, it seems sensible to employ the same low-pass filtering as in the application of Prony's method.

By comparing the two sets of results obtained with the actual natural frequencies and with each other, we can resolve a few issues that had to be left open in the discussion given until now. First, we can verify the conclusion drawn in Subsection 6.4.4 that reconstructed configurations as shown in Figures 6.4.5c and 6.4.8e,f have the same reflection coefficients as the actual configuration over the real-frequency range excited. If this conclusion is correct, we should observe good agreement between the natural frequencies of the reconstructed configurations whose imaginary parts lie in that range and their actual counterparts. Second, we can decide whether the piecewise-linear expansions employed are flexible enough. If the deviation between the indirectly obtained natural frequencies and those obtained otherwise increases considerably with their increasing imaginary parts, it may be necessary to choose a larger number of subintervals, i.e. a larger M. Third, by applying a Prony-type method, we can find out which of the natural-mode contributions are present in the filtered time-domain fields from which the reconstructed profiles are evaluated. Thus, we can confirm the argument used in Subsection 4.3.2 that the lower-order poles are most characteristic for the slab configuration. Finally, by comparing the accuracy of the natural frequencies found directly and indirectly from the reflected fields, we can assess whether the modified Prony-type procedure formulated in Section 4.3 leaves room for improvement.

In Table 6.4.1, results are presented of determining the natural frequencies as proposed above for the lossless three-layer slab specified in Figure 6.4.5. With the obvious exception of the real part of the natural frequency nearest to the cut-off of the low-pass filtering, the Prony results turn out to be slightly better than the results obtained indirectly by determining the natural frequencies of the reconstructed configuration given in Figure 6.4.5c. This suggests that a slight improvement of that reconstruction should, in principle, be possible. This would be in agreement with the rather large field error remaining in the final iteration step. In view of the relatively small reduction of the field error achieved in steps 5 and 6, however, it does not seem sensible to try and obtain this improvement by carrying out some more iteration steps. Note that the Prony results listed in Table 6.4.1, in turn, are less accurate than those computed for the four-layer slab considered in Figure 4.3.3

Table 6.4.1 Comparison of the actual natural frequencies of the slab configuration specified in Figure 6.4.5 with the ones recovered from the time-domain reflected field excited by a sine-squared incident pulse with T = 0.7.

Actual natural frequency	Approximation from profile reconstruction	Approximation from Prony-type method
-1.3636	-1.3820	-1.3639
-1.7288+i2.6023	-1.7492+i2.5926	-1.7299+i2.6045
-1.8719+i4.3320	-1.8885+i4.3154	-1.8753+i4.3269
-1.4461+i6.4973	-1.4588+i6.4814	-1.4438+i6.4902
-1.5208+i9.3500	-1.5435+i9.3289	-1.5026+i9.3373
-1.8974+i11.3444	-1.9425+i11.3084	-1.8556+i11.3863
-1.6441+i13.1536	-1.7558+i13.0785	-1.6726+i13.1590
-1.3782+i15.8968	-1.3529+i15.4588	-1.1806+i15.8341

and Table 4.3.1. This can be understood from the fact that, in the latter computation, the marching-on-in-time evaluation of the reflected fields was performed with a smaller space-time step.

In Table 6.4.2, results of repeating the computation for the lossy three-layer slab specified in Figure 6.4.8 are presented. From the actual natural frequencies listed in the first column, it can be seen that there are three poles located on the negative real s-axis. Apparently, we have the same situation as discussed in Section 2.2. Passing over from the lossless slab specified in Figure 6.4.5 to the lossy one specified in Figure 6.4.8, we will observe two poles coalescing into a double pole on the negative real s-axis, and, subsequently, splitting up into a pole pair that remains on that axis. This observation also indicates that the configuration specified in Figure 6.4.8 is already so lossy that a field reflected at one end

Table 6.4.2 Comparison of the actual natural frequencies of the slab configuration specified in Figure 6.4.8 with the ones recovered from the time-domain reflected fields excited by a sine-squared incident pulse with T = 0.7.

Actual natural frequency	Approximation from profile reconstruction	Approximation from Prony-type method
-4.819	-4.962	-4.569
-4.422	-4.659	
-2.635	-2.490	-2.631
-2.186+i3.195	-2.151+i3.144	-2.186+i3.195
-1.824+i6.145	-1.795+i6.097	-1.818+i6.142
-2.224+i9.110	-2.153+i9.092	-2.195+i9.084
-2.588+i10.960	-2.553+i10.973	-2.471+i11.049
-2.143+i12.851	-2.180+i12.818	-2.162+i12.889
-1.943+i15.797	-1.814+i15.510	-1.628+i15.822

of the slab will hardly penetrate as far as the opposite slab interface. As in the lossless case, the Prony results are slightly better than the natural frequencies found indirectly. It turns out that Prony's algorithm is not able to distinguish between the two poles located close together at the far end of the negative real s-axis. Instead, only a single pole with a residue twice as large is recovered. This is the same feature that was observed in Table 4.3.1 and Figure 4.3.4, namely that the natural frequency of a more rapidly decaying residual contribution will be obtained with less accuracy. In fact, this same phenomenon may be the cause of the overall loss of accuracy observed in comparing the results in Table 6.4.2 with those in Table 6.4.1.

Looking at both tables, we see that especially the natural frequencies nearest the origin of the complex s-plane are reconstructed well by the Prony-type method described in Section 4.3. Hence, comparing these frequencies with the ones obtained indirectly via profile reconstruction may indeed provide an additional way of judging the quality of the profile estimates obtained.

Furthermore, the conclusion that the Prony results are superior to the natural frequencies recovered indirectly from exactly the same reflected-field information plus additional a priori information indicates that there is little occasion to try and improve our version of Prony's method. In particular, this refers to the possible improvement proposed towards the end of Subsection 4.2.4, which would amount to minimizing a cost function similar to that we minimized in solving the inverse-scattering problem.

6.5 Conclusions

In this chapter, we have investigated three different ways to reconstruct one or both constitutive parameters of an inhomogeneous, lossy dielectric slab from the time-domain reflected fields caused by a sine-squared pulse incident from one or both sides of the slab. The essential difference with the solution procedures discussed in the previous chapter is that we have used time-domain integral relations to construct the successive matrix equations for the unknown coefficients in the piecewise-linear approximations employed.

In Section 6.2, we applied the time-domain version of the Born-type iterative procedure with a vacuum background medium employed previously in Section 5.4. The reconstructions obtained appear to be somewhat better than those obtained with the frequency-domain version of the procedure. The best results are obtained when the summation method is employed to construct the square matrix equations for the unknown expansion coefficients in the successive piecewise-linear approximations. A possible explanation of both these effects can be derived from the fact that, by considering only the first traverse of the pulse across the slab, we avoid all contributions due to multiple reflections at the slab's interfaces. As a result, we have an almost one-to-one correspondence between the known reflected field in a given subinterval and the unknown profile in a single cell of the piecewise-linear approximation. Nevertheless, the reconstructions obtained clearly suffer from the band limitation that is inherent in choosing a vacuum background medium. Especially when both the susceptibility profile and the conductivity profile are to be reconstructed from two-sided reflected-field information, a minor ambiguity is observed. As a consequence, the iterative procedure must be terminated as soon as the global error in the reflected fields no longer changes significantly in order to prevent the regularization terms from exerting undue influence. From a computational point of view, the method is very efficient. Per iteration step, we only need to carry out one or two marching-on-in-time computations, according to whether we use one- or two-sided reflection data.

Next, we considered, in Section 6.3, the possibility of using all the reflection data available in the time interval $0 < t < \infty$. The non-linearity of this problem requires the application of a more conventional "brute-force" optimization routine for minimizing a function. A special feature of this problem is that, due to discretization errors, the cost function is only available with a limited accuracy. Hence, we cannot estimate the derivatives constituting the gradient and the Hessian matrix by interpolating locally. As far as the gradient is concerned, this problem can be resolved by evaluating a closed-form expression, which involves solving the adjoint integral equation (6.3.20). For the Hessian matrix, such an expression has not been derived. As a consequence, the application of quasi-Newton methods, which do attempt to build up an approximation to the Hessian matrix, is ruled out. This leaves the conjugate-gradient method, in which the Hessian matrix is not used. In view of the computational effort required, we have only considered the reconstruction of a single unknown profile from a single time-domain reflected field. The reconstructions obtained are considerably better than those obtained with the Born-type iterative procedure with a vacuum background medium. Each iteration step of the conjugate-gradient method requires at least two evaluations of the cost function and its gradient. This represents a numerical effort equivalent to performing at least four marching-on-in-time computations. For two-sided reflection data, this effort would have to be doubled. In addition, the conjugate-gradient method requires more iteration steps than the Born-type procedure. From that point of view, the optimization approach does not seem to be very efficient.

Finally, in Section 6.4, we have investigated the Born-type iterative procedure with the background medium being identical to the reference medium. As argued in Subsection 5.2.5, it only makes sense to apply this procedure in the time domain. By considering the proper time interval, we may then hope to eventually avoid the band limitation in the reconstructed profiles which is inherent in the Born-type approximation employed. A disadvantage of the procedure is that, in each iteration step, we must carry out two subsequent time-domain convolutions for each space point figuring in the discretized integral relations pertaining to the reflected fields. In the present implementation, this problem has been resolved provisionally by starting from the frequency-domain solutions of the successive approximate direct-scattering problems. However, it should be possible to devise a faster method of evaluation. Both the cases of one and two unknown constitutive parameters have been investigated. The quality of the reconstructions obtained seems comparable to that attained by applying an optimization approach. The contention that the band limitation inherent in the Born-type methods is removed as the iterative procedure progresses was verified by comparing the natural frequencies of the actual and the reconstructed slab configurations. Like the Born-type procedure with a vacuum background medium, the improved scheme requires the solution of one or two approximate direct-scattering problems per iteration step for one- and two-sided reflection data, respectively.

In all three applications, we observe two common aspects that are connected to the underlying direct-scattering problem rather than to the specific reconstruction method employed. In the first place, the effective band limitation in the sine-squared incident pulse as analyzed in Subsection 4.3.2 prevents a reconstruction of the short-range behavior of the unknown susceptibility and/or conductivity profile. This means that this short-range behavior should be supplied additionally from a priori information. We have chosen to do this by representing the unknown profiles by a piecewise-linear approximation, and by including a small regularization term in the cost function. This term compensates for a possible excess in flexibility in that piecewise-linear approximation. In the second place, it should be remarked that, by choosing a sine-squared incident pulse - filtered or unfiltered -, we have disregarded the conclusion reached in Chapter 5 that we should ideally have $|F(i\omega)| = O(\omega^{-1})$ as $|\omega| \to \infty$. Moreover, an incident-pulse shape that does exhibit this behavior would be discontinuous in time, and, hence, incompatible with the marching-on-in-time method employed. Nevertheless, the iterative procedure does converge, which is contradictory to the convergence results reported in Subsection 5.4.4. The only possible explanation is the one given above, i.e. that by considering only a finite time interval, we have established a better correspondence between the known reflected fields and the unknown profiles.

To end this section, we wish to stress that the three inverse-profiling methods investigated in this and the previous chapter all seem generalizable to more complicated inverse-scattering problems. In this respect, we seem to have gained on the methods requiring a Liouville transformation as reviewed in Section 5.1. It should be pointed out, however, that, for multi-dimensional inverse-profiling problems, the efficiency of the respective methods may depend on the set-up of the experimental measurements. To demonstrate this, let us consider two extreme situations. In Situation I, we have excitation by a single point source and observation at N points away from the source point. In this situation, the Born-type iterative procedure with a vacuum background medium requires the solution of only a single approximate direct-scattering problem per iteration step. A single iteration step with the conjugate-gradient method still takes a computational effort equivalent to resolving four such problems. The relevant gradient can be obtained by repeating the analysis of Subsection 6.3.3 for the total integrated squared error in all the measured fields. The Born-type iterative procedure with the background medium being identical to the reference medium, however, requires the solution of N + 1 approximate direct-scattering problems. It seems most feasible, therefore, to start the reconstruction by determining a band-limited approximation with the aid of the original Born-type procedure. If necessary, this reconstruction can then be improved by applying either the optimization approach or the improved Born-type procedure.

In Situation II, we have excitation by N independent point sources and observation at the source points only. In this situation, a single iteration step in either of the Born-type procedures takes the solution of N approximate direct-scattering problems, while one step in the conjugate-gradient method requires an effort equivalent to solving 4N of these problems. Hence, it seems preferable to apply the improved Born-type procedure directly.

References

Bolomey, J.-Ch., Durix, Ch., and Lesselier, D. (1979), Determination of conductivity profiles by time-domain reflectometry, *IEEE Trans. Antennas Propagat. 27*, 244-248.

Brodlie, K.W. (1977), Unconstrained minimization, In: *The state of the art in numerical analysis*, Jacobs, D. (Ed.), Academic Press, London, pp. 229-268.

Coen, Sh., Mei, K.K., and Angelakos, D.J. (1981), Inverse scattering technique applied to remote sensing of layered media, *IEEE Trans. Antennas Propagat. 29*, 298-306.

Fletcher, R., and Reeves, C.M. (1964), Function minimization by conjugate gradients, *Computer J. 7*, 147-154.

Gill, P.E., and Murray, W. (1974), *Numerical methods for constrained minimization*, Academic Press, London.

Gill, P.E., and Murray, W. (1977), Linearly-constrained problems including linear and quadratic programming, In: *The state of the art in numerical analysis*, Jacobs, D. (Ed.), Academic Press, London, pp. 313-363.

Lesselier, D. (1978), Determination of index profiles by time domain reflectometry, *J. Optics 9*, 349-358.

Lesselier, D. (1982a), Optimization techniques and inverse problems: reconstruction of conductivity profiles in the time domain, *IEEE Trans. Antennas Propagat. 30*, 59-65.

Lesselier, D. (1982b), *Diagnostic optimal de la lame inhomogène en régime temporel. Applications à l'électromagnétisme et à l'acoustique*, Ph.D. Thesis, l'Université Pierre et Marie Curie, Paris.

NAG Fortran Library Mark 11 (1984), Numerical Algorithms Group, Oxford.

Tijhuis, A.G. (1981), Iterative determination of permittivity and conductivity profiles of a dielectric slab in the time domain, *IEEE Trans. Antennas Propagat. 29*, 239-245.

Widder, D.V. (1946), *The Laplace transform*, Princeton University Press, Princeton, pp. 61-63.

SUBJECT INDEX

SAMENVATTING

Dit proefschrift geeft een overzicht van het onderzoek dat door de samensteller gedurende een aantal jaren is verricht op het gebied van de directe verstrooiing van elektromagnetische golven aan een- en tweedimensionale obstakels en het gebruik daarvan bij het reconstrueren van de elektromagnetische eigenschappen van eendimensionale configuraties (eendimensionale inverse verstrooiing). Het onderzoek heeft geleid tot een reeks van resultaten, waarvan het merendeel eerder in artikelvorm in de internationale literatuur is gepubliceerd. Het betreft hier zowel de gebruikte oplosmethoden in het algemeen als de toepassing hiervan op specifieke problemen. In dit proefschrift is getracht deze oplosmethoden en hun toepassingen op systematische wijze te presenteren.

Verschillende technieken worden besproken voor het oplossen van elektromagnetische directe verstrooiingsproblemen met een willekeurige tijdsafhankelijkheid. Deze problemen worden zowel rechtstreeks in het tijddomein opgelost als indirect via een Fourier of Laplace transformatie naar het reële of complexe frequentiedomein. Van iedere oplosmethode worden de analytische en de rekentechnische aspecten besproken. Bovendien worden representatieve numerieke resultaten van de toepassing van elk van de technieken op tenminste één illustratief probleem gepresenteerd en besproken. Speciale aandacht wordt besteed aan de fysische interpretatie van de theoretische en numerieke resultaten.

Voor het eendimensionale geval wordt tevens aangegeven hoe de speciale eigenschappen van de verschillende technieken kunnen worden gebruikt bij het oplossen van inverse verstrooiingsproblemen. Hierbij worden de onbekende elektromagnetische eigenschappen van een obstakel bepaald uit het verstrooide veld. Zowel de identificatie van het obstakel uit een eindige klasse van mogelijkheden als het bepalen van de ruimtelijke distributie van materiaaleigenschappen (inverse profilering) komen aan de orde. In dit gedeelte van het proefschrift worden dezelfde analytische, numerieke en fysische aspecten benadrukt als voor het directe probleem. Speciale aandacht wordt besteed aan de bandbegrenzingen in de reconstructie die ontstaan door mogelijke beperkingen in het frequentiespectrum van het invallend veld en door de benaderingsfouten die inherent zijn aan de gebruikte Born-achtige oplosmethoden.

LEVENSBERICHT

De samensteller van dit proefschrift werd op 6 augustus 1952 geboren te Oosterhout, Noord-Brabant. Nadat hij aldaar aan het Oelbertgymnasium het diploma Gymnasium Bèta had behaald, studeerde hij van 1970 tot 1976 natuurkunde aan de Rijksuniversiteit Utrecht. Van 1973 tot 1976 was hij tevens als student-assistent bij deze instelling werkzaam. In 1976 behaalde hij het doctoraaldiploma in de theoretische natuurkunde.

Vanaf november 1976 is hij verbonden aan de Technische Universiteit Delft bij de vakgroep Elektromagnetisme, eerst als wetenschappelijk medewerker en sinds december 1985 als universitair hoofddocent. Daar heeft hij, onder leiding van Prof.dr.ir. H. Blok, wetenschappelijk onderzoek verricht en colleges verzorgd op het gebied van de theorie van de excitatie, propagatie en diffractie van elektromagnetische golven. Daarnaast heeft hij incidenteel deelgenomen aan de activiteiten op het gebied van de quantumelektronica die, onder leiding van Prof.dr. W. van Haeringen, binnen de vakgroep plaatsvonden. Het door de samensteller verrichte onderzoek heeft geleid tot een groot aantal publikaties en, uiteindelijk, tot dit proefschrift.